LED Lighting for Urban Agriculture

Toyoki Kozai • Kazuhiro Fujiwara
Erik S. Runkle
Editors

LED Lighting for Urban Agriculture

Editors
Toyoki Kozai
Japan Plant Factory Association (NPO)
Kashiwa, Chiba, Japan

Erik S. Runkle
Department of Horticulture
Michigan State University
East Lansing, Michigan, USA

Kazuhiro Fujiwara
Graduate School of Agricultural and Life
 Sciences
The University of Tokyo
Bunkyo-ku, Tokyo, Japan

ISBN 978-981-10-1846-6   ISBN 978-981-10-1848-0   (eBook)
DOI 10.1007/978-981-10-1848-0

Library of Congress Control Number: 2016957396

© Springer Science+Business Media Singapore 2016
This work is subject to copyright. All rights are reserved by the Publisher, whether the whole or part of the material is concerned, specifically the rights of translation, reprinting, reuse of illustrations, recitation, broadcasting, reproduction on microfilms or in any other physical way, and transmission or information storage and retrieval, electronic adaptation, computer software, or by similar or dissimilar methodology now known or hereafter developed.
The use of general descriptive names, registered names, trademarks, service marks, etc. in this publication does not imply, even in the absence of a specific statement, that such names are exempt from the relevant protective laws and regulations and therefore free for general use.
The publisher, the authors and the editors are safe to assume that the advice and information in this book are believed to be true and accurate at the date of publication. Neither the publisher nor the authors or the editors give a warranty, express or implied, with respect to the material contained herein or for any errors or omissions that may have been made.

Printed on acid-free paper

This Springer imprint is published by Springer Nature
The registered company is Springer Science+Business Media Singapore Pte Ltd.

Acknowledgments

We would like to thank Ms. Tokuko Takano for her editorial assistance and dedication. Thanks are extended to Professors T. Maruo, M. Takagaki, T. Yamaguchi, and Y. Shinohara of Chiba University for their academic guidance, and to K. Yamada, K. Ohshima, and S. Sakaguchi of PlantX Corporation for their technical support. We also appreciate the guidance given by Dr. Mei Hann Lee and Ms. Momoko Asawa of Springer.

Contents

Part I Perspective and Significance of LED Lighting for Urban Agriculture

1 Why LED Lighting for Urban Agriculture?............................ 3
   Toyoki Kozai

2 Integrated Urban Controlled Environment Agriculture
   Systems.. 19
   K.C. Ting, Tao Lin, and Paul C. Davidson

3 Open-Source Agriculture Initiative—Food for
   the Future?... 37
   Caleb Harper

Part II Plant Growth and Development as Affected by Light

4 Some Aspects of the Light Environment............................. 49
   Toyoki Kozai and Geng Zhang

5 Light Acts as a Signal for Regulation of Growth and
   Development... 57
   Yohei Higuchi and Tamotsu Hisamatsu

6 Factors Affecting Flowering Seasonality........................... 75
   Yohei Higuchi and Tamotsu Hisamatsu

7 Light Environment in the Cultivation Space of Plant Factory
   with LEDs... 91
   Takuji Akiyama and Toyoki Kozai

Part III Optical and Physiological Characteristics of a Plant Leaf and a Canopy

8 Optical and Physiological Properties of a Leaf 113
Keach Murakami and Ryo Matsuda

9 Optical and Physiological Properties of a Plant Canopy 125
Yasuomi Ibaraki

10 Evaluation of Spatial Light Environment and Plant Canopy Structure 137
Yasuomi Ibaraki

11 Lighting Efficiency in Plant Production Under Artificial Lighting and Plant Growth Modeling for Evaluating the Lighting Efficiency 151
Yasuomi Ibaraki

12 Effects of Physical Environment on Photosynthesis, Respiration, and Transpiration 163
Ryo Matsuda

13 Air Current Around Single Leaves and Plant Canopies and Its Effect on Transpiration, Photosynthesis, and Plant Organ Temperatures 177
Yoshiaki Kitaya

Part IV Greenhouse Crop Production with Supplemental LED Lighting

14 Control of Flowering Using Night-Interruption and Day-Extension LED Lighting 191
Qingwu Meng and Erik S. Runkle

15 Control of Morphology by Manipulating Light Quality and Daily Light Integral Using LEDs 203
Joshua K. Craver and Roberto G. Lopez

16 Supplemental Lighting for Greenhouse-Grown Fruiting Vegetables 219
Na Lu and Cary A. Mitchell

17 Recent Developments in Plant Lighting 233
Erik S. Runkle

Part V Light-Quality Effects on Plant Physiology and Morphology

18 **Effect of Light Quality on Secondary Metabolite Production in Leafy Greens and Seedlings**.......................... 239
Hiroshi Shimizu

19 **Induction of Plant Disease Resistance and Other Physiological Responses by Green Light Illumination**................... 261
Rika Kudo and Keiji Yamamoto

20 **Light Quality Effects on Intumescence (Oedema) on Plant Leaves**....................................... 275
Kimberly A. Williams, Chad T. Miller, and Joshua K. Craver

Part VI Current Status of Commercial Plant Factories with LED Lighting

21 **Business Models for Plant Factory With Artificial Lighting (PFAL) in Taiwan**...................................... 289
Wei Fang

22 **Current Status of Commercial Plant Factories with LED Lighting Market in Asia, Europe, and Other Regions**.......... 295
Eri Hayashi

23 **Current Status of Commercial Vertical Farms with LED Lighting Market in North America**....................... 309
Chris Higgins

24 **Global LED Lighting Players, Economic Analysis, and Market Creation for PFALs**........................... 317
Eri Hayashi and Chris Higgins

25 **Consumer Perception and Understanding of Vegetables Produced at Plant Factories with Artificial Lighting**.......... 347
Yuki Yano, Tetsuya Nakamura, and Atsushi Maruyama

Part VII Basics of LEDs and LED Lighting Systems for Plant Cultivation

26 **Radiometric, Photometric and Photonmetric Quantities and Their Units**..................................... 367
Kazuhiro Fujiwara

27 **Basics of LEDs for Plant Cultivation**...................... 377
Kazuhiro Fujiwara

28 **Measurement of Photonmetric and Radiometric Characteristics of LEDs for Plant Cultivation**........................... 395
Eiji Goto

29	Configuration, Function, and Operation of LED Lighting Systems..	403
	Akira Yano	
30	Energy Balance and Energy Conversion Process of LEDs and LED Lighting Systems................................	417
	Akira Yano	
31	Health Effects of Occupational Exposure to LED Light: A Special Reference to Plant Cultivation Works in Plant Factories...	429
	Motoharu Takao	
32	Moving Toward Self-Learning Closed Plant Production Systems..	445
	Toyoki Kozai and Kazuhiro Fujiwara	

Index.. 449

Part I
Perspective and Significance of LED Lighting for Urban Agriculture

Chapter 1
Why LED Lighting for Urban Agriculture?

Toyoki Kozai

Abstract The benefits of using light-emitting diodes (LEDs) in urban agriculture are discussed, along with the necessity of introducing information and communication technology (ICT). The incorporation of ICT into urban agriculture is now economically viable because the marginal costs of information processing, storage, and transfer are approaching zero. Electricity generated from renewable resources such as solar energy and biomass is also becoming cost-competitive with that generated from fossil fuel and nuclear power. Internet-connected plant factories lit with LEDs and greenhouses with LED supplemental lighting will serve as key components in urban agriculture. The potential for combined applications of ICT, artificial intelligence, and the Internet of Things in urban agriculture is described briefly. Finally, the concept of closed plant production system (CPPS) and its application in plant factory with LED lighting are described.

Keywords Controlled-environment agriculture • Greenhouse • Light-emitting diode (LED) • Plant factory with artificial lighting (PFAL) • Supplemental lighting • Urban agriculture

1.1 Introduction

Since 2007, more than half of world population are living in urban areas, and it is estimated that over 70 % of world population would live there in 2050. Then, more and more people have recently been interested in urban agriculture or vertical farming (Despommier 2010). Indoor urban agriculture includes atriums, potted plants and plant stands to create green interiors with or without supplemental artificial light, and plant factories with artificial lighting (PFALs) (Kozai et al. 2015). Outdoor urban agriculture includes community gardens (or city farms) in public spaces, home vegetable/flower gardens, orchards, and greenhouses

---

T. Kozai (✉)
Japan Plant Factory Association (NPO), Kashiwa-no-ha, Kashiwa, Chiba 277-0882, Japan
e-mail: kozai@faculty.chiba-u.jp

with or without supplemental lighting. Issues on PFALs not discussed in this book are mostly discussed in Kozai et al. (2015).

1.1.1 Benefits of Urban Agriculture

Urban agriculture has two basic functions. One is to allow individuals to enjoy environmental horticulture as a hobby, and the other is to produce food and ornamental plants locally for sale to nearby residents. Food and ornamental plant production for local consumption can (1) save fossil fuel, labor time, and packaging material and thus transportation costs; (2) reduce postharvest losses due to damage during transport; (3) increase job opportunities, which benefits those living in urban areas; and (4) allow residents to enjoy a greater variety of fresh fruit and vegetables. By consuming locally-produced fresh food with minimum loss of quality and quantity, urban dwellers also use less electricity and/or fuel for shopping, processing, and cooking.

Since land prices in urban areas are high, the annual productivity of crops for sale per unit of land area must be considerably higher in PFALs and greenhouses than that in open fields. The annual productivity of leaf lettuce plants per unit land area is about 200-fold higher in PFALs and approximately tenfold higher in controlled-environment greenhouses compared with that in open fields (Kozai et al. 2015). If the soil is not sufficiently fertile to grow plants and/or is contaminated with toxic chemicals, heavy metals, etc., hydroponic systems with artificial substrates can be utilized which are isolated from the ground soil.

1.1.2 Benefits of Using Light-Emitting Diodes

Light-emitting diodes (LEDs) are increasingly common in numerous fields due to their good cost performance, relatively high electricity-to-light energy conversion factor, varied coloration (spectra), relatively low surface temperature, long lifetime, solid-state construction without gas, etc. Luminous efficacy (lumen per watt) of white LED tips was 75 in 2010, is 150 in 2016, and will reach around 200 in 2020 (Fig. 1.1). Recent improvement in the luminous efficiency of organic LEDs has also been significant.

Applications of LEDs for horticultural research have been conducted intensively since the 1990s (Massa and Norrie 2015). Fluorescent lamps in PFALs have gradually been replaced by LEDs after the first LED-lit PFAL was built in 2005 for the commercial production of leafy greens. As of 2015, more than 10 of about 200 Japanese PFALs in operation relied on LEDs. While supplemental lighting for greenhouse crops with high-pressure sodium (HPS) lamps has remained popular mainly in the Netherlands and the northern USA since the 1990s (Lopez and Runkle 2016), the HPS lamp versions are also now being replaced by LEDs.

1 Why LED Lighting for Urban Agriculture?

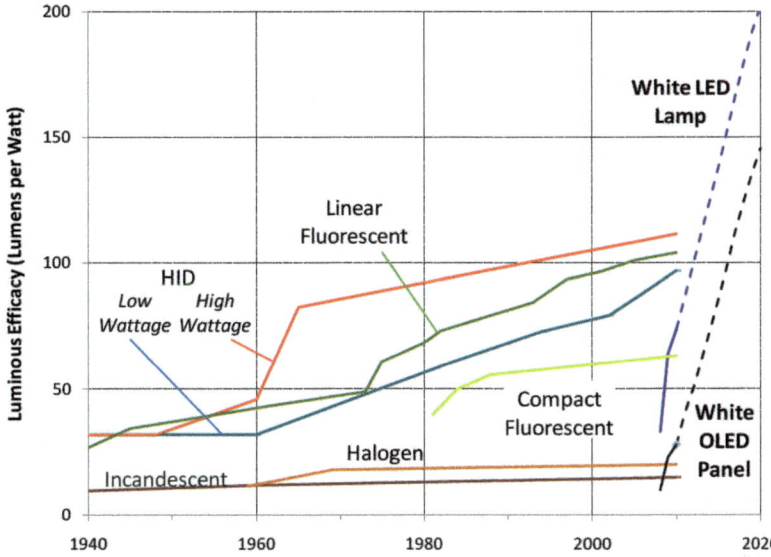

Fig. 1.1 Historical and predicted luminous efficacy of light sources (US Department of Energy 2011)

Plant growth and development are affected by the ambient light, including photosynthetic photon flux density (PPFD, sometimes called light intensity), cycle (light/dark period), ratio of diffuse to direct PPFD, angle determined by geometrical position or solar altitude and azimuth, and quality (wavelength or spectral distribution).

Plant morphology (flower bud initiation, internode length, branching, rooting, etc.) and secondary metabolite production (pigments, vitamins, etc.) are affected significantly by light quality and cycle. Therefore, LEDs with varying light qualities can be used to control morphogenesis and secondary metabolite production more efficiently, increasing the value of crops (see Parts 2 and 5 in this volume).

1.2 Scope of this Publication

This book focuses on LED lighting, mainly for the commercial production of horticultural crops in PFALs and greenhouses with controlled environments, with special attention to (1) plant growth and development as affected by light environment and (2) business and technological opportunities and challenges with regard to LEDs (Fig. 1.2). It contains 31 chapters grouped into seven parts: (1) overview of controlled-environment agriculture and its significance, (2) the effects of ambient light on plant growth and development, (3) optical and physiological characteristics of plant leaves and canopies, (4) greenhouse crop production with supplemental LED lighting, (5) effects of light quality on plant physiology and morphology, (6) current status of commercial plant factories under LED lighting, and (7) basics

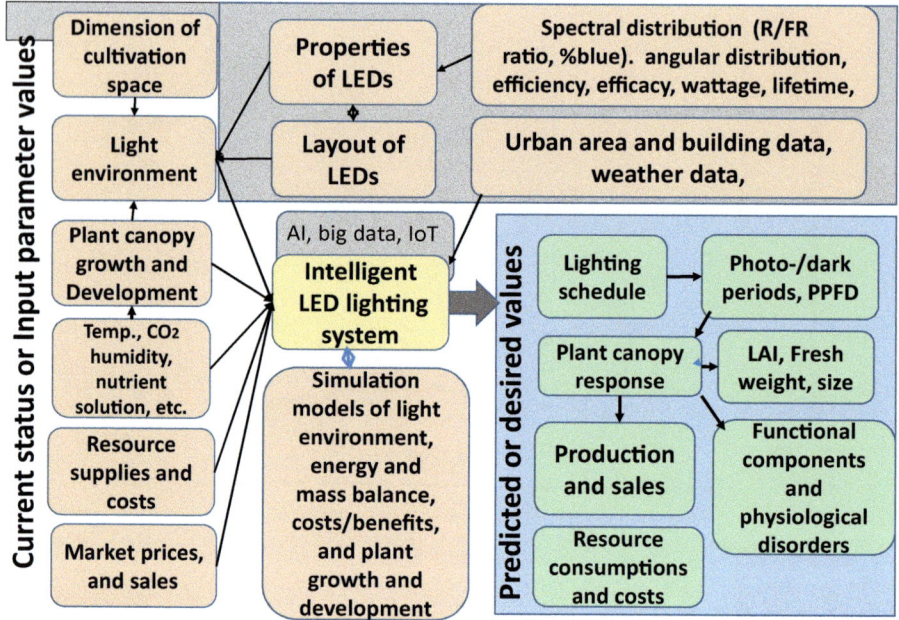

Fig. 1.2 Scientific, technological, and business key components and their structure of LED lighting for urban agriculture

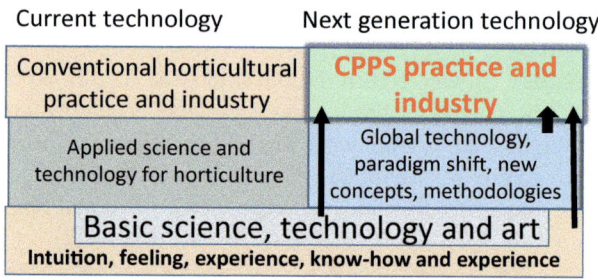

Note: CPPS denotes Closed Plant Production System (see the text for its definition)

Fig. 1.3 A scheme showing that closed plant production system (CPPS) as a major part of urban agriculture in the forthcoming decades will be largely dependent on a new paradigm, concepts, and methodologies, which are not directly related to conventional greenhouse horticulture

of LEDs and LED lighting for plant cultivation. Broader aspects of PFALs, excluding LED lighting, are described in Kozai et al. (2015).

It should be noted that "LED lighting for urban agriculture" in the forthcoming decades will not be just an advanced form of current urban agriculture. It will be largely based on two fields: One is a new paradigm and rapidly advancing new concepts, global technologies on LED, ICT, renewable energy, etc. and methodologies (Fig. 1.3); the other one is basic science and technology which should not be changed for the next several decades. Then, we need to forget about conventional horticultural technology once and to start thinking about the forthcoming urban agriculture based on the abovementioned two fields.

1.3 Technological Background to the Urban Agriculture of the Future

This section describes the technological background to the forms of urban agriculture expected to become widespread in the future. It should be noted that the marginal costs of information processing, storage, and transfer are now approaching zero, as are those of plant DNA genome sequencing (Rifkin 2015). Current fee rates for electricity generated from renewable energy sources are competitive with those generated by fossil fuel and nuclear power. These cost reductions will enable the development of sustainable, economically viable plant production systems with high yields and quality using minimal resources.

1.3.1 Local and Global Technology

Technology can be roughly divided into the local and global types (Fig. 1.4). Many local technologies were originally developed in specific agricultural and/or rural areas, influenced by the climate, soil, water availability, landscape, and other resources. The natural environment and associated agricultural practices then shaped local culture, including festivals, music and dance, cuisine, tools, and social rules. Traditional agriculture was generally sustainable, although not necessarily transferable to other regions.

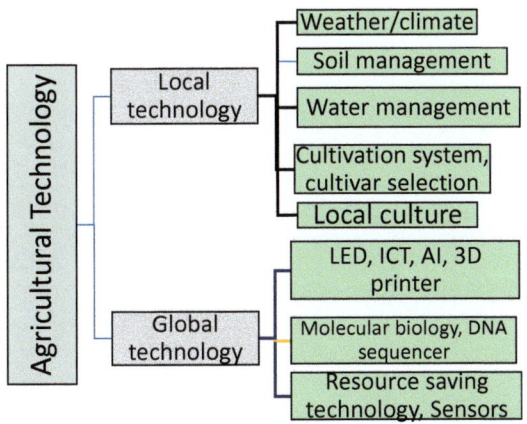

Fig. 1.4 Components of local and global technologies in urban agriculture. *ICT* (information and communication technology), *(AI* (artificial intelligence), *3D* (three dimensional) printer

Global technology, on the other hand, was developed mostly in cities by scientists, engineers, artists, and others, who created what we call "civilization." It sometimes refers to the comfort and convenience of modern life, regarded as available only in towns and cities, as cited in the *Oxford English Dictionary*. Global technology is often universal, expandable, and thus transferable to multiple regions, forming the basis of "Western" science. Typical examples of recent global technologies are computers, the Internet, LEDs, molecular biology-based DNA sequencing, and 3D (three-dimensional) printers.

1.3.2 Introducing Global Technology Locally

While ICT is a global technology, it can be customized by downloading application software, often free of charge, via the Internet and then adjusting the parameters for personal use or by a local or multinational group. It can also be used anywhere, anytime, by anyone at minimum cost. With the application of such global technologies, a sustainable plant production system can be developed as a form of "local culture" that relies on the available natural and human resources.

Any such system must be economically feasible, i.e., resulting in the maximum production of the highest-quality plants with the least possible yield variation, using minimum resources but with the highest use efficiency, and with the lowest cost and pollutant emission. Human welfare and global as well as local sustainability depend on these feasibility considerations.

Along with rapid advances in ICT, other new technology trends are being adopted in many industries (Fig. 1.5) which will affect agriculture in the near future. When local industries introduce global technology to enrich local resources, the advantage of scale enjoyed by large production units tends to decrease.

Fig. 1.5 Technological trends in agriculture

No.	from	to
1	Open (material)	Semi-closed or closed
2	Closed (information)	Open (information)
3	Centralized	Distributed & networked
4	Automatic	Intelligent & flexible
5	Market expansion	Market creation
6	Improvements	Innovations
7	Wired	Wireless, cloud computing

1.3.3 Innovative Global Technologies Influencing Next-Generation Urban Agriculture

1.3.3.1 Reductions in the Cost of Information and Bioinformatics

Information processing speeds (million instructions per second per US$1, Mips/US$) of microprocessors increased from 50 in 1990 to 4000 in 2000 and to 7 million in 2010 and will reach more than 100 million in 2015 (Fig. 1.6). In 2015, a micro-SD card measuring 11 mm wide, 15 mm high, and 1 mm thick, weighing only 1 g and priced at US$10, had a 32-GB memory (32 billion bytes; 1 byte is 8 binary digits needed to represent one alphabetic or numeric character), with a data transfer speed of 40 MB/s. Mobile phones had a data transfer speed of 9.6 kilobits per second (kbps) in 1980, which had increased to 100 Mbps in 2015.

Search engines such as Google and Yahoo and a huge number of public databases on genomes, weather, etc. are accessible free of charge. With these advances, computer networks have evolved from large central mainframes with many small terminals into distributed, autonomous, intelligent Internet-based, or networked systems. This change was enabled by steep decreases in the marginal cost of information, i.e., the increase in the total cost for adding one additional unit of new information is approaching zero.

The marginal cost of DNA sequencing is approaching zero even more rapidly than that of microprocessor information. The processing speed of DNA sequencing was 50 Mips/US$ in 2000, 7 million Mips/US$ in 2010, and 1 billion Mips/US$ in 2015 (Fig. 1.7) (Wetterstrand 2011). Fees for analyzing other bioinformatics have

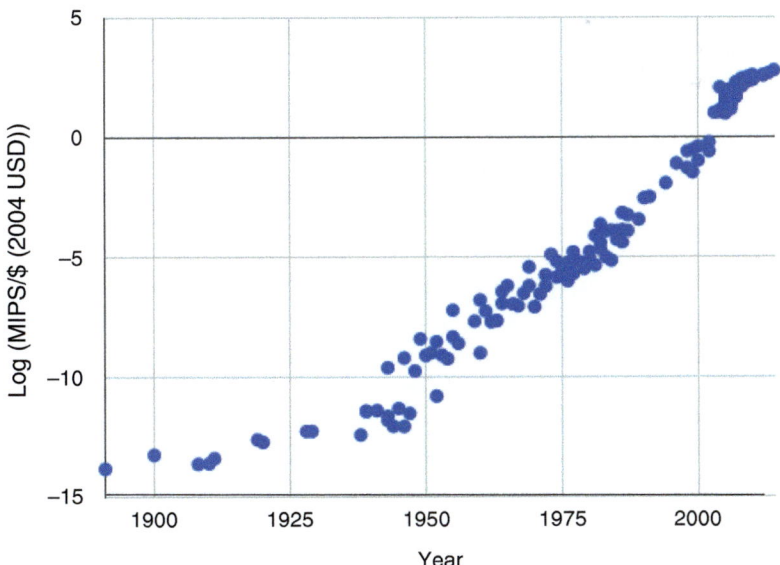

Fig. 1.6 Cost (USD) of microprocessor chip for MIPS (one million floating-point operations per second). (Trends in the cost of computing, 2014), http://aiimpacts.org/trends-in-the-cost-of-computing/

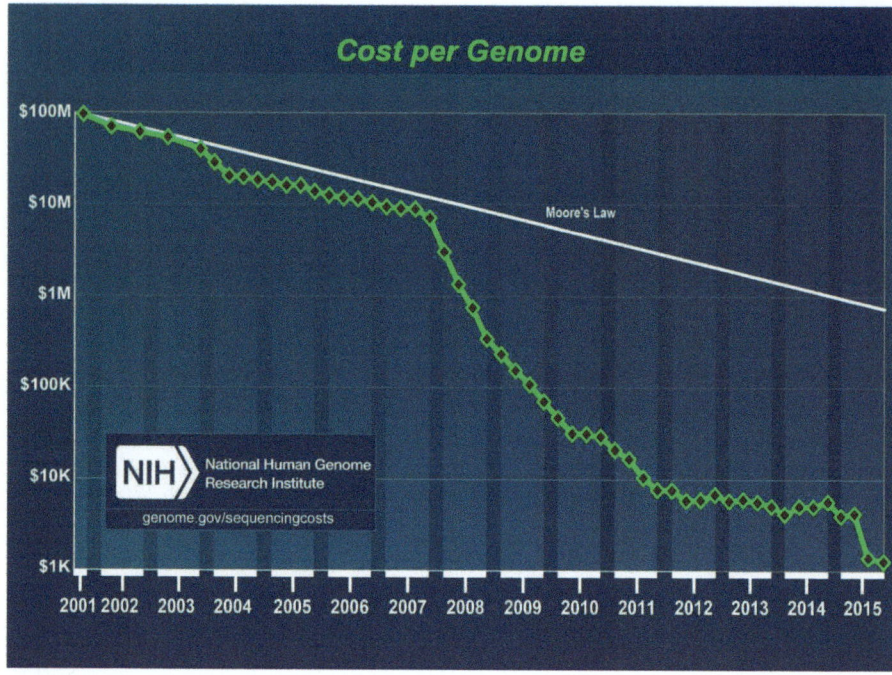

Fig. 1.7 Recent trend in cost of sequencing a human-sized genome. National Human Genome Research Institute (NHGRI) (https://www.genome.gov/images/content/costpermb2015_4.jpg)

also dropped sharply. These have allowed the new research area of "phenomics" to emerge, which involves the measurement and analysis of changes in the physical and biochemical traits of organisms in response to genetic mutations and environmental effects.

1.3.3.2 Levelized Cost of Electricity Generated from Renewable Energy Sources

The levelized cost of electricity (LCOE) is a measure of a power source to compare different methods of generation. It is an economic assessment of the average total cost of building and operating a power-generating asset over its lifetime divided by its total energy output over that lifetime. LCOE in the Organization for Economic Co-operation and Development (OECD) and non-OECD countries is shown in Fig. 1.8 (International Renewable Energy Agency 2015). Although the cost ranges for renewables are wide, reflecting varying resource quality and capital costs, the weighted average LCOE is competitive with new fossil fuel-fired generation options. For example, where oil-fired generation is the predominant power generation source (on islands, off grid, and in certain countries), a lower-cost renewable solution almost always exists today.

1 Why LED Lighting for Urban Agriculture? 11

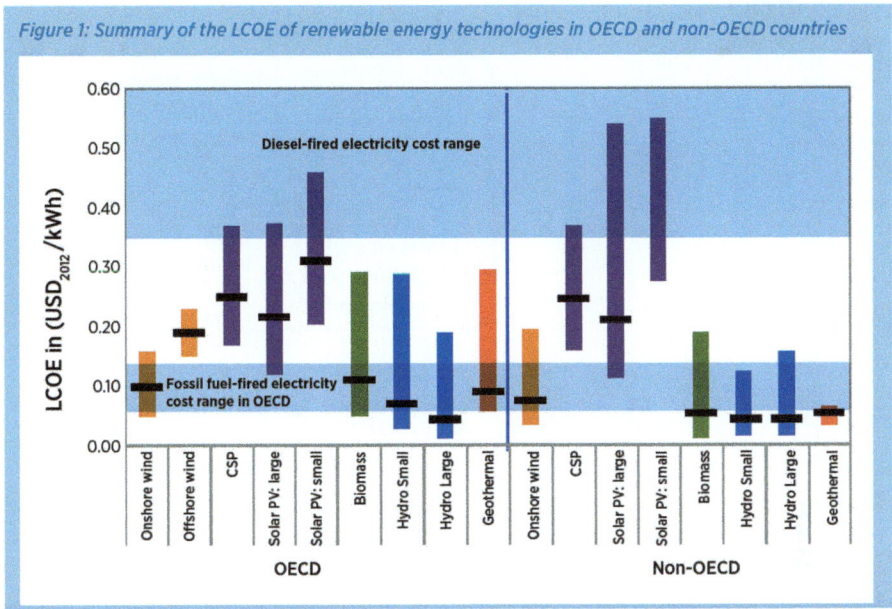

Fig. 1.8 Levelized cost of electricity (LCOE) of renewable energy technologies in the Organization for Economic Co-operation and Development (OECD) and non-OECD countries (2016). International Renewable Energy Agency, http://costing.irena.org/

The competitiveness of renewable power generation technologies continued improving in 2013 and 2014, reaching historic levels. Biomass, hydropower, geothermal, and onshore wind resources can all provide electricity competitively compared with fossil fuel-fired generation. Solar photovoltaic power has also become increasingly competitive, with its LCOE at utility scale falling by one-half in 4 years.

1.3.3.3 3D Printing

A 3D printer is a type of industrial robot, and 3D printing refers to various processes used to synthesize three-dimensional objects made of different kinds of metals, plastics and soils. According to a Wikipedia entry (2016), in 3D printing (https://en.wikipedia.org/wiki/3D_printing#Printers), successive layers of material are formed under computer control to create an object. The object can be of almost any shape or form and is produced from a 3D model or other electronic data sources. It will eventually be possible to send a blueprint of any product to any location place via the Internet for replication on a 3D printer. For example, with a 3D printer installed at home or in a nearby facility, 3D data on an object will be downloadable from the Internet to produce objects or machine parts as needed.

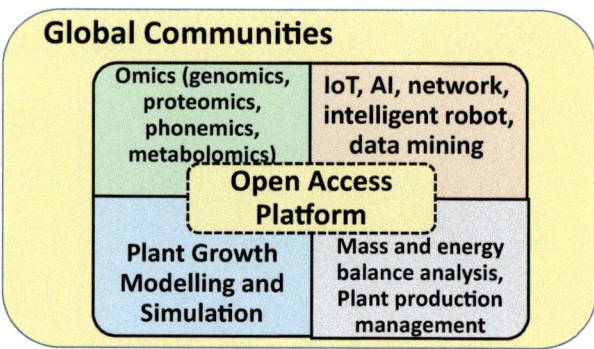

Fig. 1.9 Open-access platform for next-generation urban agriculture

1.4 Next-Generation Urban Agriculture

PFALs and greenhouses with LEDs will play a central role in urban agriculture because of the recent improvements in the electricity-to-light energy conversion factor of LEDs and reductions in the cost of electricity generated by renewable energy sources. Simultaneous decreases in information processing and bioinformatics costs are ushering in a new era of agricultural technology. Big data can be collected using ICT from many PFALs, greenhouses, and other agricultural facilities, while data in open-access databases can be analyzed by cloud computing with artificial intelligence via the Internet at minimal expense (Harper and Siller 2015). Adopting the Internet of Things and 3D printing in urban agriculture will improve the resource use efficiencies of plant production systems and/or food chains in urban areas (see Chap. 3).

On the other hand, in closed or semi-closed plant production systems such as PFALs and greenhouses, ecophysiological modeling and simulation are useful to predict plant growth and development, as well as mass and energy balances in the systems (Takakura and Son 2004; Yabuki 2004). By constructing open-access platforms consisting of databases, knowledge bases, rule bases, search engines, etc., more efficient, sustainable plant production systems are within reach (Fig. 1.9).

1.5 Closed Plant Production System (CPPS) (Kozai 2013; Kozai et al. 2015)

Requirements of a commercial plant production system in urban agriculture include high annual productivity per unit land area, high cost performance, safe and healthy produce, sustainable and stable production, high resource use efficiencies, and economic and social viabilities. The use of LEDs can contribute to meet all of these requirements, especially when used in a closed plant production system (CPPS).

The CPPS concept is relatively well introduced in commercial plant factories with artificial lighting (PFALs), while there exist few commercial "CPPSs with solar light" or "closed greenhouses," although its research, development, and commercial trial were extensively conducted during 2000–2015 in the Netherlands and other countries (De Gelder et al. 2012).

Difficulties about the commercialization of PFAL and closed greenhouse at that time period were due to its high cost and some technological problems. Even so, potential benefits of PFAL and closed greenhouse with LEDs in urban agriculture are considerable. Thus, understanding the concept, characteristics, and related methodology of CPPS is important when designing and operating PFAL and closed greenhouse with LED lighting.

1.5.1 Concept of CPPS

The CPPS is briefly defined as a plant production system covered with thermally insulated and airtight walls (e.g., Kozai 2013; Kozai et al. 2015). PFAL is one type of CPPS covered with walls which do not transmit solar radiation at all. PFAL consists of six main components: (1) tiers with lamps and hydroponic culture unit, (2) air conditioner, (3) CO_2 supply unit, (4) nutrient solution supply unit, (5) environmental control unit, and (6) thermally insulated and airtight structure to accommodate the abovementioned six units. Closed greenhouse is another type of CPPS covered with walls which transmit a large or small portion of solar radiation and heat energy through plastic or glass walls, with or without thermal/shading screen inside or outside the walls. (Ventilation fans are installed in the CPPSs for emergency use only.)

In both types of CPPS, since air exchanges (or ventilation) are negligibly small, hourly amounts of input material resources (water, CO_2, fertilizer, seeds, etc.) and of electric energy supplied can be measured relatively accurately. Those of output materials (produce, wastewater, plant residue, etc.) can also be measured relatively accurately (Fig. 1.10), while accurate estimation of heat and radiation energy exchanges between inside and outside the closed greenhouse is not so easy.

Chemical energy fixed in whole plants can be estimated relatively accurately based on the fresh weight of whole plants and its averaged percent dry matter. Hourly heat energy removed to the outside by the air conditioner and light energy emitted by the lamps can be estimated accurately in the case of PFALs. In summary, visibility and controllability of the environment and plant growth are highest in PFAL, followed by closed greenhouse, ventilated greenhouse, and open field. More importantly, in the CPPS, we can estimate or measure and control hourly values of "rate variables" (see below) in addition to those of "state variables."

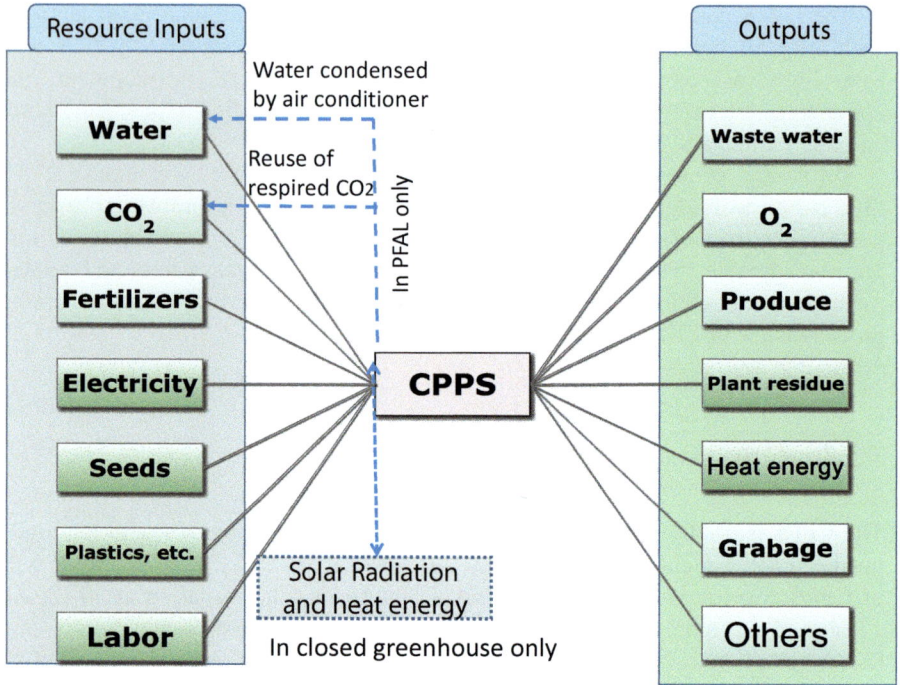

Fig. 1.10 Resource inputs into and outputs from CPPS (PFAL and closed greenhouse)

1.5.2 Estimating Rate Variable Values in the CPPS

Rate variable is a variable with unit of time (e.g., hourly change in weight; kg h^{-1}), while state variable is one without unit of time (e.g., kg). In the CPPS, we can estimate the ecophysiological responses of plants to the environments, such as hourly rates of water uptake, transpiration, net photosynthesis, dark respiration, fertilizer uptake, and hourly or daily rates of fresh weight increase and waste emission.

Uptake rate of each nutrient element can also be estimated by using the measured values of water uptake and concentration of each nutrient element in the culture bed (Kozai 2013). Also, we can measure seed germination rate and yield rate of plants (the weight ratio of salable part of plants to the whole plants) every day.

This advantage of CPPS with the use of measured rate variables will open a new era of plant production system, because hourly measurements of rate variables enable us to estimate resource use efficiencies (RUEs) online. Until now, measured values of state variables such as temperature, CO_2 concentration, water vapor pressure deficit (VPD), pH, and electric conductivity (EC) of nutrient solution only are used to control the environment for plant growth in PFAL and closed greenhouse.

1.5.3 Resource Use Efficiency (RUE) and Cost Performance (CP)

Based on the measured values of rate variables mentioned above, resource use efficiency (RUE, the ratio of resource fixed or held in products or plants to the resource input) in the PFAL can be calculated and visualized on the monitor screen for each resource component hourly, daily, and/or weekly. The RUE includes the use efficiencies of light energy, water, fertilizer, seeds, electricity, etc. Electricity use efficiencies include (1) electricity-to-light energy conversion factor (LEDs generally show higher values than other light sources), (2) conversion factor of light energy to chemical energy fixed in plants, and (3) coefficient of performance (COP) of air conditioners. Seed use efficiency and/or seedling use efficiency is also calculated. In "perfect CPPS," all the resource inputs are converted to the produce with or without plastic bags, so that no environmental pollutants, except for heat energy, are emitted to the outside.

In this way, we can monitor and analyze the RUE for each resource component at an arbitrary time interval, which enables us to improve each RUE and the total performance of PFAL systematically. This feature of PFAL is important for developing a methodology to improve the RUE and cost performance (CP) with low coefficient of variance (CV, ratio of standard deviation to the average) steadily with time. Once the methodology is developed for CPPS, the methodology can be applied for the closed greenhouse, after some modifications, and can also be applied in the ventilated greenhouse.

Cost performance (CP) for each resource input is briefly defined and calculated by the equation

$$CP = RUE \times (E/P)$$

where RUE is the resource use efficiency, E is the economic value per unit weight, and P is the production cost per unit weight. The time span can be hourly, daily, weekly, or monthly. The unit production cost includes the unit cost for processing the environmental pollutants.

Overall economic benefit is expressed as a product of overall CP and a total amount of produce. High CP needs to be associated with high RUE and low CV for sustainable food production. The above equation needs to be generalized for multi-resource inputs in its actual business application.

1.5.4 Rate Variable Control

The yield and quality of a crop cultivar are strongly affected by the changes in rate variables of photosynthesis, dark respiration, transpiration, nutrient uptake, water uptake, translocation, etc. Those rate variables are affected by the environments and ecophysiological characteristics of the crop. In turn, the rate variables affect the environments in the CPPS. These relationships in the CPPS are expressed by simultaneous differential equations.

In the PFAL, we can measure those rate variables relatively easily and also can measure the rate variables of resource inputs such as electricity, water, fertilizer, and CO_2 relatively accurately (Kozai 2013). Then, hourly and daily RUE for each resource element can be estimated relatively easily, while it is difficult and costly to measure such rate variables in the open greenhouse and open field. This is an essentially important point of the plant production in the CPPS.

1.5.5 Current Advantages of PFAL

Even though there exists only one commercial PFAL at Chiba University operated by PlantX Corp. with rate variable measurement control, there are many other advantages of PFAL shown below:

1. *Relative annual productivity* per unit land area is currently 100–200 times higher in the PFAL with high operation skills than in the open fields and 10–20 times higher than in the hydroponic greenhouse. In the case of the PFAL with ten tiers, a typical annual productivity is about 2500 leaf lettuce heads/m^2 (80 g fresh weight per head) or 200 kg/m^2.

   This annual productivity being proved by many commercially operated PFALs is mainly due to (1) multilayers (10–15 tiers), (2) plant growth promotion by environmental control, (3) no damage by pest insects and weather, (4) high planting density, and (5) transplanting on the same day as the day of harvest (360 days in cultivation at the same culture space). In addition, wholesale price per kg is about 30 % higher compared with that of greenhouse-grown vegetables, because of its cleanness (See Nos. 2 and 3 below), etc. Plant growth rate can also be promoted or slowed down to meet the variable demands, market prices, and costs with time.

   The relative annual productivity per unit land area of PFAL with 15 tiers is expected to increase up to 300 times or higher within 5–10 years by further improving the environmental control method, LED lighting system, production process management, hydroponic culture system, and other factors which will be described in the following section.

2. *Pesticide-, pest insect-, and foreign substance-free* are important characteristics of PFAL-grown leaf vegetables such as lettuce and spinach. Because of this characteristic, there is virtually no need to inspect foreign substances (fine insects, metals, plastic film pieces, etc.) in the vegetables before serving them. Also, PFAL mangers need virtually no knowledge of pathogen-originated disease and pesticides to grow plants in the PFAL.

3. *Colony-forming unit (CFU)* of PFAL-grown leaf vegetables is generally lower than 500, while the CFU of greenhouse-grown leaf vegetables is generally 10,000 or higher. Thus, there is no need to wash before eating the leafy greens

fresh. Then, we can save a large amount of water for washing. In the case of greenhouse- or field-grown vegetables, washing with tap water or water containing hypochlorous acid (HClO) is necessary to keep its salinity, by which water-soluble nutrients such as vitamin are dissolved and lost into water.

4. *Duration of life* of PFAL-grown produce is around twofold compared with that of greenhouse-grown produce when they are purchased at shops and kept in the refrigerator at home. This is mostly because the PFAL-grown leafy greens with CFU lower than 500 are packed in a plastic bag and sealed in the culture room just after harvesting and stored in a precooling room at a temperature of 2–5 °C until shipping. By doing so, we can save loss of vegetables after being purchased. Experimental data under different conditions need to be revealed.
5. *Traceability* from seeding through harvesting to delivery to customers is almost perfect with electronic and digital data. Flows and stocks of all the supplies (consumables) and products, wastes, environmental conditions, and operation hours of workers are recorded electronically, and monitor cameras are working all day.
6. *Higher labor productivity* due to light works under comfortable and safe working conditions (20–25 °C, 70–80 % relative humidity, 50 cm/s air current speed) regardless of weather. Then, labor efficiency is improved.
7. *Nighttime (often surplus) electricity* can be used (at a reduced price in many countries). Electricity cost is affected by lighting schedule under the same electricity consumption in case that the cost per kWh is dependent on the time of day and the season of year.
8. *Resource-saving* characteristics of PFALs, except for considerable electricity consumption, are significant, compared with those in the greenhouse (Kozai 2013). Roughly speaking, the following reductions in resource inputs can be realized compared with those in the open fields: pesticide by 100 %, land area by over 95 %, fertilizer by 50 % (recycling use), labor hour per production by 50 % (small land area), and plant residue by 30 % (less loss of plant parts). Water consumption for hydroponic culture is reduced by over 95 % (recycling use of condensed air at the cooling panels of air conditioners).

1.5.6 Current Disadvantages and Challenges of PFAL

Generally speaking, current technology level is much lower in PFAL industry than in Dutch greenhouse industry. PFAL technology has just been emerging and is still at the initial stage, although technological and business potentials of PFAL are very high. Current problems of PFAL business include (1) high production cost consisting of high initial, electricity, and labor costs due to poor design and management, (2) low quality and yield of produce due to poor environmental control and poor prediction of plant growth, and (3) poor production planning and process management. Also, we still do not know how to use LEDs most efficiently.

It is expected that, by 2020–2025, the production cost will be halved by improving light energy use efficiency and the productivity per floor area will be doubled by better environmental control and optimal selection of cultivars (Kozai et al. 2015). In order to achieve this goal, we need to use the global technology to develop the next-generation LED-lit PFAL with computer software of predictive modeling, simulation, and management of PFAL.

References

De Gelder A, Dieleman JA, Bot GPA, Marcelis LFM (2012) An overview of climate and crop yield in closed greenhouses. J HortScience Biotech 87(3):193–202

Despommier D (2010) The vertical farm: feeding the world in the 21st century. St Martin's Press, New York, 336 pp

Harper C, Siller M (2015) Open AG: a globally distributed network of food computing. Pervasive Comput 14(4):24–27

International Renewable Energy Agency (2015) Renewable power generation costs in 2014, 162 pp

Kozai T (2013) Resource use efficiency of closed plant production system with artificial light: concept, estimation and application to plant factory. Proc Jpn Acad Ser B 89(10):447–461

Kozai T, Niu G, Takagaki M (eds) (2015) Plant factory: an indoor vertical farming system for efficient quality food production. Academic, London, 423 pp

Lopez RG, Runkle ES (2016) Managing light in controlled-environment agriculture. Meister Media Worldwide, Ohio, USA, (in press)

Massa G, Norrie J (2015) LEDs electrifying horticultural science: proceedings from the 2014 Colloquium and Workshop. HortSci 50(9):1272–1273

Rifkin J (2015) The zero marginal cost society: the internet of things, the collaborative commons, and the eclipse of capitalism. St. Martin's Griffin, New York, 368 pp

Takakura T, Son JE (2004) Simulation of biological and environmental processes. Kyushu University Press, Fukuoka, 139 pp

US Department of Energy (2011) Solid-state lighting research and development: multi year program plan (Fig. 3.4), 130 pp

Wetterstrand K (2011) DNA sequencing costs: data from the NHGRI (Human Genome Research Institute) large-scale genome sequencing program. http://www.genome.gov/sequencingcosts

Yabuki K (2004) Photosynthetic rate and dynamic environment. Kluwer Academic Publishers, Dordrecht, 126 pp

Chapter 2
Integrated Urban Controlled Environment Agriculture Systems

K.C. Ting, Tao Lin, and Paul C. Davidson

Abstract Controlled environment agriculture (CEA) has evolved from very simple row covers in open fields to highly sophisticated facilities that project an image of factories for producing edible, ornamental, medicinal, or industrial plants. Urban farming activities have been developed and promoted as a part of the infrastructures that support residents' lives in high-population-density cities. Technology-intensive CEA is emerging as a viable form of urban farming. This type of CEA is likely to include engineering and scientific solutions for the production of plants, delivery of environmental parameters, machines for material handling and process control, and information for decision support. Therefore, the deployment of CEA for urban farming requires many components, subsystems, and other external influencing factors to be systematically considered and integrated. This chapter will describe high-tech CEA as a system, provide a systems methodology (i.e., the concept of automation-culture-environment systems or ACESys), propose a decision support platform (i.e., the concurrent science, engineering, and technology or ConSEnT, computational environment), and identify challenges and opportunities in implementing integrated urban controlled environment agriculture systems or IUCEAS.

Keywords Urban agriculture • Controlled environment agriculture • Systems integration • Systems informatics and analytics • Decision support

---

K.C. Ting (✉) • P.C. Davidson
Department of Agricultural and Biological Engineering, University of Illinois at Urbana-Champaign, 1304 W. Pennsylvania Ave., Urbana, IL 61801, USA
e-mail: kcting@illinois.edu; pdavidso@illinois.edu

T. Lin
Department of Agricultural and Biological Engineering, University of Illinois at Urbana-Champaign, 1304 W. Pennsylvania Ave., Urbana, IL 61801, USA

College of Biosystems Engineering and Food Science, Zhejiang University, 866 Yuhangtang Road, Hangzhou, Zhejiang 310058, People's Republic of China
e-mail: lintao1@zju.edu.cn

2.1 Introduction

Since the beginning of human civilization, protective structures for plant cultivation have been developed with increasing sophistication, ranging from growing plants under simple covers to producing large-quantity and high-quality crops within precisely controlled environments. The analysis, planning, design, construction, management, and operation of high-tech controlled environment agriculture (CEA) for plant production require multidisciplinary expertise. Plant science and engineering technology, as well as their interrelationships, are the foundation for technically workable and economically viable high-tech CEA. Today, there is a wealth of knowledge for designing and managing plant-based engineering systems, i.e., phytomation systems (Ting et al. 2003).

It is commonly known that plants require air, light, water, and nutrients while exposed to appropriate ranges of temperature and relative humidity, to effectively grow and develop. The extent of growth and development varies with different plants when subjected to different combinations of the factors above. Plant scientists have, for many years, investigated the fundamental phenomena of plant physiology, photosynthesis, pathology, etc. Horticulturists have explored ways to cultivate and produce plants to satisfy certain purposes. Engineers have developed methods and equipment to create and deliver growing environment, support structures, material handling devices, and logistics operations to enable plant production at various scales. As mentioned above, these expertise need to be integrated in order to result in functional (and preferably optimized) CEA systems. It is also important to consider social, economic, and surrounding environmental conditions for successful "commercial" scale CEA systems (Nelkin and Caplow 2008; Despommier 2010).

It is predicted that, by 2050, the global population will exceed nine billion people and more than 70 % will live in high-population urban areas (United Nations 2014). Food security, in the context of availability, accessibility, utilization, and stability, is expected to be a daunting challenge, especially for the constant supply of fresh vegetables. Energy security and water security are strongly linked with food security. They have to be addressed in an integrated fashion. Therefore, the nexus of food-energy-water plays a very important role in urban food systems. CEA, especially in the form of plant factories (a.k.a. vertical farming), is well positioned to be part of urban food systems and deserves to be systematically analyzed within that context.

Systems analysis is a methodology that emphasizes the interfaces among the components of a system to investigate how components should work together. It is an important task to determine whether it makes sense to integrate interrelated components to achieve predetermined overall (i.e., system level) goals. The analysis can also help identify ways to resolve the interconnectedness of components and explore ways to improve the overall performance or derive the best system design and operation scenario under various constraints (Ting 1998). Systems analyses have been carried out by CEA researchers and practitioners in various

2.2 Recent Evolution of CEA

Figure 2.1 depicts the technological and functional evolution of CEA over the past 50 years. Light, temperature, air relative humidity and composition, plant nutrition, etc. are critical environment and physiology factors that determine the plant productivity and quality. Controlled environment, from protected cultivation and greenhouses to sophisticated, environmentally controlled plant factory, aims to provide extended range of microenvironmental conditions to support plant production either during the times when the natural environments are not conducive to plant growth or throughout the year.

2.2.1 Protected Cultivation

Protected cultivation refers to simple covers over plants in the production fields without advanced environmental control systems. They are normally seen in the forms of anchored plastic mulch, floating mulch, and low tunnels (Baudoin 1999). The purpose of protected cultivation is to improve the plant microenvironment for enhanced crop productivity in open fields. The key benefit of protected cultivation is to provide relatively low-cost crop protection from direct impact by the natural elements, such as frost and freezing. It can also promote water use efficiency and reduce risks of damages from insects, weeds, and other predators. There has been a continuing expansion in crop production areas utilizing protected cultivation, as well as an increase in its application in higher-value vegetable crops and flowers/ornamental plants (Wittwer and Castilla 1995). The comparative advantages derived from protected cultivation were the driving force for researchers and farmers to explore the technical workability and economic viability of creating and investing in increasingly sophisticated operations, equipment, and facilities for plant production.

2.2.2 Greenhouses

Commercial greenhouses started to emerge when better and larger enclosing structures and more elaborate plant growing configurations and devices were added to the original concept of protected cultivation. The larger structure of greenhouses allows sufficient vertical and horizontal spaces for workers to perform

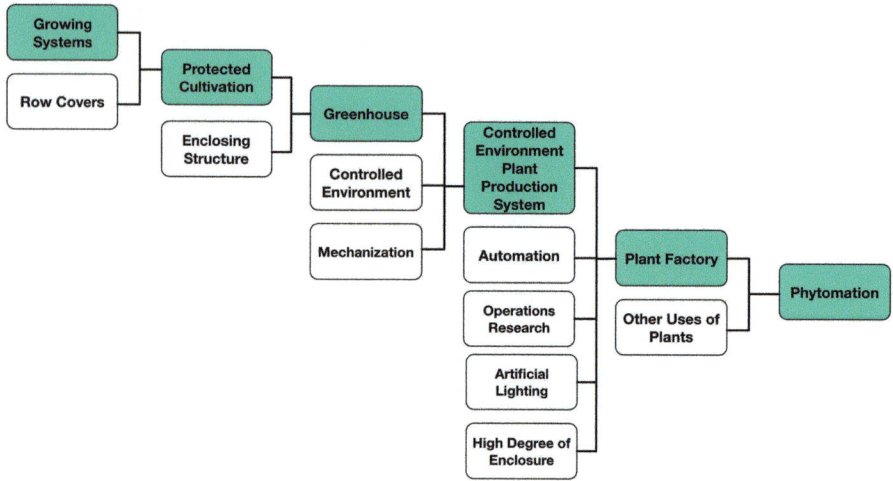

Fig. 2.1 From protected cultivation to phytomation

plant culture tasks and for taller plants to grow upright. The early form of greenhouses had a limited indoor environmental control ability; however, it was capable of providing much better modified environments for plants to produce a profitable yield during unfavorable outdoor conditions. The greenhouse's ability to control the environment under its enclosed structure allowed for an increased productivity of plants and human workers in addition to other direct benefits to plants and workers (Wittwer and Castilla 1995). Many growers started to improve upon the low-cost simple greenhouse structures that had poor environmental control and did not allow plants to reach their potential yield and quality (Baudoin 1999). Heating, cooling, ventilation, lighting, and CO_2 enrichment are key environmental control considerations within a greenhouse. Among them, better temperature controls, especially by heating, were the initial purpose for growers' adoption of greenhouses. The details of functional characteristics and design requirements of greenhouses have been reviewed by von Elsner et al. (2000a, b).

2.2.3 Controlled Environment Plant Production Systems (CEPPS)

Building on the advantages of greenhouses, additional investments were made to add more technologies, including automated indoor environmental control and mechanized plant growing and handling equipment. The impact of the entire production facility to the outdoor environment also started to attract interest. The concept of environmental friendliness of enclosed plant production operations became an important topic in the late 1980s and early 1990s. The increase in complexity of biological, physical, and chemical requirements for efficient plant growth and development, combined with the added constraints of social acceptance and government regulations, started to require plant producers to be skillful in the management aspect of their operations. The term "controlled environment plant production systems (CEPPS)" emerged as a more appropriate description for "advanced greenhouses." CEPPS is a form of controlled environment agriculture (CEA). CEA implies that it may include the production of livestock or fish. In recent years, "plant factory" and "vertical farming" have been used to represent certain forms of CEPPS that are particularly suitable as a part of urban agriculture and food systems.

2.2.4 Phytomation

CEPPS may be used for purposes other than producing plants for commercial markets. One example is phytoremediation processes for treating contaminated water within a controlled environment. This system has a similar form as a hydroponic plant growing system; however, the functional objective is reversed. Instead of supplying nutrient solutions to grow plants as a marketable product, it uses special types of plant to "rhizo-filtrate" contaminated water into clean water (Dushenkov et al. 1997; Fleisher et al. 2002). The cleaned water is the product of a phytoremediation CEPPS. Another example is the crop production unit within an advanced life support system (ALSS) for human long-duration space exploration. The crop production unit is similar in concept of a plant factory or vertical farming operation; however, the purposes are more than providing food for the crew members. It also participates in cleaning the recycled water and converting CO_2 to O_2 in the atmosphere within the ALSS (Kang et al. 2001; Rodriguez et al. 2003).

The added functional dimensions of enclosed plant production systems called for the need of an effective methodology for systems analysis and integration. An automation-culture-environment oriented systems analysis (ACESys) concept was proposed (Ting 1997). A term "phytomation" was created to capture this phase of evolution to describe all plant-based engineering systems (Ting et al. 2003).

2.2.5 Plant Factories with Artificial Light

The term "plant factory" has been used, mostly in Asia, to describe a commercial plant production facility that has similar operational principles as a typical industrial manufacturing facility. It typically has a very structured interior configuration with carefully designed processes for handling plants through their various stages of growth and development. The environmental and plant support parameters, such as temperature, relative humidity, light, CO_2, and nutrient solution conditions, in the facility are controlled within predetermined target ranges (Watanabe 2011; Goto 2012). Many plant factories are equipped with high-tech sensing, computing, process control, and automation devices for plant, nutrient solution, and environmental monitoring, as well as task planning and execution. In addition, some plant factories also have automated or mechanized systems for manipulating and transporting plants and assessing plant quality. Due to the relatively high financial investment required for plant factories, they are designed and managed to achieve very high space and resource use efficiency and to produce predictable high quantity, quality, and market value crops.

There are three options for lighting source for plants: (1) 100 % sunlight through translucent roof and/or walls, (2) 100 % "artificial" light from electricity-powered illumination devices, and (3) the combination of the two. The second type is called "plant factory with artificial light" (Kozai 2013). It allows the facility enclosure to be constructed in a way to minimize the influence of external environmental conditions. This provides the plant factory manager a better control of crop production operations. However, due to the total dependency on electricity for providing light energy, the requirements and cost-effectiveness of lighting devices for plant growth and development need to be carefully understood, selected, and operated.

The commonly used artificial light sources include high-pressure sodium lamps, fluorescent light tubes, light-emitting diodes (LEDs), etc. In recent years, LEDs have gained a significant amount of attention in the research community, CEA industry, and lighting device manufacturers. There are a number of advantages of LEDs as compared to the other forms of lighting equipment. LEDs have the potential of reducing the electricity costs from efficient conversion of electric power to targeted light wavelengths usable by plants, as well as from reducing cooling costs due to the lower thermal energy generation. The LED's compact design also allows its placement near the plants, which enables the configuration of multiple plant production layers, stacked vertically, within a plant factory facility. The high capital and operating costs of LEDs continue to be a high-priority research and development subject for future plant factories. This book will provide a comprehensive treatise of the state-of-the-art plant factories with artificial light and LEDs.

2.3 CEA's Role and Participants Within Urban Food and Agriculture Systems

Agriculture for food production as part of modern urban infrastructures has gained considerable attention in recent years (Pearson et al. 2010). There are many reasons why urban agriculture is desirable or even essential. Locally grown food, especially fresh vegetables, is not readily available in many metropolitan areas with high population densities. Fresh produce that travels a long distance to reach consumers in big cities requires high fuel and logistics costs and is prone to quantity and quality losses. Community-based food production systems are expected to contribute to the establishment of smart cities and healthy cities. CEA in or near urban areas, as part of urban food systems, can provide a reliable and safe food supply year-round (Despommier 2010). It may also enhance economic and social development of the cities. Its impact to the vegetable supply chain has been a research topic of some researchers (Hu et al. 2014).

There has been accelerated development in commercial CEA applications in the form of plant factories in East Asia, most noticeably in Japan and Taiwan. The major players include research and educational institutions, various levels of governments, real estate developers and builders, construction companies, heating/ventilation/air conditioning industry, electronics industry, supermarkets, restaurants, consumers, media, etc. This signifies the emerging opportunities for a wide range of businesses, as well as the unique challenges in how to simultaneously make things work better and make things work together.

2.4 CEA's Functional Components and Subsystems

Innovations in greenhouse engineering and horticulture have provided technical advances that have helped to bring about the state-of-the-art facilities and operations in CEA. The innovations are the results of responding to the need for improving CEA operations, as well as to the anticipated strategic changes in production systems. Operational factors that influence CEA systems include consumer preferences, market accessibility, labor availability, energy cost, logistics, etc. Factors that have strategic implications have mostly resulted from broader, regional issues such as environmental impact, product safety and consistency, and consumer demand (Giacomelli et al. 2008). A more detailed description of the controlled environment plant production system indicated in Fig. 2.1 is shown in Fig. 2.2.

The growing systems are the center stage of a CEPPS (i.e., CEA). It supports the plants (i.e., crops) that are established, grown, and harvested as the marketable product. Attention is normally paid to ensure plant quality by exposing the plants to appropriate environment, substrate, and nutrients. Plants at certain stages of production will need to be physically supported, manipulated, and transported either manually or by machines. There are many forms of external structures that can be used to enclose the entire production area and space. The most common glazing materials that allow sunlight transmission are glass panels and film and rigid plastics. Plant factories with 100 % artificial light sources normally use opaque construction materials. This presents an opportunity to have the walls and roofs well insulated for easier inside temperature control and better energy conservation. Some plant factories are situated inside a commercial building and may coexist with business and residential areas.

It is important to understand how the surrounding environmental factors impact the growth and development of the plants. In designing and operating a CEA system, it is also necessary to know how to deliver the desirable environmental

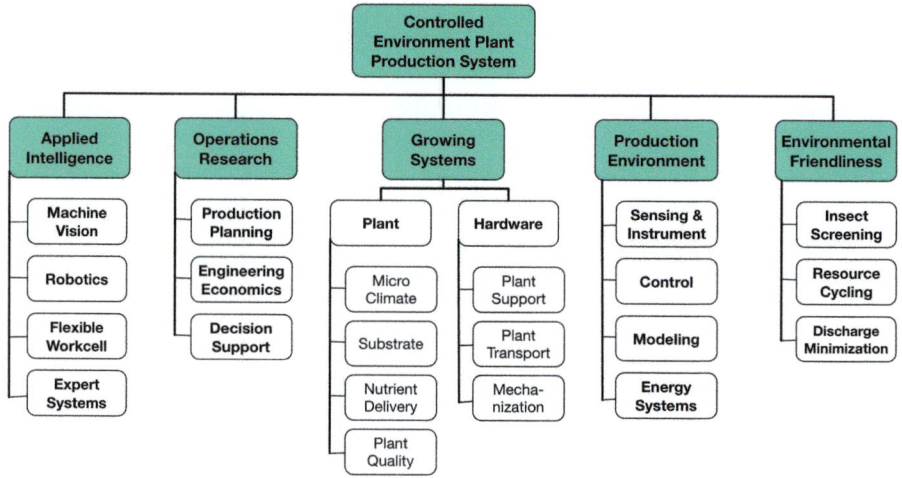

Fig. 2.2 Functional components of controlled environment plant production system

conditions. Therefore, within a CEA system, it will be ideal to have the capabilities of providing heating/cooling/ventilation, controlling relative humidity, enriching CO_2, and supplementing/adjusting lighting. The actions taken to activate these capabilities are sensing, data acquisition, and feedback/feedforward/model-based control. In many parts of the world, energy consumption for heating and/or cooling CEA environments is a significant portion of the initial and operating costs. Therefore, there have been a substantial amount of studies on utilizing energy from alternative or renewable sources to replace the heat and power derived from fossil-based fuels.

Recent developments in information technologies and mechatronics have worked their way into CEA system management and operations. Commonly seen applied intelligence includes computer vision-supported machine guidance for materials handling and watering, as well as plant quality evaluation and sorting. This added capacity also allows the automation of production planning and adaptive control of cultural tasks and environmental factors. A number of innovative ideas for minimizing impacts to the external environment have been put into practice, which include insect and disease screening, discharge minimization, and resource recycling.

2.4.1 CEA as Integrated Systems: An ACESys Model

So far, we have used the term "CEA system" without providing a systems approach for analyzing the system. We will start to bring that concept into our discussion. Within the context of this chapter, the technologies and knowledge bases needed for delivering a successful CEA system are in the areas of automation, culture,

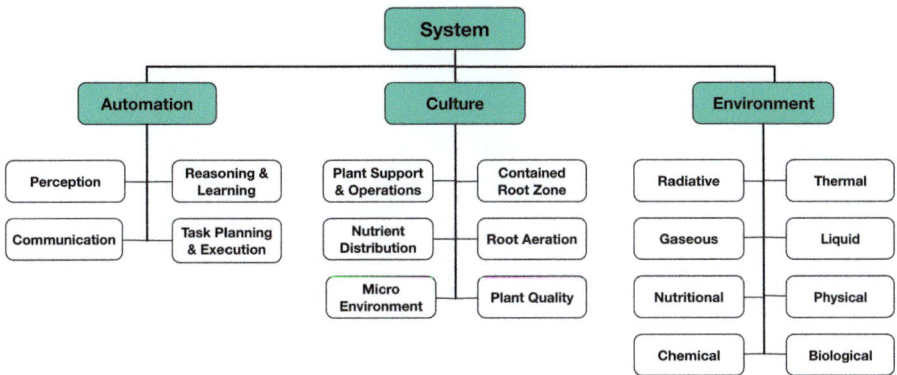

Fig. 2.3 The ACESys concept for CEA systems

environment, and systems (ACESys for short). Figure 2.3 is another version of Fig. 2.2, with an emphasis on the systems concept. Here are the brief descriptions of A, C, E, and Sys:

Automation deals with information processing and task execution related to a system's operation including the capabilities of perception, reasoning/learning, communication, and task planning/execution.

Culture includes the factors and practices that can directly describe and/or modify the biological growth and development of plants.

Environment encompasses the surroundings of plants, which consist of climatic and nutritional, as well as structural/mechanical, conditions.

Systems analysis and integration is a methodology that starts with the definition of a system and its goals and leads to the conclusion regarding the system's workability, productivity, reliability, and other performance indicators.

A plant factory with 100 % artificial light sources is a "closed" plant production system that provides a high level of control over plant production. This form of "closed" system exhibits the integration of automation, plant cultural requirements, and environmental control. An object-oriented approach guided by the ACESys concept may be taken to analyze plant production systems. The purpose was to develop a set of foundation classes that could be used to effectively describe the components of closed plant production systems. For example, eight foundation classes could be developed as the result of the object-oriented analysis, namely: Automation, Culture_Plant, Culture_Task, Culture_Facility, Environment_Rootzone, Environment_Aerial, Environment_Spatial, and Shell. Every class may contain key attributes and methods that provide appropriate systems informatics and analysis utilities for the systems under study. A computer model developed based on these classes and attributes would be capable of calculating crop yield, inedible plant material, transpiration water, power usage, automation, labor requirement, etc. over time for various crop mixes and scheduling scenarios (Fleisher et al. 1999; Kang et al. 2000; Rodriguez et al. 2003).

This modeling methodology can be modified into another example of analyzing CEA systems. Ting and Sase (2000) developed the following foundation classes, using the ACESys concept: (I) Category of Automation – Class_Perception, Class_Reasoning/Learning, Class_Communication, Class_Task_Planning, and Class_Task_Execution; (II) Category of Culture – Class_Crop, Class_Cultural_Task, and Class_Cultural_Support; (III) Category of Environment – Class_Environment_Structure and Class_Environment_Equipment; (IV) Category of System Level – Class_System_Requirement; and (V) Category of Result of Analysis – Class_Model_Output. The information flow pattern among the classes is depicted in Fig. 2.4. The arrows connecting the class objects indicate the key aspects of compatibility among the objects to be investigated. For example, perception class objects need to be capable of measuring the status and/or activities of class objects of environment_equipment, environment_structure, culture_task, and crop. Based on the signal from class object task_planning, the task_execution class objects must issue commands to activate class objects of environment_equipment and culture_task. The physical and functional compatibility among class objects is essential in ensuring the technical workability of the entire CEA system. Furthermore, the information is helpful in improving (or optimizing) the system design. The above examples represent very simple abstractions of CEA systems. The same methodology may be scaled up to represent CEA systems at a more sophisticated level.

2.5 Intelligence-Empowered CEA

Modern agriculture is an intelligence-empowered production system that requires capability for information collection/processing and decision-making, mechatronic devices for sensing, controls and actions, and ability to synergistically integrate components into functional systems. CEA is no exception. Its activities require actions taken by the growers in physical spaces, such as the core activities mentioned in Fig. 2.5. Ideally, these actions should be supported and guided by the intelligence resulted from analyses in the information space. An information system consisting of effective contents and efficient delivery methods will be very valuable in empowering growers in their decision-making.

Information technologies that can potentially provide the needed intelligence to agriculture include (1) *perception* using sensing and data acquisition/management technologies; (2) *reasoning and learning* involving mathematical, statistical, logical, and heuristic methodologies; handling of incomplete and uncertain information; and data mining; (3) *communication* by considering the contents, sources and recipients, and delivery platforms including wired, wireless, local area network, wide area network, Internet, and mobile technologies and devices; (4) *task planning and execution* that involve control logic, planning of physical tasks, intelligent machines, robotics, and flexible automation workcells; and (5) *systems integration* to provide computational resources and capabilities of systems informatics,

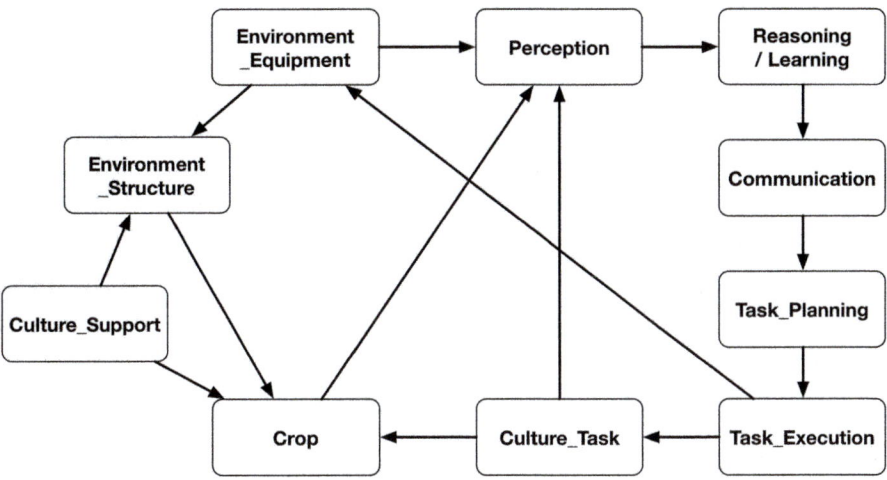

Fig. 2.4 Information flow diagram of ACESys objects for a CEA system

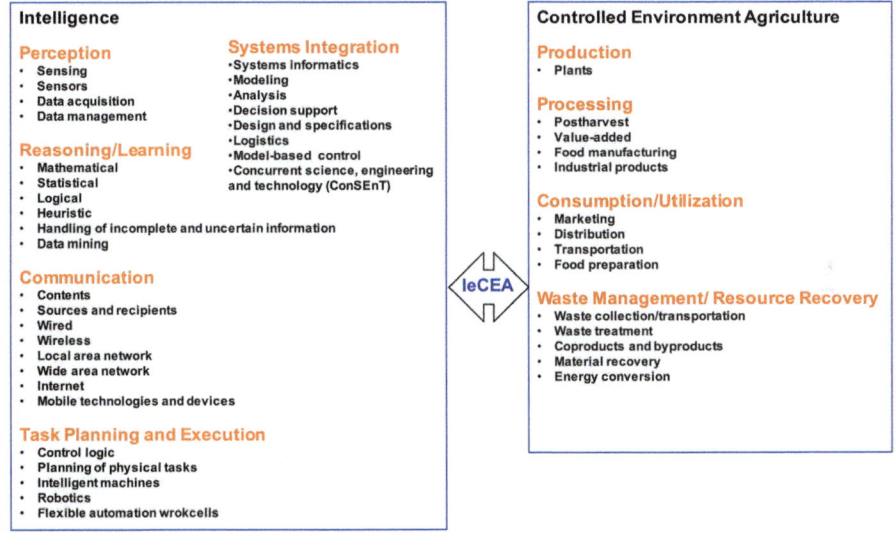

Fig. 2.5 Intelligence-empowered controlled environment agriculture (IeCEA)

modeling, analysis (simulation, trade studies, parametric analysis, life cycle analysis, optimization, etc.), decision support, design and specifications, logic and model-based control, and concurrent science, engineering, and technology (ConSEnT).

A substantial amount of work has been done in adding certain kind of intelligence to specific CEA tasks. Commonly seen studies have been in applying reasoning algorithms to develop control logic for controlling production environment within CEA. Related work may include sensing and communication technologies. Kolokotsa's team and Park's team have developed methodologies to monitor and control key environmental parameters, such as temperature, relative humidity, CO_2 concentration, and illumination based on Zigbee and Bluetooth wireless communication technologies (Kolokosta et al. 2010; Park and Park 2011). Mathematical models have been developed for predictive simulation and/or controlling CEA environment (Fitz-Rodriguez et al. 2010; Ishigami et al. 2013). Some models were developed specifically for controlling the cooling (Villarreal-Guerrero et al. 2012) and heating (Reiss et al. 2007) of greenhouses. van Straten's team developed a greenhouse environmental control strategy focusing on an optimization principle of making efficient use of the resource to maximize the economic return or minimize the cost of production (van Straten et al. 2000).

Another area of empowerment of CEA production operation has been the use of machine vision capabilities for quality sorting of plant materials and/or task guidance of automated machines (Tai et al. 1994; Kondo and Ting 1998).

2.6 CEA Systems Informatics and Analytics

As described above, opportunities exist for creating intelligent CEA systems enabled by systems informatics and analytics (SIA) concepts. The positive driving forces are the wealth of CEA-related domain knowledge; higher technology readiness level; available information technology, mechanization, and computer modeling capabilities; effective communication systems and computational platforms; improved economic picture; better market acceptance; potential spin-off technologies; ability to implement emerging technologies; etc.

The intelligence resources needed are informatics, computer modeling, systems analysis (in the forms of methodologies and tools for computation, simulation, and optimization), and actionable decision support (in the forms of analytics and consultative advices). All of these need to take into consideration of content, reasoning, audience, delivery, and action.

Systems analysis is a well-studied science that includes problem-focused and conclusion-targeted analytical algorithms and computational tools. One proven procedure for carrying out systems analysis is as follows:

1. Define system's scope and objectives.
2. Identify system constraints.
3. Establish indicators of success.
4. Conduct system abstraction.
5. Obtain data and information.
6. Handle uncertainty and incomplete information.
7. Incorporate heuristic and fuzzy reasoning.
8. Develop system model.

9. Verify and validate model.
10. Investigate what-ifs.
11. Draw conclusions.
12. Plan and execute actions.
13. Communicate outcomes.
14. Continuous monitoring and improvement.

The 14 steps are normally done in a sequential manner with the possibility of some steps being omitted or repeated depending on the nature of the system being analyzed and purpose of the analysis.

A concurrent science, engineering, and technology (ConSEnT) guiding concept implemented in a cyber-based platform may be used to facilitate the implementation of SIA (Ting et al. 2003; Liao 2011).

2.6.1 ConSEnT for CEA Decision Support

The core analytical activity in a ConSEnT cyber environment is to support decisions on the actions in physical space by carrying out analysis in the information space, i.e., the concept of cyber-physical systems (Chen et al. 2015). The processes contained in the centered circle in Fig. 2.6 depict the transformation of a system in physical space to its representation in information space, as well as the implementation of actions decided in the system's information space back to its physical space. The 14-step procedure described above may be used to facilitate these processes. There are at least four additional factors to be considered in a concurrent fashion in this analysis:

Systems requirements – What are the required functionality and desirable design considerations of the CEA system under study? How critical is each requirement or consideration?
Mission scenarios – What are the site-specific conditions that would influence the design and operation of the CEA system?
Candidate technologies – What are the available technologies or resources that could be implemented or utilized to satisfy the systems requirements? What is the technology readiness level of each technology?
Systems configuration and design – How would the necessary hardware, software, and other components be integrated into a functional system?

Figure 2.7 shows the computational and functional components of ConSEnT. It also emphasizes the importance of modularity, interrelationships, and concurrency of the components. The starting point of analysis is the system scope and objectives. The outcome of analysis is to support decisions and provide actionable analytics at the strategic, tactical, and operational levels. The computational resources needed are (1) informatics for managing the capture and flow of data, information, knowledge, and wisdom; (2) modeling and analysis tools for processing and interpreting

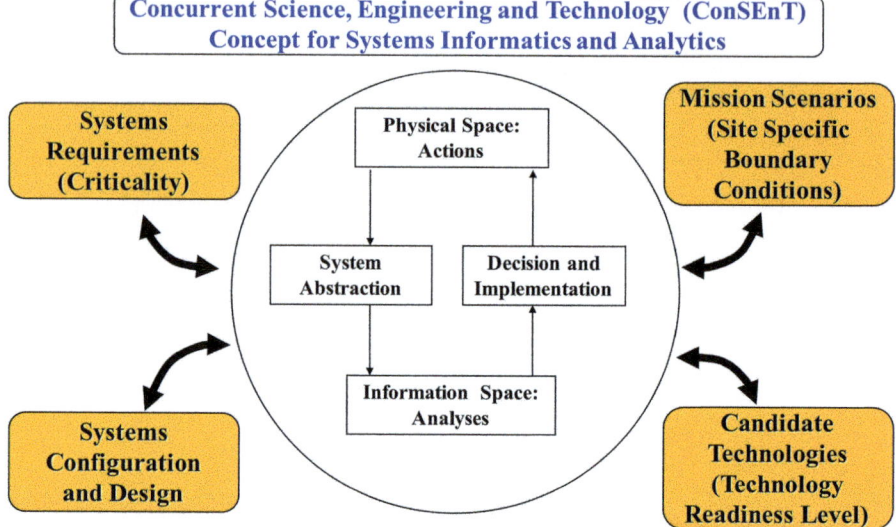

Fig. 2.6 ConSEnT concept for systems informatics and analytics

the information; and (3) decision support to connect the outcome of analysis to actions. The purpose of the ConSEnT cyber environment is to make all the resources available to the users in a way that facilitates broad and near real-time participation.

2.6.2 Decision Support and Analytics

Analytics is a way to discover and communicate key and meaningful characteristics in information. It has become an effective way of evaluating the performance of systems. It can be used to support decisions in various fashions. All available and emerging data processing and visualization tools have been used to produce and present analytics; many are custom designed for target audiences. For CEA systems, useful analytics are as follows:

- Technical workability
- Maintainability
- Controllability
- Reliability
- Productivity
- Economic competitiveness
- Energy efficiency
- Resource requirement
- Environmental impact

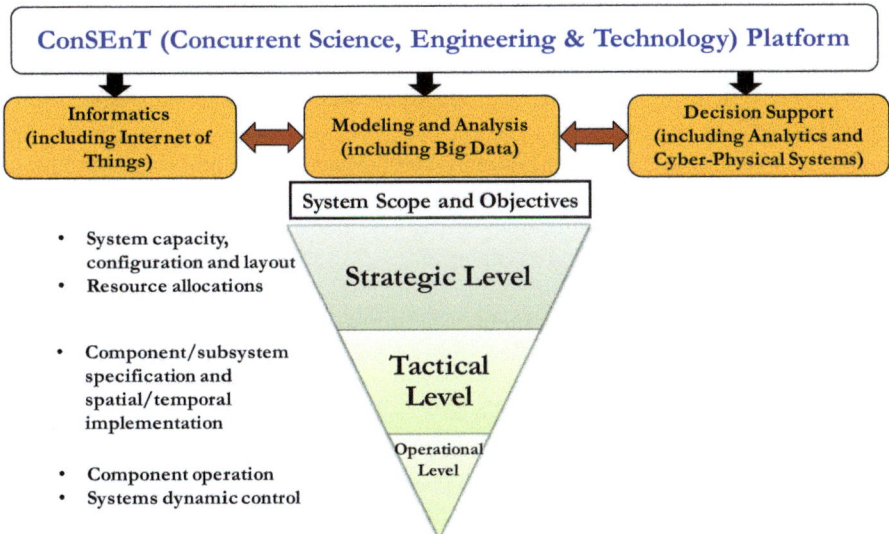

Fig. 2.7 Computational and functional components of ConSEnT

- Ecological harmony
- Social acceptance
- Optimization ability
- Operation and management capability
- Sustainability

Some of the above analytics may need to be expressed by a number of sub-analytics. For example, economic competitiveness may include capital investment, operating cost, revenue, return on investment, etc. Productivity may be expressed by sellable harvest per unit production area or revenue per unit input.

2.7 Current and Future CEA Challenges and Opportunities

The concept and practice of CEA are not new. However, there have been continued development, and the possibilities are plentiful. Just like other economic sectors, CEA, especially as part of urban food systems, has some unique challenges and opportunities. Some of the challenges and opportunities related to automation and systems informatics and analytics are discussed in this section.

2.7.1 Challenges

Automation
- Make return on investment attractive.
- Achieve systems optimization by proper integration of automation, plant culture, and controlled environment.
- Balance fixed automation and flexible automation (i.e., identifying appropriate level of necessary machine intelligence).
- Explore multiple uses of machines or parts of machines.
- Improve market demand and acceptance.
- Emphasize safety in operation.
- Enhance research and development capabilities.

Systems Informatics and Analytics
- Consider top level vs. process level.
- Make analytical tools expandable, compatible, adaptable, and transferable.
- Conduct effective system abstraction processes.
- Understand target participants and audiences.
- Validate systems models.
- Handle heuristic, uncertain, and incomplete information.
- Produce meaningful and useful deliverables from analysis outcomes.
- Coordinate multidisciplinary and multi-objective activities (e.g., ConSEnT).

2.7.2 Opportunities

Automation
- Take the advantage of improved technology readiness level and economic viability of automated information gathering/processing and materials handling.
- Build on past success of agricultural mechanization and modeling capabilities.
- Utilize effective communication systems and computational platforms.
- Enhance market acceptance.
- Increase the potential of spin-off technologies.
- Facilitate implementation of emerging technologies.

Systems Informatics and Analytics
- Establish information protocols and analysis algorithms for CEA.
- Develop a computerized environment for real-time information integration and analysis.
- Produce unified and robust models of CEA components and entire system.
- Perform studies at the system level to aid in design, operation, and research recommendations of CEA systems.

- Implement the systems informatics and analysis environment in a concurrent computational platform (e.g., ConSEnT), i.e., make things work better and together.

2.8 Concluding Remarks

It is probably an understatement to say that intelligent integration and optimized operation of urban CEA systems for achieving sustainability and competitiveness are a complicated task. Systematic approaches by involving multidisciplinary experts in evaluating and integrating available resources and candidate technologies will prove to be very productive endeavors. A concurrent science, engineering, and technology (ConSEnT) cyber environment may be created to enable real-time analysis, integration, design, management, and operation of urban CEA systems.

References

Baudoin WO (1999) Protected cultivation in the mediterranean region. Acta Hortic 491:23–30

Chen N, Zhang X, Wang C (2015) Integrated open geospatial web service enabled cyber-physical information infrastructure for precision agriculture monitoring. Comput Electron Agric 111:78–91

Despommier PD (2010) The vertical farm: controlled environment agriculture carried out in tall buildings would create greater food safety and security for large urban populations. J Verbr Lebensm 6(2):233–236

Dushenkov S, Vasudev D, Kapulnik Y et al (1997) Removal of uranium from water using terrestrial plants. Environ Sci Technol 31(12):3468–3474

Fitz-Rodriguez E, Kubota C, Giacomelli GA et al (2010) Dynamic modeling and simulation of greenhouse environments under several scenarios: a web-based application. Comput Electron Agric 70(1):105–116

Fleisher DH, Ting KC, Hill M et al (1999) Top level modeling of biomass production component of ALSS. In: The 29th international conference on environmental systems. SAE, Warendale, Technical Paper No.1999-01-2041

Fleisher DH, Ting KC, Giacomelli GA (2002) Decision support software for phytoremediation systems using rhizofiltration processes. Trans CSAE 18:210–215

Giacomelli G, Castilla N, Van Henten E et al (2008) Innovation in greenhouse engineering. Acta Hortic 801:75–88

Goto E (2012) Plant production in a closed plant factory with artificial lighting. Acta Hortic 956:37–49

Hu MC, Chen YH, Huang LC (2014) A sustainable vegetable supply chain using plant factories in Taiwanese markets: a Nash–Cournot model. Int J Prod Econ 152:49–56

Ishigami Y, Goto E, Watanabe M et al (2013) Development of a simulation model to evaluate environmental controls in a tomato greenhouse. In: International symposium on new technologies for environment control, energy saving and crop production in greenhouse and plant 1037:93–98

Kang S, Ozaki Y, Ting KC et al (2000) Identification of appropriate level of automation for biomass production systems within an advanced life support system. ASAE annual international meeting, Milwaukee, Wisconsin, July 9–12, 2000, Paper No. 003075

Kang S, Ting KC, Both AJ (2001) Systems studies and modeling of advanced life support systems. Agric Biosyst Eng 2(2):41–49

Kolokotsa D, Saridakis G, Dalamagkidis K et al (2010) Development of an intelligent indoor environment and energy management system for greenhouses. Energy Convers Manag 51(1):155–168

Kondo N, Ting KC (1998) Robotics for bioproduction systems. An ASAE monograph. ASABE, St Joseph, 325 pp

Kozai T (2013) Resource use efficiency of closed plant production system with artificial light: concept, estimation and application to plant factory. Proc Jpn Acad Ser B Phys Biol Sci 89(10):447–461

Liao YC (2011) Decision support for biomass feedstock production enabled by concurrent science, engineering, and technology (ConSEnT). M.S. thesis, Department of Agricultural and Biological Engineering, University of Illinois at Urbana-Champaign, 151 pp

Nelkin J, Caplow T (2008) Sustainable controlled environment agriculture for urban areas. Acta Hortic 801:449–456

Park DH, Park JW (2011) Wireless sensor network-based greenhouse environment monitoring and automatic control system for dew condensation prevention. Sensors 11(4):3640–3651

Pearson LJ, Pearson L, Pearson CJ (2010) Sustainable urban agriculture: stocktake and opportunities. Int J Agric Sustain 8(1–2):7–19

Reiss E, Mears DR, Manning TO et al (2007) Numerical modeling of greenhouse floor heating. Trans ASABE 50(1):275–284

Rodriguez LF, Kang S, Ting KC (2003) Top-level modeling of an ALS system utilizing object-oriented techniques. Adv Space Res 31(7):1811–1822

Tai YW, Ling PP, Ting KC (1994) Machine vision assisted robotic seedling transplanting. Trans ASAE 37(2):661–667

Ting KC (1997) Automation and systems analysis. In: Plant production in closed ecosystems. Springer, Dordrecht, pp 171–187. Retrieved from http://link.springer.com/chapter/10.1007/978-94-015-8889-8_11

Ting KC (1998) Systems analysis, integration, and economic feasibility. In: Robotics for bioproduction systems. ASABE, St. Joseph, pp 287–320

Ting KC, Sase S (2000) Object-oriented analysis for controlled environment agriculture. In: Environmentally friendly high-tech controlled environment agriculture. National Research Institute of Agricultural Engineering, Tsukuba, pp 101–109

Ting KC, Fleisher DH, Rodriguez LF (2003) Concurrent science and engineering for phytomation systems. J Agric Meteorol 59(2):93–101

United Nations (2014) World urbanization prospects: the 2014 revision. United Nations Publications, New York

van Straten G, Challa H, Buwalda F (2000) Towards user accepted optimal control of greenhouse climate. Comput Electron Agric 26(3):221–238

Villarreal-Guerrero F, Kacira M, Fitz-Rodríguez E et al (2012) Implementation of a greenhouse cooling strategy with natural ventilation and variable fogging rates. Trans ASABE 56(1):295–304

von Elsner B, Briassoulis D, Waaijenberg D et al (2000a) Review of structural and functional characteristics of greenhouses in European Union countries: Part I, Design requirements. J Agric Eng Res 75(1):1–16

von Elsner B, Briassoulis D, Waaijenberg D et al (2000b) Review of structural and functional characteristics of greenhouses in European Union countries: part II, typical designs. J Agric Eng Res 75(2):111–126

Watanabe H (2011) Light-controlled plant cultivation system in Japan-development of a vegetable factory using LEDs as a light source for plants. Acta Hortic 907:37–44

Wittwer SH, Castilla N (1995) Protected cultivation of horticultural crops worldwide. HortTechnology 5(1):6–23

Chapter 3
Open-Source Agriculture Initiative—Food for the Future?

Caleb Harper

Abstract In the 10,000 years of agriculture's history, advancements have enabled three society-altering revolutions. We believe that food computing, an alternative, distributed farming system based on new methods of communication, sensing, data collection, and automation, will enable network-effect advantages in the next generation of food production and give rise to the Internet of Food and the next agricultural revolution. At the MIT OpenAG Initiative, we are working on building this new digital-plant-recipe-centric network, a database of "climate recipes" for achieving the desired phenotypic expression of the plant in question. The food computer, or FC, that we are developing is a term for an agricultural technology platform that creates a controlled environment using robotic control systems and actuated climate, energy, and plant sensing mechanisms, designed to optimize agricultural production by monitoring and actuating a desired climate inside of a growing chamber that in turn creates desired phenotypic traits in plants. With iterative experimentation, we could hypothetically map the entire phenome of a selected plant and correlate certain phenotypic traits with specific environmental stimuli—our Open Phenome Project, a catalog of the epigenetic expression of plant life. Currently, the commercially available systems are being developed as closed proprietary systems noncompatible with other platforms of the same scale or across scales. We imagine a very different future, with open and cross-compatible technology platforms underlying a distributed network of FCs of various scales using digital plant recipes and the controlled environment climate as the scaling factor.

Keywords Food computing • Networked experimentation • Climate recipes • Plant phenomics • Controlled environment agriculture

In the 10,000 years of agriculture's history, advancements have enabled three society-altering revolutions. First came the domestication of plants and the resulting first human settlements in 8000 BC, followed by the horse and plow and the rise of technology-based societies in 600 AD, and finally the vertical integration

---

C. Harper (✉)
MIT Media Lab, 75 Amherst Street, Cambridge, MA 02139, USA
e-mail: calebh@media.mit.edu

of farming brought on by the mechanization, chemical fertilization, and biotechnology of today (Baker 1996). Agricultural revolution has been a major driving force behind humanity's societal progress. The current industrialized food system feeds 7.3 billion people (United Nations, World Population Prospects 2015), of whom more than half live in cities (United Nations, World Urbanization Prospects 2015) and very few of whom are involved in the production of their own food. The backbone of this system is comprised of large, centralized, chemically intensive single-crop farms. With natural-resource scarcity, flattening yields, loss of biodiversity, changing climates, environmental degradation, and booming urban populations, our current food system is rapidly approaching its natural limit. What will define the fourth agricultural revolution, and how will it impact and shape global societies?

This is the central research question of the Open Agriculture (OpenAG) Initiative at the MIT Media Lab. As we at the OpenAG Initiative and our collaborators develop a greater understanding of the unintended ecological and nutritional consequences of industrialized agriculture—including its contribution to global warming from CO_2 emissions from farming, shipping, and storing food, the pollution of oceans from agricultural runoff, nutrient-depleted produce and resulting malnutrition, "food deserts," and obesity and diet-related illnesses such as type II diabetes—we envision an alternative, distributed farming system based on new methods of communication, sensing, data collection, and automation that will enable network-effect advantages in the next generation of food production.

The Internet was built to compute and share information using interconnected open systems and networks. In the same way, this next agricultural revolution will be based on interconnected open food production platforms (food computers) to increase production either by scaling up or scaling out and sharing data to form a new kind of network: the Internet of Food (IoF). This new Internet is a digital-plant-recipe-centric network, a database of "climates" created for achieving the desired phenotypic expression of the plant in question. One can imagine digital plant recipes as the equivalent of "html" in computer networks. The recipes are the logical structured containers of exchanged information within the IoF. We believe that food computing, open data platforms, and networked production communities will each play a pivotal role in the next agricultural revolution.

3.1 Food Computing

Donald Baker, a distinguished fellow in the American Society of Agronomy and the American Association for the Advancement of Science, suggests the following:

> The third revolution may run its course or it may receive a boost from biotechnology. But with or without the application of a new technology, a fourth method of yield measurement may be used in the near future. It is the ratio of yield to a critical factor other than land. As the critical factor in the past has gone from human effort, to the amount of seed sown, to the amount of land used, it may soon change, for example, to the nitrogen, the phosphorus, or

the energy expended. Perhaps the best one would be an economic one, since it also requires a superior bookkeeping system. Thus, the next yield expression might become yield per dollar spent. (Baker 1996)

It is the premise of the OpenAG Initiative that the "superior bookkeeping system" to which Baker refers could be realized through leveraging the networked and computational power of "food computing" in the fourth agricultural revolution (Harper and Siller 2015).

The food computer, or FC, is a term for an agricultural technology platform that creates a controlled environment using robotic control systems and actuated climate, energy, and plant sensing mechanisms. Not unlike climate-controlled datacenters optimized for rows of servers, FCs are designed to optimize agricultural production by monitoring and actuating a desired climate inside of a growing chamber that in turn creates desired phenotypic traits in plants. Climate variables—such as carbon dioxide, air temperature, photosynthetically active radiation levels, leaf surface humidity, dissolved oxygen, potential hydrogen, electrical conductivity, and root-zone temperature—are among the many potential points of actuation within the controlled environment.

These points of actuation, coupled with the plant machine interface (PMI), are the drivers of plant-based morphologic and physiologic expressions. For example, FCs can program biotic and abiotic stresses, such as an induced drought, to create desired plant-based expressions of color, texture, taste, and nutrient density—vintners often apply the method of inciting a strategic vineyard drought, implemented by analog means, to sweeten grapes, but this technique has not been applied to the much broader variety of food that can be grown in FCs, with exponentially greater control. Operational energy, water, and mineral consumption are monitored (and adjusted) through electrical meters, flow sensors, and controllable mineral dosers throughout the growth period. When a plant is harvested from the FC, a digital plant recipe is created based on the corresponding data.

Digital plant recipes are composed of layered data that includes operational consumption data, plant morphology and physiology data, and a series of climate set points. Such points read like machine code and include a time stamp, an environmental control code, and a value associated with that environmental control. For example, a single climate set point reading of "00:00:00 SAHU 60" would set the air humidity to 60 % at time 00:00:00 h or 12:00 am. All of these layers of data are collated to form a repeatable digital plant recipe with known inputs and outputs. The goal is to create a genuine cyber-physical system in which each attribute has a closed feedback loop from sensor, to actuator, to biologic expression. With iterative experimentation, we could hypothetically map the entire phenome of a selected plant and correlate certain phenotypic traits with specific environmental stimuli.

With sophisticated correlation, we anticipate building digital plant models based on real-time data that can rapidly carry through iterative experimentation and suggest new digital plant recipes without having to grow each individual permutation. We are interested in a paradigm shift from simple controlled environments to adaptive environments where knowledge gained in process is feedback continuously based on the desired attributes of the product.

The human plant interface (HPI), a combination of the user experience and user interface, is a software layer that lets a human operator monitor sensors and actuator systems; browse and predict inventory; and load, override, or create derivative digital plant recipes. The HPI abstracts the operator from the mechatronics and reduces the biological or engineering expertise required to operate the system. Most importantly, the software is being designed as hardware agnostic enabling the development of a modular hardware ecosystem. This is crucial for enabling FCs to integrate easily with rapid advancements currently being made in the control environment agriculture space (sensors, LEDs, actuators, etc.).

There are currently three scales of control environment platforms being developed globally. The consumer electronic scale is a product scale (2–10 ft^2 or 0.18–0.93 m^2) designed for an at-home user, hobbyist, or student (see Fig. 3.1). The boutique production scale is a shipping container scale (200–500 ft^2 or 18.6–46.5 m^2) FC designed for owners/operators or franchisees to sell small amounts of high-value produce into local markets, restaurants, or cafeterias (see Fig. 3.2). The factory farm is a light industrial scale (+10,000 ft^2 or 929 m^2) designed to operate in urban or peri-urban environments and distribute fresh produce into a regional supply chain or produce a large quantity of a very high-value crop (see Fig. 3.3).

Currently, the commercially available systems are being developed as unique, closed, and proprietary systems that are noncompatible with other platforms of the same scale or across scales. CAE environments are being designed as static control systems, set points are created once, and an agricultural product is created. Knowledge is being developed locally on closed platforms, and questions remain unanswered regarding the functionality, scalability, economic viability, safety, and environmental sustainability of these systems.

Fig. 3.1 The new FC (food computer) prototype

3 Open-Source Agriculture Initiative—Food for the Future?

Fig. 3.2 The MIT OpenAG FC (food computer), built against the façade of the Media Lab and designed to be shipped anywhere in the world

Fig. 3.3 The view from within a farm factory scale FC, with multilayer design

3.2 Open Platforms and Open Data

We, at the MIT Media Lab OpenAG Initiative and our collaborators, imagine a very different future. This future is one where open and cross-compatible technology platforms underlie a distributed network of FCs of various scales using digital plant recipes and the controlled environment climate as the scaling factor. Conventional agricultural data has been difficult to export and use to replicate plant growth and output, because the growing conditions are dependent on the idiosyncratic variables created by the time of year, regional climate, and local resource availability.

FCs operate autonomously of local climate. Therefore, creating an agricultural product in one FC can easily be shared as a digital plant recipe and recreated, almost identically, in another compatible FC anywhere in the world, greatly expanding the concept of agricultural exports. We have begun piloting this concept through collaborations among boutique FCs at MIT, Guadalajara, and India.

This cross-platform compatibility would create the framework for rapid scalability of valuable discoveries. For example, innovations made at the personal scale could be quickly tested and verified across a network of compatible FCs and then deployed at the boutique or light industrial scales. FCs, then, could be imagined as networked cores of agricultural experimentation and production, capable of responding to local or global environmental, cultural, or market demands.

As the global network of FCs begins to create, iterate, and deploy digital plant recipes, we imagine these recipes would be open-source licensed and hosted on a public forum, modeled after Wikipedia, and downloadable as an executable file. Similar to the Human Genome Project, we envision the Open Phenome Project to be a crowdsourced cataloging of plants and their phenotypic traits correlated with the causal environmental variable. Over time, recipes would be optimized to decrease water, energy, and mineral use, while increasing nutrient density, taste, and other desirable characteristics. This database of functional plant phenomics would be the basis for scientific discovery, interdisciplinary collaboration, and new methods of efficient and distributed food production.

We believe this database could eventually impact conventional farmers by simulating a change in future climates to simulate future productivity and by using FCs to "climate prospect" which areas in the world would be best for a particular crop and its desired attributes—thereby rendering climate as a useful tool, by providing access to a catalog of climates and allowing farmers to make more informed, articulate choices about where and what they plant globally.

3.3 Integrating Artificial Intelligence Experimentation

The ultimate goal of the Open Phenome Project is to serve as a comprehensive catalog of epigenetic plant data, but the compilation thereof will rely entirely on crowdsourced and iterative experimentation. We have recently taken that idea further with experimentations in artificial intelligence (AI) and machine learning applications. To that end, we are collaborating with Rene Redzepi, renowned chef and owner of Noma, the molecular gastronomy restaurant in Copenhagen, and Sentient Technologies, an artificial intelligence program most commonly used for stock prediction.

Though our plants are already two to three times more nutrient-dense than those conventionally grown on farms, grow three to five times faster, and use 50–90 % less water, we are not limited to optimizing nutrition—we can also use our climates to enhance or change the phenotypic expression that results in flavor. For example, much of what humans perceive as flavor in herb plants like basil derives from molecules classified as secondary metabolites. These are referred to as "metabolites" in the sense that they are the result of living cellular processes and as "secondary" in the sense that they are not directly involved in the growth, development, or reproduction of the plant, but serve as enhancements to one or more of these, or to survival more generally. Secondary metabolites are involved in protecting the plant against environmental stresses, such as drought, UV light, and predation from herbivores. When exposed to different combinations of these stressors, basil has been shown to increase the concentration of these volatile secondary metabolites by up to an order of magnitude as compared to unstressed plants.

This means that it is possible to grow more intensely flavorful basil by stressing the plant in specific ways. With Noma's flavor chemist, Arielle Johnson, we have designed an experiment relating input data on basil-growing conditions (including main effects such as light source, chitosan addition, and UV supplementation and secondary effects like temperature, nutrition, water EC, and pH) to output data on markers for basil flavor and quality, for various volatile-increasing stress conditions.

The experiment involves the testing of three common types of light fixtures (fluorescent, Philips LED, and Illumitex LED), as well as supplementation with UV-B light and chitosan. UV-B has been shown to increase production of basil trichomes, the gland-like structures on the surface of the leaf that store most of the volatiles, and boost essential oil/volatile output. Chitosan, a polymer modified from the exoskeletons of crustaceans, confuses the plant into thinking that insects are in its vicinity or eating it and has likewise been shown to increase volatile output. This plant defense mechanism renders basil offensive to insect predators by yielding increased pungency, which makes such stressed basil ideal for culinary purposes.

We are using the MIT OpenAG boutique FC, partitioned into three rooms, each with three trays, such that potential interactions between light source and stress can be evaluated. Each tray within a room will receive one of the three light sources (fluorescent, Philips, or Illumitex), and each room will feature one of three stress treatments—UV-B addition, chitosan addition, or a control (no added stress). Within the nine trays, we will grow the basil to at least five leaf pairs—about 5–6 weeks total, including germination—and then analyze for dry weight, fresh weight, and volatile profile. The GC-MS volatile analysis is performed by pooling the first four leaf pairs on a basil plant, grinding them with liquid nitrogen and extracting the volatiles from this homogenate.

Once we have the nine experimental conditions' volatile profiles and other metrics, we will run the information through Sentient Technology's neural network AI algorithm, which will begin to generate suggestions about optimal growing conditions for essential oil yield. From there, we plan to design further experiments to vary nutrition, levels of different stressors (such as chitosan concentration, the time course for adding chitosan, or other "volatile elicitors" such as methyl jasmonate, UV light exposure times, and others), microbial profiles, temperature, and water availability.

Other experiments we have already completed include culturing the water within the MIT OpenAG boutique FC in order to study the microbiome of our plants. Ultimately, the goal is to understand the microbiota of root structures, stalks, and leaves peculiar to each plant such that we can mimic microterroirs from all over the world by introducing different cultures of bacteria to the water. Soon, we will also have the benefit of an additional experimental space in Middleton, Massachusetts, currently under construction; a 5000 foot^2 or 464.5 m^2 shipping container which we will divide into four boutique FCs; and one factory farm FC—the largest open-source vertical multilayer farm in the world.

3.4 Building the IoF and Enabling Communities

As the IoF develops, we look to the development of the Internet for useful guiding principles for adaptation—specifically flexibility, diversity, and openness. These guiding principles enable user-driven development, which has characterized the Internet's history (Abbate 1999). The IoF design is based on the premise of abstracting the information we share and its logical distribution algorithms from how we implement and interconnect physical things. In other words, the logical definition of the information is conceived in a different plane from the physical implementation. Therefore, the physical implementation is left open and enables the desired versatility, flexibility, and easy management of the IoF.

This openness would also allow for FCs customized in size and design, according to their application. For example, a research lab design (such as the MIT OpenAG boutique FC) might be different from a restaurant implementation. Imagine retailers in different locations designing a custom "case," similar to a custom iPhone case, for their FCs to coordinate with local décor, marketing, and regional aesthetics, while still not altering the core functionality of the FC. FCs might also have customized HPIs according to farming expertise: new, amateur, expert, or research farmer.

Because we recognize that due to lack of expertise, some farmers will not be able to construct their own PCs using the OpenAG schematics, instructional videos, and readily available materials—and might therefore not become involved due to the daunting sense of frontloaded effort, while they might otherwise have become active contributors to the IoF—we are working on the design of the next generation of in-home FC kits, which will be available for market purchase in 1–2 years.

The combination of open-sourced digital plant recipes, open technology platforms, and the IoF will lead to the democratization of food production enabled by massive communities of users. Social communities will be formed by users according to interests, preferences, levels of expertise, and so on. These communities will be comprised of the usual features: chats, forums, wikis, social networks, and blogs. We have recently launched a community at forum.openag.media.mit.edu that now has over 100 users building platforms on six continents—all but Antarctica—in more than 20 countries. Recipes will be quickly shared, validated, and customized. The recipes and FC hardware and software customization will become the center of the socialization process. The sharing of recipes (data) will allow a knowledge base to develop new plant data and technological setups, creating more accessible food production methods and meeting the everlasting demand for optimized food growth.

Open communities of food innovators, drawn together by collaborative and readily accessible technology platforms, will form the foundation of the next agricultural revolution. These communities will yield a diversity of thought and solutions and will nurture new connections between people and their food. The more ubiquitous the tools and knowledge of production systems become, the more informed, innovative, and empowered the average person can be in contributing to the global future of food. The accessibility of data, hardware, software, and, most importantly, food and nutrition for the projected nine billion people of 2050 hinges on fostering a creative, interdisciplinary forum of thinkers and doers on collaborative platforms today.

3.5 A Platform for Expression

Such collaboration on the IoF also creates a natural platform for expression, not necessarily or exclusively food-centric, on the personal, corporate, educational, governmental, and global scales. On a personal level, the in-home FC will allow hobbyist farmers to grow their own nutraceuticals, geared toward the specific requirements of their own genetics and physiology—or even for aesthetic, decorative effect. With resources like the Svalbard Global Seed Vault, we can resurrect more nutritious and flavorful cultivars that fell out of favor due to the difficulty of their production and inhospitable climates, providing the individual with greater diversity in both palate preference and personalized health—and without the attenuation of nutrition brought about by degradation of produce shipped great distances and stored for long periods of time.

From a corporate standpoint, the shift from an emphasis on abundant cheap food to better, more sustainable food is driving corporations toward seeking tools for providing transparency. With the accessibility of the Internet, the consumer has come to demand more transparency and food education from retailers and food producers, and companies must respond by transitioning to open, objectively healthier offerings.

Not only can the FC be turned to increasing adult literacy in food, it can also be used to bolster educational curriculums with an emphasis on science, technology, engineering, and mathematics (STEM) for children. We already have seven FCs in schools all over Boson, with children using an interactive, three-dimensional interface, generating modifications, and emailing recipes to each other. Teachers can use their classroom FCs to teach coding, chemistry, biology, or data science. Of course, every experiment they run will also serve as an additional core of processing for the Open Phenome Project as students upload their recipes.

Alternatively, some corporations and restaurateurs simply wish to integrate FCs into existing or imminent operations. Finally, on a governmental or global level, FCs can also be used to boost the economy by creating jobs in a clean, high-tech field. Ultimately, we are aiming to build an interdisciplinary, agile platform able to empower the collective of its users by capitalizing on their particular skills, lending them instant access to the aggregate wisdom of experts in other fields and meeting their unique needs.

Acknowledgment The author is thankful to Lana Popovic for her editorial support.

References

Abbate J (1999) Inventing the internet. MIT Press, Cambridge, MA, 275 p.p

Baker D (1996) A brief excursion into three agricultural revolutions. Keuhnast Lecture, University of Minnesota. http://climate.umn.edu/doc/journal/kuehnast_lecture/l4-txt.htm

Harper C, Siller M (2015) OpenAG: a globally distributed network of food computing. Pervasive Comput 14(4):24–27

World Urbanization Prospects: The 2014 Revision. United Nations, Dept. of Economic and Social Affairs, Population Division, ST/ESA/SER.A/366, 2015; http://esa.un.org/unpd/wup/Highlights/WUP2014-Highlights.pdf

World Population Prospects: The 2015 Revision, Key Findings and Advance Tables. United Nations, Dept. of Economic and Social Affairs, Population Division, working paper no. ESA/P/ WP.241, 2015; http://esa.un.org/unpd/wpp/Publications/Files/Key_Findings_WPP_2015.pdf

Part II
Plant Growth and Development as Affected by Light

Chapter 4
Some Aspects of the Light Environment

Toyoki Kozai and Geng Zhang

Abstract Roles of light as energy and signal sources for growing plants are discussed mainly with respect to photosynthesis and photomorphogenesis. Components of the light environment above and within a plant canopy, such as spectral distribution, photosynthetic photon flux density (PPFD), lighting cycle, and lighting direction, are shown. Characteristics of LED arrays as light source and characteristics of light environment in plant factory with artificial lighting (PFAL) are discussed. Concepts and significances of supplemental upward lighting in the PFAL and supplemental lighting in greenhouse are described. Advantages of supplemental upward lighting in the PFAL to prevent senescence of lower leaves and increase marketable fresh weight of leaf lettuce plants are shown. Strategy of environmental control under artificial lighting is discussed.

Keywords Energy source • Light environment • Photomorphogenesis • Photosynthesis • Signal source

4.1 Light as an Energy and Signal Source

Light is a source of energy for photosynthesis as well as a source of signals or information activating photomorphogenesis and other physiological processes such as secondary metabolite production in plants (Fig. 4.1). Light is also a source of information for human eyes. The wavelengths of photosynthetically active light (400–700 nm) and physiologically active light (300–800 nm) overlap, so that photosynthesis and photomorphogenesis are often concurrent. Both are photochemical reactions, and the amount of light received by plants is measured in units of moles (1 mol = 6.03×10^{23} photons), not in joules (energy). In addition, light affects human color and shape perception, along with health (Chap. 31). (More detailed explanations

T. Kozai (✉)
Japan Plant Factory Association (NPO), Kashiwa-no-ha, Kashiwa, Chiba 277-0882, Japan
e-mail: kozai@faculty.chiba-u.jp

G. Zhang
Graduate School of Horticulture, Chiba University, Matsudo-shi 271-8510, Japan

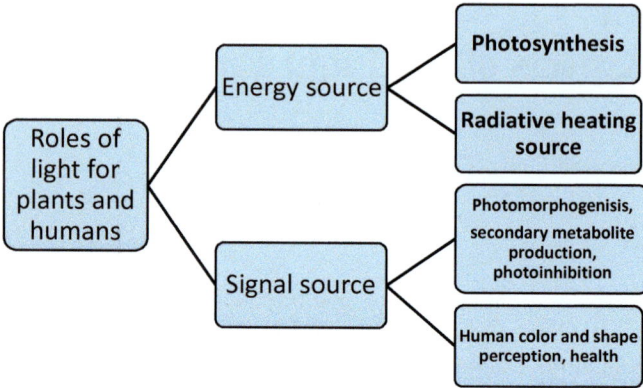

Fig. 4.1 Roles of light for plants and humans

of light and its units are given in Chaps. 26 and 28). Active light for human eyes ranges between 380 and 780 nm with a peak wave length of 555 nm.

Light is absorbed, reflected, or transmitted when it reaches plants. Some photosynthetic photons are captured and converted into chemical energy as carbohydrates in plants, while the remaining light absorbed is ultimately converted into heat (Kozai 2013; Chap. 9). In this sense, photosynthesis is a process of energy conversion from light or photosynthetically active radiation (PAR) to chemical and heat energy as measured in joules. The conversion from joules to moles is determined once the wavelength is specified (Chap. 26). This chapter describes the light environment in plant factories with artificial lighting (PFALs) and greenhouses with supplemental lighting, with details not discussed in other chapters.

4.2 Components of the Light Environment

Components of the light environment above and within a plant canopy (or community) are shown in Fig. 4.2. The photosynthetic photon flux density (PPFD) or PAR flux density on the horizontal plane is the most important variable in the light environment. The average PPFD or PAR flux density decreases exponentially with increasing plant canopy depth (referred to as the cumulative leaf area index [LAI] (Chaps. 10 and 11).

4.2.1 Spectral Distribution of Light Within the Plant Canopy

The light quality or spectral distribution of light over the plant canopy differs significantly from that within it. Green light generally penetrates the plant canopy more easily than blue and red light, because its transmissivity, reflectivity, and

4 Some Aspects of the Light Environment

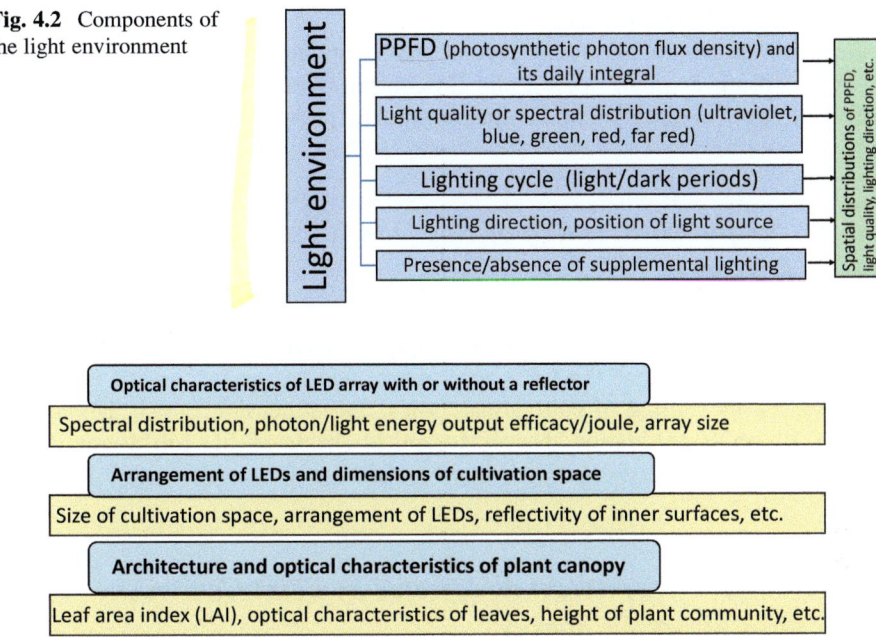

Fig. 4.2 Components of the light environment

Fig. 4.3 Factors affecting the light environment in the cultivation space of a PFAL

absorptivity by green leaves are about 30 %, 20 %, and 50 %, respectively (Kozai et al. 2015), while those of blue and red light are about 0 %, 100 % and 100 %, respectively. Blue and red light is mostly absorbed by the uppermost layer of the plant canopy. A detailed description of the optical and physiological properties of leaves and canopies is given in Part III (Chaps. 8, 9, 10, 11, 12, and 13).

4.3 Light Environment in PFALs

In PFALs, the light environment over plant canopies in cultivation spaces is significantly affected not only by the optical characteristics of light-emitting diode (LED) arrays and their layout but also by the optical characteristics of each cultivation space and plant canopy (Fig. 4.3).

4.3.1 Characteristics of LED Arrays as Light Source

LEDs as a light source are characterized by the: (1) dimension and structure of the LED array; (2) spectral distribution (ultraviolet, blue, green, red, and far red) of photons emitted from the LED array; (3) directional (spatial) distribution of photons emitted from the LED array; (4) ratio of light energy emitted to electric energy consumed (joule/joule) or ratio of photosynthetic photons emitted to electric energy consumed (μmol /joule); and (5) control of the lighting cycle (light/dark periods), spectral distribution, and photosynthetic photon flux (see Part VII).

4.3.2 Spatial Distribution of PPFD in Empty Cultivation Spaces in PFALs

The spatial distribution of PPFD in an empty cultivation space is affected by the optical characteristics of the LED array mentioned above, layout of LED arrays installed on the ceiling of cultivation spaces, and dimension (width, height, and length) of empty cultivation spaces and optical characteristics (diffused/mirrored reflection, reflectivity, etc.) of their inner surfaces.

4.3.3 Light Environment as Affected by Plant Canopies in Cultivation Spaces

The light environment in a cultivation space with a plant canopy differs significantly from that in an empty cultivation space. In a PFAL, the PPFD and light quality above and within a plant canopy change markedly with time as the canopy grows for two main reasons. First, the top layer of a canopy approaches the light source with upward growth; second, immediately after transplanting seedlings onto a white culture panel, most (about 80 %) of downward-directed light is reflected back to the ceiling. As the plant canopy grows, the downward blue and red light is mostly absorbed by plant leaves, while about 50 % only of green light, if any, is absorbed by plant leaves; about 20 % is reflected back to the ceiling and about 30 % is transmitted through leaves (Fig. 4.4). Therefore, in a PFAL lit with red and blue

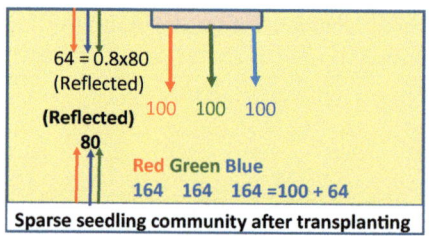
Reflectivity of white culture panel for red, green and blue light: 0.8: 08: 0.8.

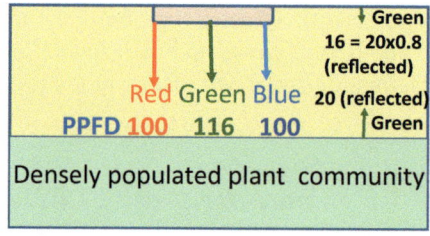
Reflectivity of plant community for red, green, and blue light: 0.0: 015: 0.0.

Fig. 4.4 Scheme showing the effect of reflectivity of the plant canopy/culture panels on the PPFD above the plant canopy in a PFAL. The PPFD immediately before harvesting is 67 % of that immediately after transplanting

LEDs alone, the PPFD above a densely populated plant canopy is decreased by about 40 % (=100–100/(100 + 100 × 0.8 × 0.8), compared with that immediately after seedling transplant onto white culture panels.

4.4 Supplemental Upward Lighting

High planting density in a PFAL tends to accelerate leaf senescence in the lower (or outer) leaves as a result of shading by the upper (or inner) leaves and neighboring plants, which decreases yields and increases labor costs for trimming senescent leaves. Thus, the development of cultivation systems to retard outer leaf senescence is an important research goal to improve the yield and profitability of PFALs.

Zhang et al. (2015) developed an upward-directed LED lighting system to improve the light environment for PFAL cultivation of leafy vegetables (Fig. 4.5). Romaine lettuce (*Lactuca sativa* L.) was grown hydroponically under downward-directed white LEDs at a PPFD of 200 μmol $m^{-2} s^{-1}$, with or without supplemental upward white LED lighting (PPFD facing the culture panel of 40 μmol $m^{-2} s^{-1}$ 4 cm above the culture panel). They showed that the supplemental lighting retarded the senescence of outer leaves and decreased the percent waste (i.e., dead or low-quality senescent leaves), leading to an improvement in the marketable leaf fresh weight (Zhang et al. 2015). After 16 days, romaine lettuce grown with downward lighting plus supplemental upward white LED lighting had a leaf fresh weight and marketable leaf fresh weight 11 % and 18 % greater, respectively, and the percent waste was 6 % less than in lettuce grown without supplemental lighting (Table 4.1). In addition, the average chlorophyll (a+b) content of the first to sixth leaves from the lowest leaf and net photosynthetic rate of the third leaf from the lowest leaf were significantly higher with supplemental white LED upward lighting (Table 4.1).

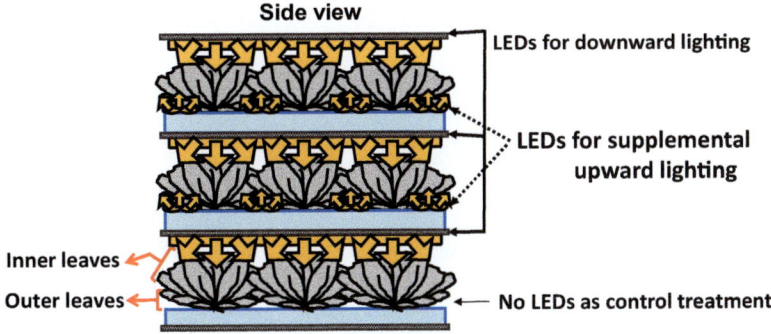

Fig. 4.5 Schematic diagram of supplemental upward LED lighting in the cultivation space of a PFAL (Zhang et al.,2015)

Table 4.1 Growth parameters of romaine lettuce grown under downward-directed white LED lighting with or without supplemental upward-directed white LED lighting

	Supplemental upward lighting (A)	No supplemental lighting (B)	A/B
Fresh leaf weight (g/plant)	170	154	1.11
Marketable fresh leaf weight (g/plant)	158	134	1.18
Percent waste	7.2	12.8	0.56
Average chlorophyll (a+b) content (g m^{-2}) of 1st to 6th leaves from the lowest	231	185	1.28
Net photosynthetic rate (μmol m^{-2} s^{-1}) of 3rd leaf from the lowest	1.08	−0.19	–

4.5 Supplemental Lighting in Greenhouses

4.5.1 Purpose of Supplemental Lighting in Greenhouses

Supplemental lighting in greenhouses is used to promote photosynthesis by adding artificial light as an energy and/or to control photomorphogenesis or other physiological processes such as secondary metabolite production, since the artificial light acts as a signal to the photoreceptors in leaves. When used to promote photosynthesis, the spectral distribution of light emitted by LEDs is not critical physiologically as long as it ranges between 400 and 700 nm. In reality, however, red-rich LEDs are currently used in most cases to save on the initial and operating (electricity) costs for lighting. On the other hand, the initial cost of white LEDs has been decreasing since 2015.

LEDs are often placed within plant communities at an LAI of 2 or 3 and higher in greenhouses for cultivating tomatoes, cucumbers, and roses. The supplemental light is directed to the sides of plants and/or upward (Chap. 19), mainly because lower leaves receive less natural light than upper ones. When photomorphogenesis is controlled by changing the light/dark periods, LEDs emitting specific wavelengths (e.g., red and/or far red) are used (Chap. 6).

4.5.2 Environmental Control for Efficient Supplemental Lighting

Supplemental lighting for the promotion of greenhouse crop photosynthesis is significantly affected by other environmental factors. Figure 4.6 shows the effects of PPFD on greenhouse crop net photosynthetic rate under optimal temperature, vapor pressure deficit, CO_2 concentration, air current speed, water irrigation, and nutrient composition/strength conditions. Since the cost of electricity for supplemental lighting to promote photosynthesis is generally the highest among all costs

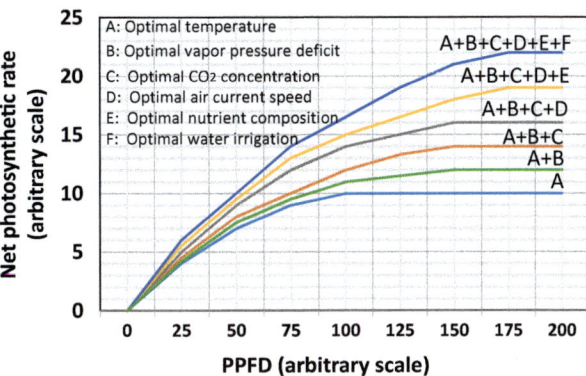

Fig. 4.6 A scheme showing the effects of PPFD on the net photosynthetic rate of plants under optimal temperature, vapor pressure deficit, CO_2 concentration, air current speed, nutrient composition/strength, and water irrigation conditions

for controlling environmental factors, those other factors need to be optimized. Costs for optimizing air current speed and water irrigation systems are generally relatively low, for example.

Supplemental lighting for photosynthesis in greenhouses can affect spatial distributions of temperature, relative humidity or water vapor pressure deficit, air movement, and CO_2 concentration. This is because light absorbed by plants is converted into sensible heat energy, latent heat energy, and/or thermal radiation energy, causing changes in air temperature, water vapor pressure deficit, etc.

References

Kozai T (2013) Resource use efficiency of closed plant production system with artificial light: concept, estimation and application to plant factory. Proc Jpn Acad Ser B 89(10):447–461

Kozai T, Niu G, Takagaki M (eds) (2015) Plant factory: an indoor vertical farming system for efficient quality food production. Academic, London, 423 pp

Zhang G, Shen S, Takagaki M, Kozai T, Yamori W (2015) Supplemental upward lighting from underneath to obtain higher marketable lettuce (*Lactuca sativa*) leaf fresh weight by retarding senescence of outer leaves. Front Plant Sci 16:1–9

Chapter 5
Light Acts as a Signal for Regulation of Growth and Development

Yohei Higuchi and Tamotsu Hisamatsu

Abstract Plants utilise light not only for photosynthesis but also as a signal to regulate optimal growth and development throughout their life cycle. The light quality (spectral composition), amount, direction and duration change depending on the season, latitude and local conditions. Therefore, to adapt to diverse light conditions, plants have evolved unique photoreceptor systems to mediate light responses to a broad range of wavelengths from ultraviolet-B to far-red light. Light signals can regulate changes in structure and form, such as seed germination, de-etiolation, leaf expansion, phototropism, neighbour avoidance, stem elongation, flower initiation and pigment synthesis. Plant hormones and transcriptional factors play an important role in the internal signalling that mediates light-regulated processes of development. Plants rely on their circadian clock to modify their growth and development in anticipation of predictable changes in environmental light and temperature conditions. The light signals perceived by photoreceptors affect the circadian clock and directly activate the induction of the light responses.

Keywords Circadian rhythm • De-etiolation • Gating effect • Photoreceptor • Phototropism • Seed germination • Shade avoidance response

5.1 Photoreceptors and Their Function

As sessile and photosynthetic organisms, plants monitor ambient light conditions and regulate numerous developmental switches to adapt to continually changing environments. A recent molecular genetic approach in the model plant *Arabidopsis* revealed that multiple photoreceptors act as light sensors for perceiving different light wavelengths (Fig. 5.1). These include phytochromes (phy), cryptochromes

Y. Higuchi
Graduate School of Agricultural and Life Sciences, The University of Tokyo,
Yayoi, Bunkyo-ku, Tokyo 113-8657, Japan

T. Hisamatsu (✉)
Institute of Vegetable and Floriculture Science, National Agriculture and Food Research Organization (NARO), Fujimoto, Tsukuba, Ibaraki 305-0852, Japan
e-mail: tamotsu@affrc.go.jp

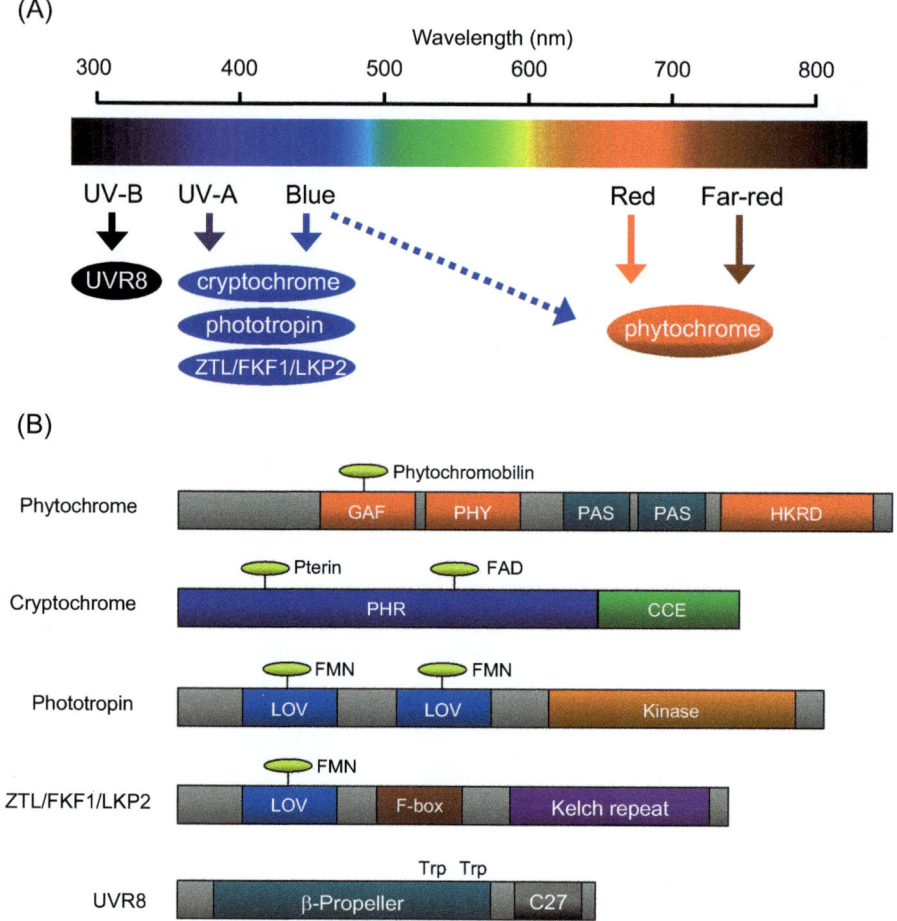

Fig. 5.1 Photoreceptors in higher plants. (**a**) Photoreceptors perceiving different parts of the light spectrum. (**b**) Structure of photoreceptor proteins. Domain structure and binding chromophores are shown. *GAF* cGMP-stimulated phosphodiesterase; *Anabaena* adenylate cyclases and *Escherichia coli* FhlA; *PAS* Per (period circadian protein), Arn (Ah receptor nuclear translocator protein) and Sim (single-minded protein); *HKRD* histidine kinase-related domain; *PHR* photolyase-homologous region; *CCE* cry C-terminal extension; *LOV* light, oxygen and voltage; *FAD* flavin adenine dinucleotide; *FMN* flavin mononucleotide

(cry), phototropins (phot), ZEITLUPE (ZTL)/FLAVIN-BINDING, KELCH REPEAT, F-BOX 1 (FKF1)/LOV KELCH PROTEIN2 (LKP2) family proteins and UV RESISTANCE LOCUS 8 (UVR8). Here, we summarise the physiological responses, light perception mechanisms and light signal transduction mechanisms regulated by multiple photoreceptors.

5.1.1 Phytochromes (Phy)

In the 1950s, the effect of exposure to different spectra of light on seed germination in lettuce was examined; the red/far-red (R:FR) reversibility was analysed, where R light exposure induced lettuce seed germination, but subsequent FR light exposure reversed the effect of R light (Borthwick et al. 1952). The photoreversible proteinous pigment, phytochrome, was extracted and analysed (Butler et al. 1959). Phytochromes are soluble proteins that bind phytochromobilin as chromophores and convert between two different photoreversible forms in vivo: R light (650–670 nm)-absorbing (Pr) and FR light (705–740 nm)-absorbing (Pfr) forms. In general, Pr absorbs R light and is converted to its biologically active form, Pfr, which induces various physiological responses; Pfr absorbs FR light and is converted to an inactive form of Pr (Fig. 5.2). This R:FR reversible response, which is a typical phytochrome reaction, is classified as a low-fluence response (LFR) that occurs in seed germination and night break (NB) responses with short light pulses. In addition to LFR, phytochrome responses include high-irradiance responses (HIR) and very-low-fluence responses (VLFR) (Casal et al. 1998). HIR include de-etiolation (inhibition of hypocotyl elongation and promotion of cotyledon expansion) and anthocyanin accumulation responses. VLFR is triggered by extremely low light intensities of all wavelengths, which is observed in light-induced seed germination. In contrast to LFR, HIR and VLFR do not show R:FR reversibility. It should be noted that in addition to R and FR regions of the spectrum, phy can also weakly absorb blue light (Figs. 5.1a and 5.3).

Since the absorption spectrum between Pr and Pfr partially overlaps (Fig. 5.3a), the phytochrome photoequilibrium (Pfr/P; where P = Pr + Pfr) under saturated light intensity changes depending on the light quality (Sager et al. 1988). A high R:FR ratio establishes a high Pfr/P, whereas low R:FR ratio creates a low Pfr/P (Fig. 5.3b). Under a vegetation canopy, shading by other plants creates a low Pfr/P that induces stem/petiole elongation and early flowering, which is a shade-avoidance response (Casal 2013). It is possible to estimate the effectiveness of light treatment by calculating Pfr/P under different light wavelengths; however, screening by other pigments such as chlorophylls, flavonoids and carotenoids could occur. For example, flowering inhibition by NB in chrysanthemum is mediated by

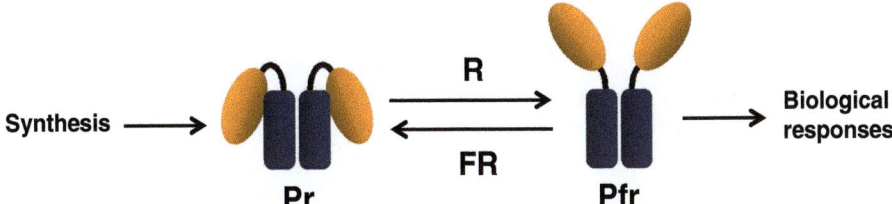

Fig. 5.2 Photoconversion of phytochrome. Phytochromes are synthesised in the Pr form. Pr absorbs R light and is converted to its biologically active form, Pfr, which induces various physiological responses. Pfr absorbs FR light and is converted it its inactive Pr form

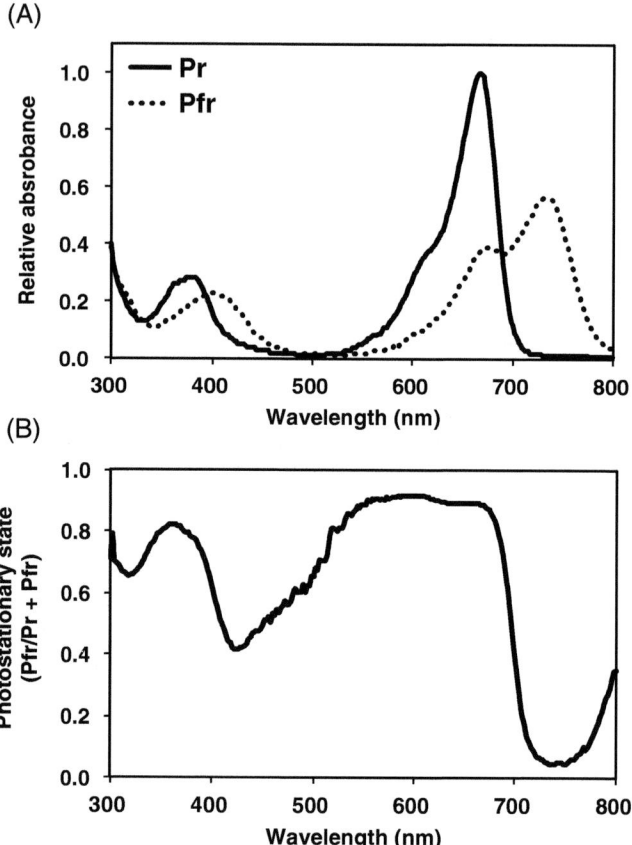

Fig. 5.3 Relative absorption spectra of purified rye phytochrome in its Pr and Pfr forms (**a**) and calculated photostationary state (**b**) (Data derived from Sager et al. 1988)

phytochromes (Higuchi et al. 2013), but spectral sensitivity to NB shifted towards shorter wavelengths (around 600 nm) than expected (Sumitomo et al. 2012). Similarly, distortion in the spectral sensitivity to flowering has been reported in *Lemna* (Ohtani and Kumagai 1980). In these cases, effects of yellow to red light have been distorted by the screening effect of chlorophyll in green leaves.

In the 1980s, molecular genetic studies identified five phytochrome genes (*PHYA, B, C, D* and *E*) in *Arabidopsis* (Clack et al. 1994). Phytochromes are classified into two groups (type I and II) according to their protein stability in light. PhyA is classified into type I, which accumulates under dark conditions and rapidly degrades when exposed to light. phyA mediates VLFR with a broad range of light and HIR with FR light. PhyB to phyE are light stable type II phytochromes that accumulate relatively constantly under light or dark conditions (Sharrock and Clack 2002). PhyB, phyD and phyE mediate R:FR reversible LFR and/or R:FR ratio response, which is the shade-avoidance response (Li et al. 2011). phyC mediates R light-induced HIR in seedling de-etiolation (Franklin et al. 2003; Monte et al. 2003). In the photoperiodic control of flowering, phyA mediates the blue- and FR-light promotion of flowering, whereas phyB mediates R-light inhibition of flowering (Goto et al. 1991; Johnson et al. 1994; Mockler et al. 2003; Franklin and Quail 2010).

5.1.2 Cryptochromes (Crys)

Cryptochromes are FAD- and pterin-containing chromoproteins that share considerable homology with DNA photolyases but lack photolyase activity (Ahmad and Cashmore 1993). Cryptochromes have two domains, the N-terminal phytolyase homology region (PHR) domain that binds chromophore and C-terminal cryptochrome C-terminus (CCT) domain, which is necessary for signal transduction (Fig. 5.1). In *Arabidopsis*, two cryptochromes (cry1 and cry2) are present as blue (B)/UV-A photoreceptors that are involved in many biological responses such as inhibition of hypocotyl elongation, entrainment of the circadian clock, stomata opening, pigment biosynthesis and photoperiodic flowering (Yu et al. 2010). Cry1 protein is light stable, but cry2 is light labile. The cry2 protein is accumulated in the dark and degraded upon exposure to B light, showing diurnal rhythms (Lin et al. 1998; Mockler et al. 2003). Cry2 promotes flowering by stabilising the CONSTANS (CO) protein, a positive regulator of florigen in the long day (LD) evening (Valverde et al. 2004).

5.1.3 Phototropins (Phots)

Phototropin was first identified as a photoreceptor mediating a blue-light-induced phototropic response in *Arabidopsis*, but its structure is different from that of cryptochromes (Huala et al. 1997). Phototropins harbour two LOV domains (LOV1 and LOV2) at their N-terminus that bind FMN as chromophores and the Ser/Thr kinase domain at their C-terminus (Fig. 5.1). *Arabidopsis* contains two phototropins (phot1 and phot2) (Huala et al. 1997; Kagawa et al. 2001) that regulate numerous blue/UV-A-induced responses, maximising photosynthetic activity such as phototropism, chloroplast relocation, leaf flattening and stomatal opening (Briggs and Christie 2002; Christie 2007). Phot1 acts over a wide range of light intensities, whereas phot2 functions predominantly at high light intensities (Christie et al. 2015).

5.1.4 Zeitlupe Family Proteins (ZTL/FKF1/LKP2)

ZEITLUPE (ZTL), FLAVIN-BINDING KELCH REPEAT F-BOX 1 (FKF1) and LOV KELCH REPEAT PROTEIN2 (LKP2) are a recently identified novel class of blue-light receptor proteins. They regulate circadian rhythms and photoperiodic flowering. The ZTL family proteins possess one LOV domain at the N-terminus, which binds FMN as chromophores. They possess an F-box and six kelch repeats at their C-terminus (Fig. 5.1) and regulate target protein degradation via the ubiquitin-proteasome system (Ito et al. 2012). ZTL forms a complex with a clock-related protein GIGANTEA (GI) in a blue-light-dependent manner and regulate protein degradation of TIMING OF CAB EXPRESSION I (TOC1), a core clock component factor, to generate circadian rhythms (Más et al. 2003; Kim et al. 2007). FKF1 also interacts with GI in a blue-light-dependent manner and controls protein degradation of CYCLING DOF FACTOR 1 (CDF1), a negative regulator of flowering, to promote flowering (Sawa et al. 2007).

5.1.5 UV-B Receptor (UVR8)

More recently, the ultraviolet-B radiation (UV-B: 280–315 nm) photoreceptor UV RESISTANCE LOCUS 8 (UVR8) has been identified in *Arabidopsis* (Rizzini et al. 2011). UVR8 is a 440 amino acid protein that has beta-propeller structures (Fig. 5.1). UVR8 exists as an inactive homodimer under UV-B-deficient light conditions, but rapidly monomerises upon UV-B irradiation, which triggers numerous UV-B responses. Unlike other photoreceptors, UVR8 does not bind subsidiary chromophores, but specific intrinsic tryptophans function as chromophores for UV-B perception (Rizzini et al. 2011). The monomerised active UVR8 forms a complex with CONSTITUTIVE PHOTOMORPHOGENIC 1 (COP1) and acts as a positive regulator of UV-B signalling via the regulation of downstream gene expressions. UVR8 mediates a number of UV-B-induced responses such as photomorphogenesis, pigment biosynthesis and pathogen resistance induction (Tilbrook et al. 2013).

5.2 Light-Dependent Seed Germination

Seed germination is the first step for seed plants to initiate a new life cycle. In several species, such as lettuce, tobacco and *Arabidopsis*, light is an important regulator of seed germination. Borthwick et al. (1952) demonstrated that red light (R; 600–700 nm) induced germination in a lettuce seed variety (cv. Grand Rapids) that was pre-soaked in water in darkness and showed that far-red light (FR; 700–800 nm) could reverse this induction (Fig. 5.4a). This photomorphogenic response ultimately led to the identification and purification of the R:FR-absorbing photoreceptors and phytochromes. Phytochromes are the major class of photoreceptors responsible for germination. Classical physiological studies have suggested the involvement of plant hormones, gibberellin (GA) and abscisic acid (ABA) as critical regulators of seed germination. R light that induces germination can be substituted by application of GA to lettuce seeds, whereas an application of ABA inhibits germination. Thus, endogenous levels of GA and ABA might be controlled by light. In fact, the endogenous levels of GA and ABA are oppositely modulated in a light-dependent manner (Seo et al. 2009). Phytochromes regulate GA biosynthesis in germinating lettuce and *Arabidopsis* seeds. Using the *phyA* and *phyB* mutants of *Arabidopsis*, phyB is the dominant phytochrome involved in the light-induced germination with the typical R:FR photoreversible response. In *Arabidopsis*, upregulation of two biosynthetic genes, *GA3ox1* and *GA3ox2*, catalyses the conversion of precursor GAs to their bioactive forms, and expression in the hypocotyl of embryos following exposure to R-light, in a phyB-mediated process, is associated with germination, whereas a GA catabolic gene, *GA2ox2*, is repressed (Fig. 5.4b). After a long period of imbibition in the dark, phyA plays a role in the irreversible response to extremely low levels of light over a wide range of wavelengths (Shinomura et al. 1996). PHYTOCHROME-INTERACTING FACTOR 3-LIKE 5 (PIL5) regulates seed germination negatively through GA (Oh et al. 2006). Expression analysis revealed that PIL5 represses the expression of *GA3ox1* and *GA3ox2* and activates the expression of *GA2ox* in both PHYA and PHYB dependent. ABA accumulates in seeds to promote dormancy and prevent premature

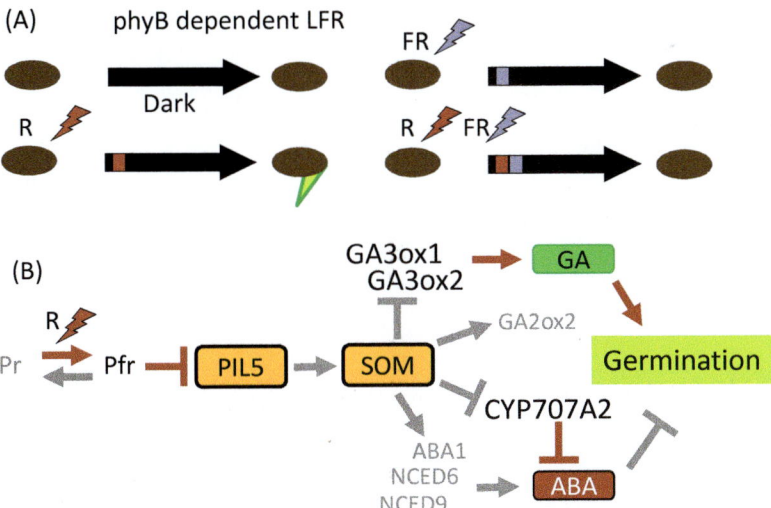

Fig. 5.4 Phytochrome B-mediated seed germination in *Arabidopsis*. (**a**) PhyB-mediated seed germination shows typical R:FR photoreversible response (LFR). (**b**) PhyB Pfr form regulates GA and ABA levels through negative regulators of seed germination, PIL5 and SOM. Active regulations after R-light exposures are shown in *Red arrows* or T-bars

germination. Consistent with the change in ABA levels, the ABA biosynthetic genes, *ABA-DEFICIENT 1* (*ABA1*), *NINE-CIS-EPOXYCAROTENOID DEOXYGENASE 6* (*NCED6*) and *NCED9*, are repressed, whereas an ABA catabolic gene, *CYP707A2*, which encodes an ABA 8′-hydroxylase, is induced by R-light exposure. PIL5 regulates both GA and ABA metabolic genes partly through SOMNUS (SOM) (Kim et al. 2008).

5.3 De-etiolation

In the dark, a seedling adopts skotomorphogenesis where it develops a long hypocotyl, an apical hook and closed cotyledons. Skotomorphogenesis is achieved by the active repression of genes that would lead to photomorphogenic development. When exposed to light, the seedling starts the de-etiolation process and switches rapidly to photomorphogenesis and inhibition of the hypocotyl elongation, promoting cotyledon development, opening the apical hook and cotyledons and initiating chlorophyll and anthocyanin biosynthesis, and true leaves begin to develop. Several classes of photoreceptors, phytochromes, cryptochromes and phototropins are involved in the photomorphogenic development (Fig. 5.5). The COP1-SUPPRESSOR OF PHYA-105 (SPA) complexes function as an E3 ubiquitin ligase and repress photomorphogenesis. COP1-SPA complexes control the light-regulated abundance of LONG HYPOCOTYL5 (HY5). HY5 is a basic leucine zipper transcription factor that binds to the promoters of numerous light-regulated genes to regulate photomorphogenic development. In the dark, COP1-SPA

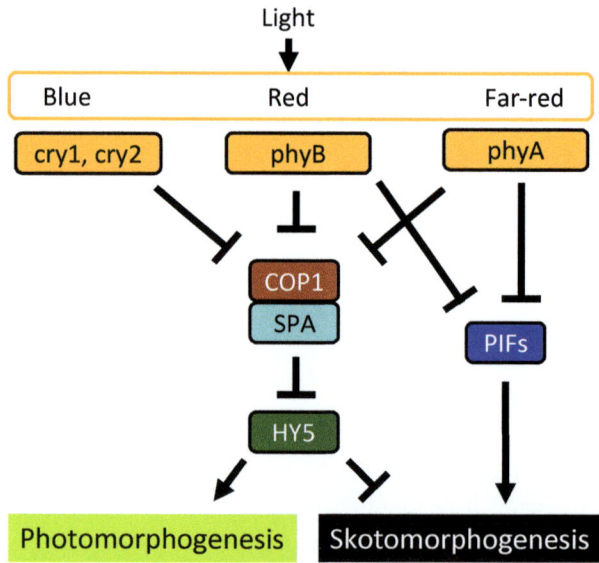

Fig. 5.5 A simplified model of the light regulation. Light sensed by photoreceptors acts to suppress two main light signalling pathways, through COP1/SPA-HY5 and PIFs

complexes target HY5 for ubiquitination, inducing proteasomal degradation. Light inactivates COP1-SPA complexes, so that HY5 accumulates. In addition to the COP1, a group of PHYTOCHROME-INTERACTING FACTORs (PIFs), basic helix-loop-helix transcription factors act to promote skotomorphogenesis (Leivar and Quail 2011). In the dark, PIFs are active and regulate gene expression to promote skotomorphogenesis. In the light, a nuclear-localised phytochrome (light-activated Pfr form) binds to PIFs and results in phosphorylation and subsequent degradation. The degradation of PIFs induces photomorphogenic development.

5.4 Phototropism

The growth of a plant towards any stimulus is called tropism, and the growth of a plant towards a light stimulus is called phototropism (Liscum et al. 2014). Phototropism is an important adaptive response where plants optimise their exposure to light. Blue wavelengths of light are more effective at orienting plant growth, which involves blue-light perception and asymmetric distribution of a plant hormone, auxin. The shoot bends towards the light because of differences in cell elongation on the two sides of the shoot. The side of the shoot that is in the shade has more auxin, and its cells therefore elongate more than those on the lighted side. The phototropic movement of plants is initiated by a blue-light receptor, phototropin (Whippo and Hangarter 2006). In *Arabidopsis*, two phototropins, phot1 and phot2, exhibit overlapping functions. The central importance of polar auxin transport and auxin signalling in phototropism has been demonstrated. Auxin efflux carrier PIN-FORMED (PIN) proteins possibly have central roles in regulating asymmetrical auxin translocation during tropic responses, including gravitropism and phototropism, in plants. When several of the PIN and kinase components were missing, plant growth was completely unresponsive to the light signals that trigger phototropism. A recent detailed analysis of various PIN gene mutants found that the contributions of PIN1, PIN3 and PIN7 to phototropic hypocotyl bending become relatively obvious when dark-grown seedlings are exposed to a short blue-light pulse (pulse-induced first positive phototropism). Strikingly, these phototropism defects become much weaker when seedlings are exposed to long-term blue-light treatments (second positive phototropism) (Haga and Sakai 2012). Blue light perceived by phototropin contributes the polar relocation of PIN proteins. Auxin streams and asymmetric growth are also regulated by AGCVIII kinases that are able to phosphorylate PINs (Barbosa et al. 2014). D6 PROTEIN KINASE (D6PK) subfamily of AGCVIII kinase-dependent PIN regulation promotes auxin transport in the hypocotyl that is a prerequisite for phot1-dependent hypocotyl bending (Willige et al. 2013). Since phytochrome-cryptochrome double mutants show a reduced phototropic response, the phototropins are not the only photoreceptors involved in phototropism (Whippo and Hangarter 2006).

5.5 Shade-Avoidance Response

The shade-avoidance response (SAR), which allows plants to escape from neighbour competitors, is an adopted response to the optimal acquisition of light energy to drive photosynthesis. The SAR is characterised by increased extension growth of the hypocotyl, stem and petiole, a more erect leaf position, increased apical dominance and early flowering (Franklin 2008). Photosynthetic pigments, such as chlorophylls and carotenoids, in the leaves absorb light over the 400–700 nm spectrum. The FR region (700–800 nm) of the spectrum is poorly absorbed by the photosynthetic pigments; consequently, sunlight reflected from or transmitted through leaves is enriched with FR light. Changes in light quality, low R:FR, are sensed by multiple light-stable phytochromes (phyB, phyD, phyE). A particular R:FR ratio is reflected in the Pfr:Pr ratio of phytochromes, thus determining the relative activity of phytochromes. Of these, phyB is the dominant phytochrome involved in the SAR (Fig. 5.6). The unique properties of phyA, a light-labile phytochrome, as an effective FR sensor in the HIR are important in natural light environments by 'antagonising' shade avoidance (Martinez-Garcia et al. 2014).

The end-of-day FR light (EOD-FR) treatment consists of a pulse of FR given at subjective dusk (Kasperbauer 1971). EOD-FR treatments result in a minimal pool of active Pfr during the dark period (Fankhauser and Casal 2004). In plants grown under day/night cycles, EOD-FR treatment mimics growth in low R:FR light conditions. The treatment is a useful method for experimentally inducing the SAR. Involvement of plant hormones, such as GA, auxin, ABA, cytokinin, ethylene and brassinosteroid, in light-regulated developments has been suggested. The perception of shade (EOD-FR) by the leaf blade induces petiole elongation in *Arabidopsis* (Kozuka et al. 2010), where it is speculated that newly synthesised auxin in the leaf blade accumulates in the petiole to induce responses. The cotyledons perceived shade (low R:FR) signal and generate auxin to regulate hypocotyl

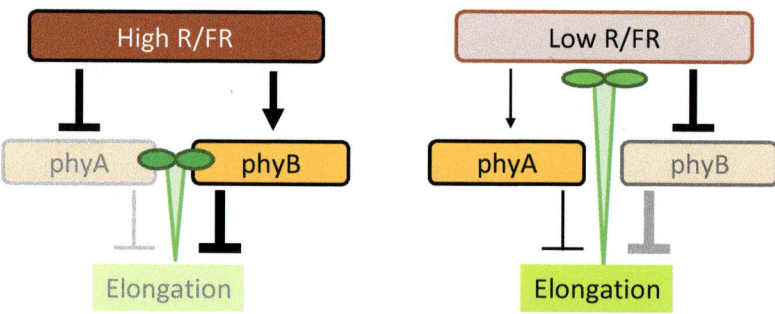

Fig. 5.6 A simplified model of the shade-avoidance response (SAR) on hypocotyl elongation. PhyB is the dominant phytochrome involved in the SAR. Light-activated phyB suppresses elongation. SAR induced by phyB deactivation is gradually antagonised by phyA, an HIR-FR response

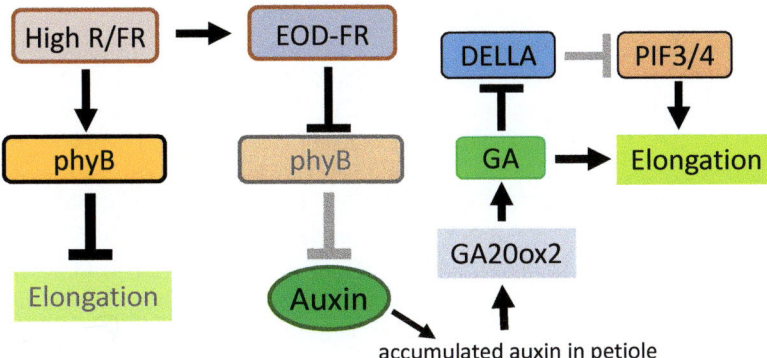

Fig. 5.7 A simplified model of the EOD-FR-induced auxin/GA cooperative petiole elongation in *Arabidopsis*. PhyB is the dominant phytochrome involved in EOD-FR. The synthesised auxin in the leaf blade by EOD-FR acts on the petiole. The *GA20ox2* transcript is upregulated in the petiole by the accumulated auxin. Increased GA promotes petiole elongation

elongation in *Brassica rapa* (Procko et al. 2014). These studies demonstrate the importance of inter tissue/organ communication in the SAR.

Transcriptomic analyses revealed that the expression of many genes related to plant hormones is regulated in response to EOD-FR (Kozuka et al. 2010). In *Arabidopsis*, upregulation of *GA20ox2* expression in the petiole following exposure to EOD-FR or low R:FR light, in a phyB-mediated process, is associated with enhanced petiole elongation (Hisamatsu et al. 2005) and floral induction (Hisamatsu and King 2008). Auxins act on the GA biosynthesis by specifically regulating the expression of two genes, *GA20ox1* and *GA20ox2* (Frigerio et al. 2006). Together, the auxin/GA cooperative response induced by EOD-FR that synthesised auxin in the leaf blade accumulated in the petiole and induced the expression of *GA20ox2*, and the increased GA induced petiole elongation (Fig. 5.7). PIFs, key transcription factors for photomorphogenesis, are growth-promoting factors (Leivar and Quail 2011), which integrate GA signalling and phytochrome-mediated SAR. The DELLA family proteins that repress GA-regulated growth are mediators of GA signalling. GA promotes DELLA degradation by binding to a GA receptor, GID1 (Hedden and Thomas 2012). DELLAs interact with PIF3/4, which blocks transcriptional activity and inhibits PIF function (Sun 2011). When plants are subjected to low R:FR from high R:FR conditions, GA levels may increase. The increased GA would promote DELLA degradation, relieving DELLA-mediated inhibition of PIF function and enhancing extension growth (Lorrain et al. 2008).

5.6 Circadian Rhythms and Biological Responses

Many circadian rhythm-related genes have been identified during flowering time mutant analyses. The time measurement mechanism is composed of an input pathway, central oscillator and output pathway. The light signals perceived by photoreceptors such as phytochromes and cryptochromes entrain the clock. The central oscillator is composed of an interlocked transcriptional and post-transcriptional feedback loop that generates an approximately 24-h free running rhythm. The clock components include CIRCADIAN CLOCK-ASSOCIATED 1 (CCA1), LATE ELONGATED HYPOCOTYL (LHY), TIMING OF CAB EXPRESSION 1 [TOC1, also called PSEUDO-RESPONSE REGULATOR 1 (PRR1)], PRR5/7/9, REVEILLE 8 (RVE8), EARLY FLOWERING 3 (ELF3), ELF4, LUXARRHYTHMO [LUX, also known as PHYTOCLOCK 1 (PCL1)], GIGANTEA (GI) and ZEITLUPE (ZTL) (Hsu and Harmer 2014; Greenham and McClung 2015) (Fig. 5.8). The loss or gain of function of these factors results in aberrant rhythmic phenotypes. The output pathway includes regulatory factors that directly involve biological processes, such as extension growth and flowering.

PIFs are important as growth-promoting factors. PIF4 and PIF5 rhythmically express over a diurnal cycle with maximal mRNA abundance either at dawn or early morning, regardless of photoperiod conditions. The circadian clock directly controls this transcript oscillation pattern. The evening clock components ELF3, ELF4 and LUX functionally repress PIF transcription at dusk. During the day, PIFs are degraded at the post-translational level by interacting with light-activated phytochromes. Therefore, PIFs accumulate during the night, when plant growth rate is highest (Nozue et al. 2007; Nusinow et al. 2011).

Correct entrainment of circadian clocks is essential because the phase of circadian rhythms relative to the day/night cycle affects flowering time. The *toc1-1* mutant, a short-period clock mutant, shows an early flowering phenotype under SD conditions. The early flowering phenotype of the *toc1-1* mutant can be explained by their short-period phenotype and phase advance in *CONSTANS* (*CO*) mRNA expression (Yanovsky and Kay 2002). *CO* rhythmically expresses over a diurnal

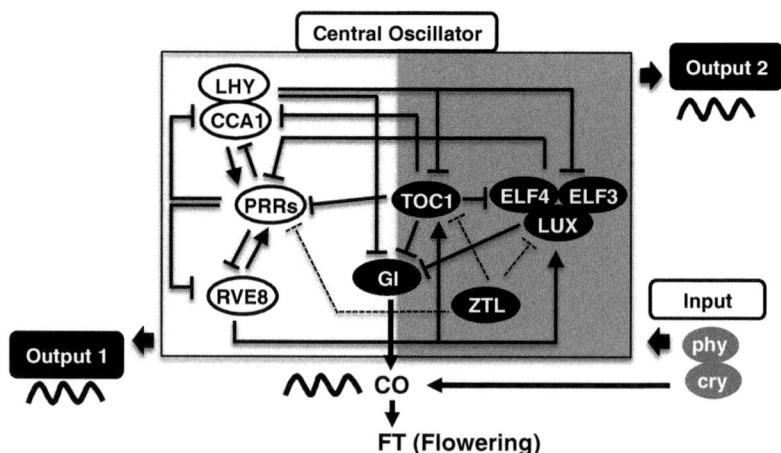

Fig. 5.8 A model for the circadian clock in *Arabidopsis*. The circadian clock is composed of multiple interlocked transcriptional and post-transcriptional feedback loops (The model was adapted from Greenham and McClung (2015))

cycle under controlling circadian clock, only when *CO* is expressed at high levels in the light phase, and the CO protein is stabilised by interacting with light signals and induces *FLOWERING LOCUS T* (*FT*) transcription, a gene encoding florigen. In wild-type plants, the rhythm of *CO* expression creates a light-sensitive phase starting from about 8 h after dawn, so there is little *CO* expression during the day in 8-h SD conditions. In *toc1-1* mutants grown in 8-h SD conditions, the phase of *CO* expression was significantly advanced, leading to a coincidence between relatively high levels of *CO* transcription during the day at dusk. The interaction between transcriptional regulation of CO by endogenous circadian clocks and external light signals at a particular phase is essential for day-length recognition for flowering.

5.7 The Gating Effects of Circadian Clocks

The phenomenon where any effects of external stimuli are limited to certain phase of the circadian clock is referred to as 'the gating effect'. For example, the inhibitory effect of NB on the floral initiation of SDP is restricted to certain time of night, and if plants were kept in continuous darkness, the photosensitive phase would appear every 24 h (Hamner and Takimoto 1964, Fig. 5.9). In *Arabidopsis*, *CHLOROPHYLL a/b-BINDING PROTEIN* (*CAB*) expression showed diurnal rhythms peaking during the day. Although a light pulse given during the day strongly induced *CAB* expression, the light given at night did not (Millar and Kay 1996), indicating that the induction of *CAB* expression by light is gated by a circadian clock. ELF3 mediates circadian gating of light responses. In the *elf3* mutant, which shows photoperiod-insensitive early flowering, the circadian

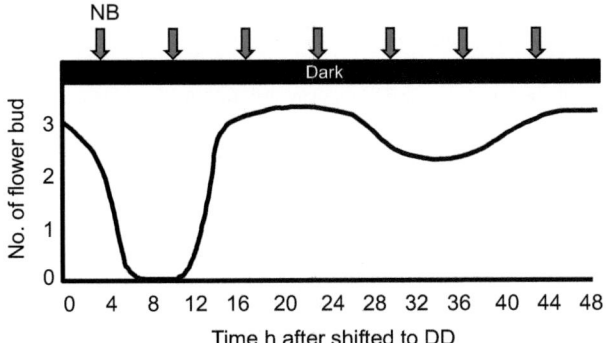

Fig. 5.9 The gated NB response in SDP. The effect of a short light pulse given at different times of night on the flowering response of *Pharbitis*. The inhibitory effect of NB on flowering occurred periodically at 8–10 h and 32–36 h after dusk under continuous darkness (Redrawn from Hamner and Takimoto (1964))

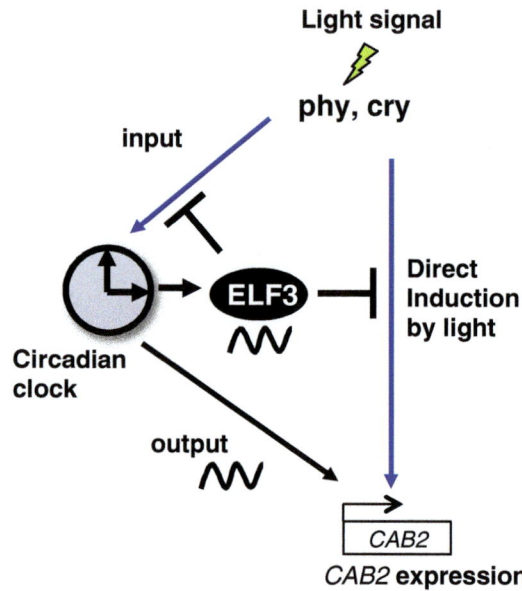

Fig. 5.10 The gating mechanism of *CAB2* expression by ELF3 and the circadian clock. *CAB2* expression shows robust diurnal rhythms peaking during the day. ELF3 acts to suppress the light input to the clock and inhibits the acute induction of *CAB2* by light at subjective night

rhythmicity of *CAB* expression disappeared under continuous light (LL), but not continuous dark (DD) (Hicks et al. 1996;). ELF3 suppressed light input to the circadian clock at a particular time of day (McWatters et al. 2000, Fig. 5.10).

References

Ahmad M, Cashmore A (1993) *HY4* gene of *A. thaliana* encodes a protein with characteristics of a blue-light photoreceptor. Nature 366:162–166
Barbosa I, Zourelidu M, Willige B et al (2014) D6 PROTEIN KINASE activates auxin transport-dependent growth and PIN-FORMED phosphorylation at the plasma membrane. Dev Cell 29:674–685
Borthwick H, Hendricks S, Parker M et al (1952) A reversible photoreaction controlling seed germination. Proc Natl Acad Sci U S A 38:662–666
Briggs W, Christie J (2002) Phototropins 1 and 2: versatile plant blue-light receptor. Trends Plant Sci 7:204–210
Butler W, Norris K, Siegelman H et al (1959) Detection, assay, and preliminary purification of the pigment controlling photoresponsive development of plants. Proc Natl Acad Sci U S A 45:1703–1708
Casal J (2013) Photoreceptor signaling networks in plant responses to shade. Ann Rev Plant Biol 64:403–427
Casal J, Sanchez R, Botto J (1998) Modes of action of phytochromes. J Exp Bot 49:127–138
Christie J (2007) Phototropin blue-light receptors. Annu Rev Plant Biol 58:21–45
Christie J, Blackwood L, Petersen J et al (2015) Plant flavoprotein photoreceptors. Plant Cell Physiol 56:401–413

Clack T, Mathews S, Sharrock R (1994) The phytochrome apoprotein family in *Arabidopsis* is encoded by five genes: the sequences and expression of *PHYD* and *PHYE*. Plant Mol Biol 25:413–427

Fankhauser C, Casal J (2004) Phenotypic characterization of a photomorphogenic mutant. Plant J 39:747–760

Franklin K (2008) Shade avoidance. New Phytol 179:930–944

Franklin K, Quail P (2010) Phytochrome functions in Arabidopsis development. J Exp Bot 61:11–24

Franklin K, Davis S, Stoddart W et al (2003) Mutant analyses define multiple roles for phytochrome C in Arabidopsis photomorphogenesis. Plant Cell 15:1981–1989

Frigerio M, Alabadi D, Pérez-Gómez J et al (2006) Transcriptional regulation of gibberellin metabolism genes by auxin signaling in Arabidopsis. Plant Physiol 142:553–563

Goto N, Kumagai T, Koornneef M (1991) Flowering responses to light-breaks in photomorphogenic mutants of Arabidopsis thaliana, a long day plant. Physiol Plant 83:209–215

Greenham K, McClung C (2015) Integrating circadian dynamics with physiological processes in plants. Nat Rev Genet 16:598–610

Haga K, Sakai T (2012) PIN auxin efflux carriers are necessary for pulse-induced but not continuous light-induced phototropism in Arabidopsis. Plant Physiol 160:763–776

Hamner K, Takimoto A (1964) Circadian rhythms and plant photoperiodism. Am Nat (Am Nat) 98:295–322

Hedden P, Thomas S (2012) Gibberellin biosynthesis and its regulation. Biochem J 444:11–25

Hicks K, Millar A, Carré I et al (1996) Conditional circadian dysfunction of the *Arabidopsis early-flowering 3* mutant. Science 274:790–792

Higuchi Y, Narumi T, Oda A et al (2013) The gated induction system of a systemic floral inhibitor, antiflorigen, determines obligate short-day flowering in chrysanthemums. Proc Natl Acad Sci U S A 110:17137–17142

Hisamatsu T, King R (2008) The nature of floral signals in Arabidopsis. II. Roles for FLOWERING LOCUS T (FT) and gibberellin. J Exp Bot 59:3821–3829

Hisamatsu T, King R, Helliwell C et al (2005) The involvement of gibberellin 20-oxidase genes in phytochrome-regulated petiole elongation of Arabidopsis. Plant Physiol 138:1106–1116

Hsu P, Harmer S (2014) Wheels within wheels: the plant circadian system. Trends Plant Sci 19:240–249

Huala E, Oeller P, Liscum E et al (1997) Arabidopsis NPH1: a protein kinase with a putative redox-sensing domain. Science 278:2120–2123

Ito S, Song Y, Imaizumi T (2012) LOV domain-containing F-box proteins: light-dependent protein degradation modules in Arabidopsis. Mol Plant 5:573–582

Johnson E, Bradley M, Harberd N et al (1994) Photoresponses of light-grown phyA mutants of Arabidopsis (phytochrome A is required for the perception of daylength extensions). Plant Physiol 105:141–149

Kagawa T, Sakai T, Suetsugu N et al (2001) Arabidopsis NPL1: a phototropin homolog controlling the chloroplast high-light avoidance response. Science 291:2138–2141

Kasperbauer M (1971) Spectral distribution of light in a tobacco canopy and effects of end-of-day light quality on growth and development. Plant Physiol 47:775–558

Kim W, Fujiwara S, Suh S et al (2007) ZEITLUPE is a circadian photoreceptor stabilized by GIGANTEA in blue light. Nature 449:356–360

Kim D, Yamaguchi S, Lim S et al (2008) SOMNUS, a CCCH-type zinc finger protein in Arabidopsis, negatively regulates light-dependent seed germination downstream of PIL5. Plant Cell 20:1260–1277

Kozuka T, Kobayashi J, Horiguchi G et al (2010) Involvement of auxin and brassinosteroid in the regulation of petiole elongation under the shade. Plant Physol 153:1608–1618

Leivar P, Quail P (2011) PIFs: pivotal components in a cellular signaling hub. Trends Plant Sci 16:19–28

Li J, Li G, Wang H et al (2011) Phytochrome signaling mechanisms. Arabidopsis Book 9:e0148

Lin C, Yang H, Guo H et al (1998) Enhancement of blue-light sensitivity of *Arabidopsis* seedlings by a blue light receptor cryptochrome 2. Proc Natl Acad Sci U S A 95:2686–2690

Liscum E, Askinosie S, Leuchtman D et al (2014) Phototropism: growing towards an understanding of plant movement. Plant Cell 26:38–55

Lorrain S, Allen T, Duek P et al (2008) Phytochrome-mediated inhibition of shade avoidance involved degradation of growth-promoting bHLH transcription factors. Plant J 53:312–323

Martinez-Garcia J, Gallemi M, Molina-Contreras M et al (2014) The shade avoidance syndrome in Arabidopsis: the antagonistic role of phytochrome A and B differentiates vegetation proximity and canopy shade. PLoS One 9(10):e109275

Más P, Kim W, Somers D et al (2003) Targeted degradation of TOC1 by ZTL modulates circadian function in *Arabidopsis thaliana*. Nature 426:567–570

McWatters H, Bastow R, Hall A et al (2000) The ELF3 zeitnehmer regulates light signaling to the circadian clock. Nature 408:716–720

Millar A, Kay S (1996) Integration of circadian and phototransduction pathways in the network controlling CAB gene transcription in Arabidopsis. Proc Natl Acad Sci U S A 93:15491–15496

Mockler T, Yang H, Yu X et al (2003) Regulation of photoperiodic flowering by *Arabidopsis* photoreceptors. Proc Natl Acad Sci U S A 100:2140–2145

Monte E, Alonso J, Ecker J et al (2003) Isolation and characterization of *phyC* mutants in Arabidopsis reveals complex cross-talk between phytochrome signalling pathways. Plant Cell 15:1962–1980

Nozue K, Covington M, Duek P et al (2007) Rhythmic growth explained by coincidence between internal and external cues. Nature 448:358–361

Nusinow D, Helfer A, Hamilton E et al (2011) The ELF4–ELF3–LUX complex links the circadian clock to diurnal control of hypocotyl growth. Nature 475:398–402

Oh E, Yamaguchi S, Kamiya Y et al (2006) Light activates the degradation of PIL5 protein to promote seed germination through gibberellin in Arabidopsis. Plant J 47:124–139

Ohtani T, Kumagai T (1980) Spectral sensitivity of the flowering response in green and etiolated *Lemna paucicostata* T-101. Plant Cell Physiol 21:1335–1338

Procko C, Crenshaw C, Ljung K et al (2014) Cotyledon-generated auxin is required for shade-induced hypocotyl growth in Brassica rapa. Plant Physiol 165:1285–1301

Rizzini L, Favory J, Cloix C et al (2011) Perception of UV-B by the Arabidopsis UVR8 protein. Science 332:103–106

Sager J, Smith W, Edwards J et al (1988) Photosynthetic efficiency and phytochrome photoequilibria determination using spectral data. Trans Am Soc Agric Eng 31:1882–1889

Sawa M, Nusinow D, Kay S et al (2007) FKF1 and GIGANTEA complex formation is required for day-length measurement in *Arabidopsis*. Science 318:261–265

Seo M, Nambara E, Choi G et al (2009) Interaction of light and hormone signals in germinating seeds. Plant Mol Biol 69:463–472

Sharrock R, Clack T (2002) Patterns of expression and normalized levels of the five Arabidopsis phytochromes. Plant Physiol 130:442–456

Shinomura T, Nagatani A, Hanzawa H et al (1996) Action spectra for phytochrome A- and B-specific photoinduction of seed germination in *Arabidopsis thaliana*. Proc Natl Acad Sci U S A 93:8129–8133

Sumitomo K, Higuchi Y, Aoki K et al (2012) Spectral sensitivity of flowering and *FT*-like gene expression in response to a night break treatment in the chrysanthemum cultivar 'Reagan'. J HortScience Biotech 87:461–469

Sun T-p (2011) The molecular mechanism and evolution of the GA–GID1–DELLA signaling module in plants. Curr Biol 21:338–345

Tilbrook K, Arongaus A, Binkert M et al (2013) The UVR8 UV-B photoreceptor: perception, signaling and response. Arabidopsis Book 11:e0164

Valverde F, Mouradov A, Soppe W et al (2004) Photoreceptor regulation of CONSTANS protein in photoperiodic flowering. Science 303:965–966

Whippo C, Hangarter R (2006) Phototropism: bending towards enlightenment. Plant Cell 18:1110–1119

Willige B, Ahiers S, Zourelidu M et al (2013) D6PK AGCVIII kinases are required for auxin transport and phototropic hypocotyl bending in Arabidopsis. Plant Cell 25:1674–1688

Yanovsky M, Kay S (2002) Molecular basis of seasonal time measurement in *Arabidopsis*. Nature 419:308–312

Yu X, Liu H, Klejnot J et al (2010) The cryptochrome blue light receptors. Arabidopsis Book 8: e0135

Chapter 6
Factors Affecting Flowering Seasonality

Yohei Higuchi and Tamotsu Hisamatsu

Abstract Environmental regulation of flowering seasonality and set seed is critical for this survival as it allows seeds to develop in the most favourable conditions. Recent genetic and molecular approaches provide a basis for understanding how plants use seasonal changes in natural daylight duration and temperature to achieve reproducible timing of flowering. Recent studies have led to the identification of members of the FLOWERING LOCUS T (FT) in *Arabidopsis*, and its orthologs in several plant species act as florigen. In addition to the floral inducer florigen, the systemic floral inhibitor anti-florigen, anti-florigenic FT/TFL1 family protein (AFT), has been identified from a wild chrysanthemum and plays a predominant role in the obligate photoperiodic response. In *Arabidopsis*, the molecular basis for vernalization process has been revealed. The key factor in the vernalization pathway is a repressor of flowering, FLOWERING LOCUS C (FLC). In temperate cereals that require vernalization to flower, three genes possibly participate in a regulatory loop to control the timing of flowering, namely, VRN1, VRN2, and VRN3. VRN2 is a key factor for flowering repression in winter varieties.

Keywords Anti-florigen • Florigen • Flowering • Seasonality • Photoperiodism • Vernalization

6.1 Photoperiodic Flowering

Many plants sense gradual change in day length (photoperiod), which is the most reliable seasonal cue at high latitude, to determine when to produce flowers. This phenomenon, photoperiodism, anticipates environmental conditions and enables plants to maximise their survival and reproduction at a suitable time of the year.

Y. Higuchi
Graduate School of Agricultural and Life Sciences, The University of Tokyo, Yayoi, Bunkyo-ku, Tokyo 113-8657, Japan

T. Hisamatsu (✉)
Institute of Vegetable and Floriculture Science, National Agriculture and Food Research Organization (NARO), Fujimoto, Tsukuba, Ibaraki 305-0852, Japan
e-mail: tamotsu@affrc.go.jp

Photoperiodism was first described in detail by Garner and Allard (1920). They demonstrated that several plant species flower in response to changes in day length, not light intensity or temperature. Flowering plants are classified into three categories based on their photoperiodic responses: short-day plants (SDP), in which flowering occurs when the night length is longer than a critical minimum; long-day plants (LDP), in which flowering occurs when the day becomes longer; and photoperiod-insensitive day-neutral plants (DNP) (Thomas and Vince-Prue 1997). Photoperiodism has had a considerable impact on the agricultural and horticultural industries, because it has enabled plant breeders and growers to control flowering time by manipulating day length. Photoperiod is perceived in the leaves, where the flower-inducing signal is synthesised under appropriate photoperiods and transmitted to the shoot apex to initiate the flower bud. In 1936, based on the grafting experiment in light-sensitive plants, Chailakhyan proposed the concept of the flowering hormone "florigen" (flower former), which is produced in the leaves and transmitted to the shoot apex to induce flowering (Chailakhyan 1936).

6.2 Florigen and Anti-florigen

Despite numerous attempts to extract florigen, the molecular structure has remained unknown for almost 70 years (Zeevaart 2008). Recently, molecular-genetic studies have demonstrated that FLOWERING LOCUS T (FT) in *Arabidopsis* and its orthologs in several plant species act as florigen (Lifschitz et al. 2006; Corbesier et al. 2007; Tamaki et al. 2007; Lin et al. 2007). *FT* was first identified as a gene responsible for the late flowering mutant of *Arabidopsis*, a facultative LDP (Kardailsky et al. 1999; Kobayashi et al. 1999). FT is expressed in the vasculature tissues of leaves under a flower-promoting LD photoperiod and forms a complex with a bZIP-type transcription factor, FD, to induce floral-meristem identity genes, such as *APETALA1* (*AP1*) and *FRUITFULL* (*FUL*) (Abe et al. 2005; Wigge et al. 2005) (Fig. 6.1). Interestingly, although *FT* is induced in leaves, *FD* expression was limited to the shoot apical meristem. In 2007, the long-distance transmission of the FT protein and its rice homolog Heading date 3a (Hd3a) from the leaves to shoot apex was determined (Corbesier et al. 2007; Tamaki et al. 2007), and the FT/Hd3a protein was demonstrated to be a molecular entity of the systemic floral stimulus florigen. *FT/Hd3a* encodes a small globular protein similar to phosphatidylethanolamine-binding protein (PEBP). Hd3a forms a complex with 14-3-3 adaptor proteins and OsFD1, which is called florigen activation complex (FAC), and then induces *OsMADS15*, a rice *AP1* homolog, transcription to induce flowering (Taoka et al. 2011). The FT/Hd3a family protein acts as universal flowering hormone "florigen" in many plant species (Wickland and Hanzawa 2015; Matsoukas 2015).

In addition to the floral inducer florigen, the systemic floral inhibitor produced in non-induced leaves can inhibit flowering. The concept of a floral inhibitor (anti-florigen) was proposed almost as early as that of florigen (Lang and Melchers

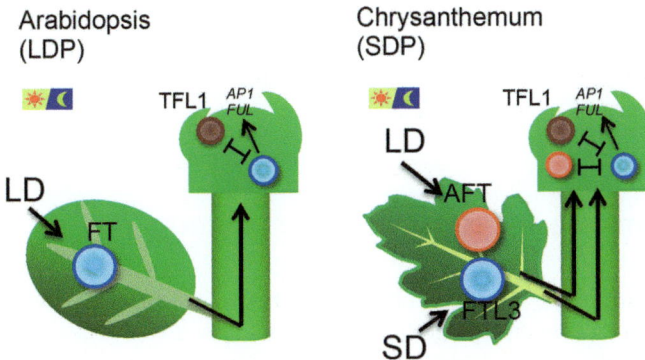

Fig. 6.1 Flowering time regulation by florigen and anti-florigen. In *Arabidopsis*, FT is synthesised in leaves under a flower-inducing LD photoperiod that moves to the shoot apex to induce floral-meristem identity genes. TFL1 is expressed in the shoot apex and suppresses flowering by antagonising FT function. In chrysanthemums, FTL3 is synthesised in the leaves under flower-inducing SD and TFL1 antagonises FTL3 function at the shoot apex. In addition, a systemic floral inhibitor (*AFT*) is synthesised in leaves under flower-inhibiting LD or NB, which antagonise florigenic activity of FTL3 at the shoot apex

1943); the appropriate photoperiod leads to the removal of the floral inhibitor, and consequently, flowering occurs. Many physiological observations such as defoliation, grafting, and localised photoperiodic treatment in *Hyoscyamus*, strawberry, *Lolium*, chrysanthemum, tobacco, and *Pharbitis* (Lang and Melchers 1943; Guttridge 1959; Evans 1960; Tanaka 1967; Lang et al. 1977; Ogawa and King 1990) suggested the existence of the systemic floral inhibitor, anti-florigen. A grafting experiment of tobacco plants with different photoperiodic responses strongly supported this hypothesis. The day-neutral (DN) tobacco normally flowered, even under an SD photoperiod, but when the LD flowering *Nicotiana sylvestris* was grafted, flowering of DN tobacco was delayed under SD (Lang et al. 1977). This result clearly indicates that a floral inhibitor produced in the leaves of *N. sylvestris* under SD systemically inhibited the flowering of DN tobacco plants. In the 1990s, molecular-genetic studies in *Arabidopsis* revealed that TERMINAL FLOWER 1 (TFL1), a member of the PEBP family protein, suppressed flowering (Bradley et al. 1997). *TFL1* is exclusively expressed in the shoot apex and maintains indeterminate inflorescence (Ratcliffe et al. 1999; Conti and Bradley 2007; Jaeger et al. 2013). Since TFL1 also formed a complex with FD, an interacting partner of FT, TFL1 suppressed flowering by antagonising florigenic activity of the FT-FD complex (Abe et al. 2005) (Fig. 6.1). Although TFL1 acts as a floral inhibitor, it possibly acts over short (cell-to-cell) distances within the meristematic zone (Conti and Bradley 2007). A recent study in a wild diploid chrysanthemum (*C. seticuspe*) identified a floral inhibitor, anti-florigenic FT/TFL1 family protein (CsAFT), which moves long distances (Higuchi et al. 2013). *CsAFT* was induced in leaves under flower-non-inductive LD or night-break (NB) photoperiods and was suppressed at very low levels under inductive SD. CsAFT proteins move long distances from leaves to the shoot apex and inhibit flowering by directly antagonising the flower-inductive activity of the FT-FD complex of *C. seticuspe* (CsFTL3-CsFDL1) (Fig. 6.1). These findings suggest that the balance between floral inducers (florigens) and inhibitors (anti-florigens) determines flowering time variations of many plant species.

6.3 Flowering and Seasonal Time Measurement

A major factor in the seasonal control of flowering time is the photoperiod. Plants flower in response to changing photoperiod, but how do they measure the length of day and night? Classical studies in plants and animals have provided several physiological models for explaining photoperiodic responses (Nelson et al. 2010). The hourglass model proposes that day length is measured simply through some regulatory product, the accumulation of which is light dependent (Lees 1973). In this model, photoperiodic responses are triggered when the amount of this product exceeds a certain threshold level (e.g. the amount of phy-Pfr has been a candidate for the sand of an hourglass). However, the external coincidence model proposes that day length is measured through a circadian oscillator that controls the expression of some regulatory product, the activity of which is modulated by light. Photoperiodic responses are triggered only when external (light) signals coincide with the light-sensitive phase of circadian rhythms (Pittendrigh and Minis 1964). The internal coincidence model proposes that light signals set two different circadian rhythms, and a response is triggered only when these rhythms are synchronised under certain photoperiods (Pittendrigh 1972). Recent studies in *Arabidopsis* and rice demonstrated that both external and internal coincidence models are consistent with the physiological and molecular-genetic evidence in plants (Greenham and McClung 2015; Song et al. 2015) (Fig. 6.2).

The external coincidence model was proposed by Pittendrigh in the 1960s based on Bunning's hypothesis (Pittendrigh and Minis 1964; Büning 1936). In this model, light has two different roles. One is to reset the circadian clock, which is a set phase of the clock. The other is to simply transfer the presence or absence of external light that triggers photoperiodic reactions. The circadian clock entrained by the light/dark cycle sets a photosensitive phase to occur at particular time of the day. The photoperiodic reaction is triggered only when the photosensitive phase "coincided" with the external light signal. In a facultative LDP *Arabidopsis*, circadian rhythms entrained by a light signal at dawn set the expression of *CONSTANS* (*CO*), a positive regulator of *FT*. Under an LD photoperiod, the peak phase of CO in the evening interacted with light signals mediated by phyA or cry2 and stabilised the CO protein. However, the peak phase of CO expression occurred after dusk under SD, and the CO protein was degraded during darkness. Therefore, the CO protein was stabilised and activated only under LD evening conditions, when it induced *FT* expression to promote flowering (Yanovsky and Kay 2002; Valverde et al. 2004). Light signals mediated by phytochromes and cryptochromes act in the input to the circadian clock and as an external light signal that directly activates the induction of

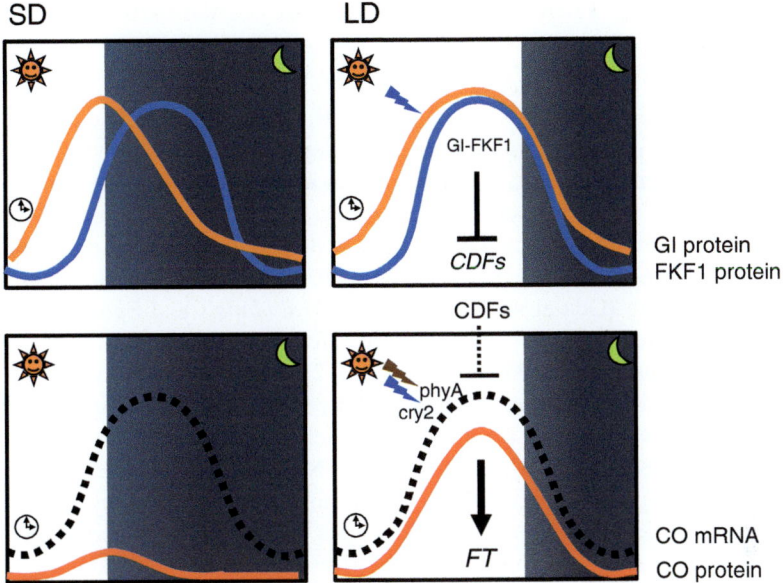

Fig. 6.2 Photoperiodic regulation of *FT* by internal and external coincidence. Accumulation of GI and FKF1 proteins synchronises in the late afternoon under LD and forms a complex (GI-FKF1) in a *blue* light-dependent manner. The activated GI-FKF1 then degrades CDF proteins, negative regulators of *CO* transcription. *CO* mRNA expression is regulated by a circadian clock to peak in the evening. Under LD, light signals mediated phyA, and cry2 stabilised CO proteins in the evening that activated *FT* transcription

florigen genes (Fig. 7.2). The evening-phased expression of *CO* under LD photoperiods is regulated by the coordinated action of a blue light receptor FLAVIN-BINDING, KELCH REPEAT, F-BOX 1 (FKF1), and a clock component GIGANTEA (GI). Transcription of both FKF1 and GI is regulated by the circadian clock. Under SD conditions, the peak phase of FKF1 and GI protein accumulation occurred at different times of the day. However, under LD, the peak phase of these proteins coincided in late afternoon and formed a complex (FKF1-GI) in a blue light-dependent manner. The FKF1-GI complex then degraded CYCLING DOF FACTORs (CDFs), negative regulators of *CO* transcription (Imaizumi et al. 2005; Sawa et al. 2007; Fornara et al. 2009). The LD-specific interaction of two different rhythms (FKF1 and GI) fitted well with the internal coincidence model (Fig. 6.2). In a facultative SDP rice, expression of *Heading date 1* (*Hd1*), an ortholog of *CO*, was regulated by a circadian clock peaking in the evening. The coincidence of Hd1 with the phytochrome signal suppressed flowering by negatively regulating the expression of the *FT* ortholog, *Heading date 3a* (*Hd3a*) (Izawa et al. 2002; Hayama et al. 2003).

6.4 Flowering Time Regulation in Chrysanthemum

Chrysanthemum (*Chrysanthemum morifolium*) is one of the most important horticultural crops worldwide. It is an obligate SDP, which flowers when the nights are longer than a critical minimum, and flowering is strictly inhibited under LD or NB. Chrysanthemum growers use blackouts or artificial lighting (day-length extension or NB) to meet the demand for marketable flowers throughout the year. Recently, molecular-genetic studies in a wild diploid *C. seticuspe* identified *FLOWERING LOCUS T-like 3* (*CsFTL3*), which encodes a systemic floral inducer in chrysanthemum (Oda et al. 2012). Unlike *Arabidopsis* and *Pharbitis*, chrysanthemums require repeated cycles of SD for successful anthesis (Corbesier et al. 2007; Hayama et al. 2007; Oda et al. 2012). Consistent with this requirement, *CsFTL3* expression is gradually increased by repeating the SD cycles (Nakano et al. 2013). However, *CsAFT* expression, which encodes systemic anti-florigen, is induced in leaves under non-inductive LD or NB, and it rapidly decreased after a shift to SD (Higuchi et al. 2013). Under non-inductive photoperiods, CsAFT produced in the leaves moved to the shoot apex and inhibited flowering by directly antagonising the florigen complex activity (CsFTL3-CsFDL1). In addition, a TFL1 homolog (CsTFL1) is constitutively expressed in shoot tips regardless of the photoperiodic conditions and shows strong floral inhibitor activity (Higuchi et al. 2013). In chrysanthemums, strict maintenance of a vegetative state under non-inductive photoperiod is achieved by a dual regulatory system: one is AFT, a systemic floral inhibitor produced in non-inductive leaves, and another is TFL1, a local inhibitor constitutively expressed at the shoot apex (Higuchi and Hisamatsu 2015; Fig. 6.3).

Light quality during NB and daytime affects chrysanthemum flowering. NB with red light effectively inhibits flowering, which is partially reversed by subsequent exposure to FR light, suggesting the involvement of light-stable-type phys in this response (Cathey and Borthwick 1957; Sumitomo et al. 2012). Interestingly, NB with blue and FR light effectively inhibit chrysanthemum flowering when grown under a daily photoperiod with monochromatic blue light, but not white (blue + red) light (Higuchi et al. 2012). This suggested that light quality during the daily photoperiod affects the sensitivity to NB at midnight, and at least two distinct phy-mediated regulation systems might exist. The knock-down of *CsPHYB* by RNAi resulted in some insensitivity to NB with red light and developed capitulum (Fig. 6.4). In *CsPHYB*-RNAi plants, *CsFTL3* was up-regulated, whereas *CsAFT* was down-regulated under NB. These results indicated that CsPHYB acts as primary photoreceptor mediating NB response and inhibits flowering by repressing *CsFTL3* and inducing *CsAFT* (Higuchi et al. 2013). Interestingly, *CsAFT* expression was strongly induced by red light given at 8–10 h after dusk under both SD and LD. Thus, induction of *CsAFT* by phy signalling is gated by the clock system, and the gate for maximal induction of *CsAFT* opens at a constant time after dusk, regardless of the entrained photoperiod (Fig. 6.4b). Moreover, if long nights (14 h) were given, flowering was successfully induced, even under non-24-h light/dark

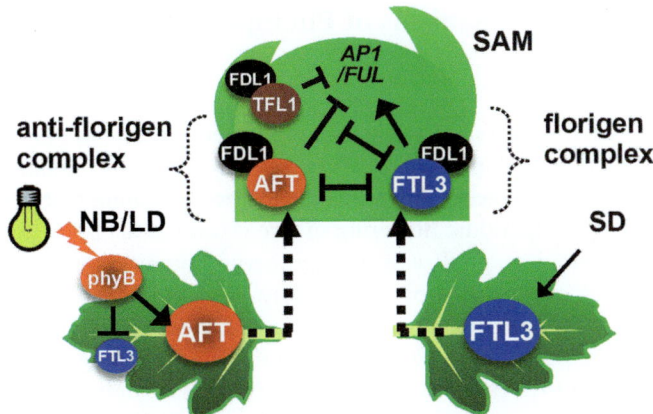

Fig. 6.3 Photoperiodic regulation of flowering in chrysanthemums. Under flower-inductive SD, FTL3 is produced in leaves to systemically induce flowering. Under non-inductive NB or LD, AFT is synthesised in leaves to systemically inhibit flowering. NB with *red* light was perceived by phyB that induces *AFT* but suppressed *FTL3* expression. *TFL1* is constantly expressed in shoot tips regardless of the photoperiodic conditions. Both AFT and TFL1 suppressed flowering by directly competing with FTL3 for binding to FDL1

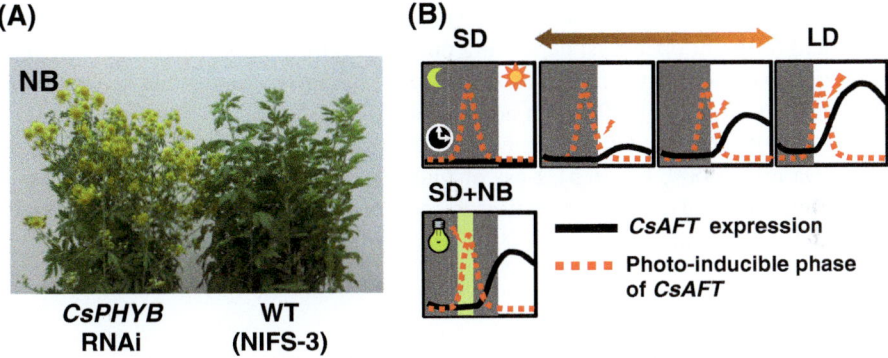

Fig. 6.4 PHYB-mediated and gated induction of *AFT*. (**a**) Flowering response of WT and *PHYB*-RNAi *Chrysanthemum seticuspe* plants under NB with red light. *PHYB*-RNAi plants are almost insensitive to NB. (**b**) Model for the induction of *AFT* in response to natural day-length extension and artificial lighting. The gate for *AFT* induction opens at a constant time after dusk regardless of the day length. As the night becomes shorter, the photo-inducible phase of *AFT* interacts with *red* light in the morning and inhibits flowering. Under NB, midnight illumination coincides with the photo-inducible phase of *AFT*

cycles. Therefore, as in the case of *Pharbitis*, day-length recognition of chrysanthemums relies on the absolute duration of darkness rather than on the photoperiodic response rhythm set by the dawn signal. Chrysanthemums measure the length of night by a timekeeping component, which is initiated from the dusk signal.

6.5 Molecular Mechanisms of Photoperiodic Flowering in Rice

Rice (*Oryza sativa*) is a facultative short-day plant that accelerates flowering under SD. Loss of function of all phytochromes (*se5* or *phyABC* triple mutant) resulted in a photoperiod-insensitive early flowering phenotype, indicating that phytochromes are required for photoperiodic flowering in rice (Izawa et al. 2000, 2002; Takano et al. 2009). In addition, phyB acts as a primary photoreceptor mediating light-induced inhibition of flowering by NB (Ishikawa et al. 2005). Compared to phys, little is known about the significance of blue light receptors such as crys and ZTL/FKF1 on the flowering time regulation in SDPs, including rice. A circadian clock output *GI-CO-FT* pathway in *Arabidopsis* is also conserved in rice (*OsGI-Hd1-Hd3a*), but the regulation of *FT* (*Hd3a*) by *CO* (*Hd1*) is reversed (Hayama et al. 2003). Rice contains an alternative and unique pathway that functions independently of *Hd1*. *Early heading date 1* (*Ehd1*) encoding a B-type response regulator promotes flowering by up-regulating *Hd3a* expression independently of *Hd1* (Doi et al. 2004). *Grain number, plant height, and heading date 7* (*Ghd7*), a CCT domain protein is induced under LD and suppressed flowering by down-regulating *Ehd1* expression (Xue et al. 2008). Interestingly, induction of both *Ehd1* and *Ghd7* by light was gated by a circadian clock. The gate for *Ehd1* induction by blue light always opened around dawn, but the gate for *Ghd7* induction with red light had different openings, depending on day length. Acute induction of *Hd3a* in response to critical day length was achieved by the interaction of these two gating mechanisms (Itoh et al. 2010). In addition to *Hd3a*, rice has another florigen gene *RICE FLOWERING LOCUS T1* (*RFT1*) that functions under LD photoperiods (Komiya et al. 2009; Fig. 6.5). Loss of function of *RFT1* results in extremely late flowering under LD, which is similar to the flowering response of absolute SDPs (Ogiso-Tanaka et al. 2013).

6.6 Flowering Time Regulation in Other Plant Species

Tomato (*Solanum lycopersicum*) is one of the most important horticultural crops worldwide and is a DNP that flowers independently of photoperiod. *SINGLE-FLOWER TRUSS* (*SFT*), a tomato homolog of *FT*, is expressed in expanded mature leaves and systemically promotes flowering (Lifschitz et al. 2006; Shalit et al. 2009). In contrast, *SELF PRUNING* (*SP*, homolog of *TFL1*) is expressed in young leaves and the shoot apex and suppresses flowering (Shalit et al. 2009). The balance between SFT and SP regulates flowering time and determinate or indeterminate shoot architecture. Weak alleles of *SFT* and mutations in *SUPPRESSOR OF SP* (*SSP*, *FD* homolog) weakened the activity of the florigen activation complex (FAC), resulting in partially determinate architecture that provided maximum yields (Park et al. 2014).

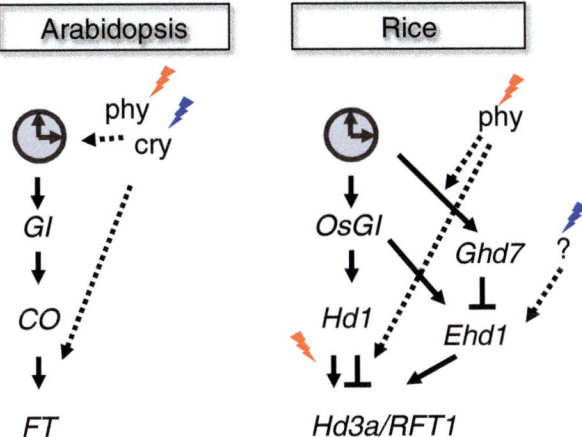

Fig. 6.5 Comparison of photoperiodic flowering pathways in *Arabidopsis* and rice. The circadian clock output *GI-CO-FT* pathway is conserved in *Arabidopsis* and rice, but the regulation of *FT* (*Hd3a*) by *CO* (*Hd1*) is reversed. Rice has an alternative pathway (*Ghd7-Ehd1*) that functions independently of *Hd1*. Rice has two florigen genes, *Hd3a* and *RFT1*. Hd3a induces flowering under inductive SD photoperiod, and RFT1 functions under non-inductive LD photoperiod

Strawberry (*Fragaria* x *ananassa*) is a perennial plant, and flowering is induced by low temperature and SD photoperiod (Heide et al. 2013). Recent studies in rose and woodland strawberry (*F. vesca*) revealed that a mutation in the *TFL1* ortholog is the principle cause of the continuous flowering phenotype in these species (Iwata et al. 2012; Koskela et al. 2012). *FvTFL1* expression is induced in shoot tips under LD but suppressed under SD. In the continuous flowering cultivar, loss of function of a strong floral repressor *FvTFL1* resulted in the derepression of flowering under LD (Koskela et al. 2012). Unlike *Arabidopsis*, *F. vesca* homologs of *FT* (*FvFT1*) and *SOC1* (*FvSOC1*) acted as floral repressors in SD flowering cultivars, because they were up-regulated under LD to activate expression of *FvTFL1* (Mouhu et al. 2013; Rantanen et al. 2014). Moreover, *FvTFL1* was regulated by a temperature-dependent pathway, independently of the regulation of *FvFT1-FvSOC1* by photoperiod (Rantanen et al. 2015) As in *F. vesca*, *F.* x *ananassa* floral inhibition pathways depend on *FaTFL1* regulation by day length via *FaFT1* and temperature, whereas the factors involved in its promotion remain unclear. A putative floral promoter, *FaFT3*, was up-regulated in the shoot tip under SD and/or low growth temperature, in accordance with the promotion of flowering in *F.* x *ananassa* (Nakano et al. 2015).

Pharbitis [*Pharbitis* (*Ipomoea*) *nil*] is an obligate SDP that initiates flowering by a single exposure to a long night (Imamura 1967). To date, *Pharbitis* homologs of *GI*, *CO*, and *FT* (*PnGI*, *PnCO*, *PnFT1/2*) have been identified (Liu et al. 2001; Hayama et al. 2007; Higuchi et al. 2011). PnFT1 has strong florigenic activity, and its expression is induced by a single SD treatment but is completely suppressed under LD or NB (Hayama et al. 2007). However, *PnCO* expression shows diurnal rhythms, but is not affected by NB (Liu et al. 2001). Interestingly, *PnFT1* induction occurs at a constant time (12–16 h) after lights off, regardless of the day length preceding the dark period, and its expression showed circadian rhythms under continuous darkness (Hayama et al. 2007). In addition, the constitutive expression of *PnGI* resulted in a longer period length and reduced the amplitude in *PnFT1* rhythmic expression and suppressed flowering (Higuchi et al. 2011). Therefore, *Pharbitis* measured the absolute duration of night through circadian clocks that were initiated on light-to-dark transition at dusk. The dark-dominant flowering of *Pharbitis* is very similar to that of chrysanthemums.

6.7 Vernalization

Temperature is also a major seasonal cue. Plants have evolved the ability to measure a complete winter season and to remember the prior cold exposure in the spring. Winter annuals and biennials typically require prolonged exposure to the cold of winter to flower rapidly in the spring. This process where flowering is promoted by cold exposure is known as vernalization.

In *Arabidopsis*, the molecular basis for this memory has been revealed. The key factor in the vernalization pathway is a repressor of flowering, FLOWERING LOCUS C (FLC), a MADS-box transcription factor (Hepworth and Dean 2015). FLC directly represses *FT* expression (Fig. 6.6a). FLC expression is high before winter but is repressed during the cold. *FLC* expression is down-regulated within 2 weeks of experiencing cold and is epigenetically silenced by polycomb repressive complex2 (PRC2) complex that contains VERNALIZATION2 (VRN2). VRN2 is constitutively expressed; its activity is boosted through the association with plant-homeodomain zinc-finger (PHD) proteins, a VRN5/VIN3-like family. Epigenetic silencing is dependent on the cold-induced *VERNALIZATION INSENSITIVE 3* (*VIN3*), a PHD gene (Sung and Amasino 2004). The VIN3-PRC2 complex, a protein complex possessing H3K27 methyltransferase activity, established the enrichment of a series of repressive chromatin modifications at the *FLC* locus to

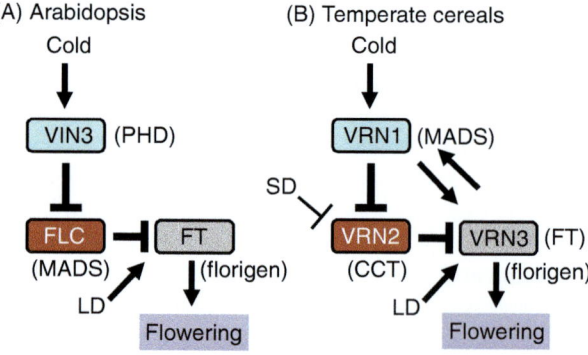

Fig. 6.6 A simplified model of the vernalization response on the florigen production in *Arabidopsis* (**a**) and temperate cereals (**b**)

keep it in a repressed state. Marking the chromatin in this way is what provides the cellular memory of winter.

In the winter, in varieties of wheat and barley that require vernalization to flower, three genes possibly participate in a regulatory loop to control the timing of flowering, namely, VRN1, VRN2, and VRN3 (Trevaskis et al. 2007; Fig. 6.6b). VRN2 is a key factor for flowering repression in winter wheat and barley. *VRN2* encodes a protein containing a putative zinc-finger and a CCT domain protein (Yan et al. 2004). Prior to cold exposure, high levels of *VRN2* act as a repressor of *VRN3* (an ortholog of *FT*) to prevent flowering (Dubcovsky et al. 2006). During cold exposure in wheat and barley, *VRN1*, a MADS-box transcription factor homologous to the floral-meristem identity gene *AP1* of *Arabidopsis*, is induced by vernalization, with the level of expression dependent on the length of cold exposure (Trevaskis et al. 2006). The induction of *VRN1* in the leaves during winter prevents the up-regulation of *VRN2* (Chen and Dubcovsky 2012). In the absence of VRN2, *VRN3* is up-regulated by LD, further enhancing an increasing *VRN1* expression and closing a positive feedback loop that leads to an acceleration of flowering. At the shoot apical meristem, *VRN1* activation by vernalization (Oliver et al. 2009) or by *VRN3* (Li and Dubcovsky 2008) accelerates the transition to the reproductive phase. GA application can substitute for vernalization in a number of biennial species as reported by Lang (1957). The substitution depends on the species. In the cold-requiring LD grass species *Lolium perenne*, exogenous GA allowed flowering in non-inductive SD conditions only in vernalized plants, whereas non-vernalized plants were unable to respond to GA either by stem elongation or flowering (MacMillan et al. 2005). The LD/GA inductive pathway is blocked unless plants are vernalized.

In sugar beet, two *FT*-like genes, *BvFT1* and *BvFT2*, have important roles in the vernalization-induced bolting and flowering (Pin et al. 2010). BvFT1 contributes to the vernalization response as a repressor. BvFT2 is essential for flowering as a promoter of flowering, whereas BvFT1 acts antagonistically and represses flowering, partly through the transcriptional repression of BvFT2.

Although VRN2 of temperate cereals and BvFT1 of sugar beet act similarly to FLC, in that it is a floral repressor, they are unrelated to the *FLC* gene. The different genes are involved in establishing vernalization in these species, indicating that vernalization systems possibly evolved after these groups of plants diverged.

References

Abe M, Kobayashi Y, Yamamoto S et al (2005) FD, a bZIP protein mediating signals from the floral pathway integrator FT at the shoot apex. Science 309:1052–1056

Bradley D, Ratcliffe O, Vincent C et al (1997) Inflorescence commitment and architecture in *Arabidopsis*. Science 275:80–83

Büning E (1936) Die endogene tagesrhythmik als grundlage der photoperiodischen reaktion. Ber Dtsch Bot Ges 54:590–607

Cathey H, Borthwick H (1957) Photoreversibility of floral initiation in Chrysanthemum. Bot Gaz 119:71–76

Chailakhyan M (1936) New facts in support of the hormonal theory of plant development. Dokl Akad Nauk SSSR 13:79–83

Chen A, Dubcovsky J (2012) Wheat TILLING mutants show that the vernalization gene VRN1 down-regulates the flowering repressor VRN2 in leaves but is not essential for flowering. PLoS Genet 8(12):e1003134

Conti L, Bradley D (2007) TERMINAL FLOWER 1 is a mobile signal controlling *Arabidopsis* architecture. Plant Cell 19:767–778

Corbesier L, Vincent C, Jang S et al (2007) FT protein movement contributes to long-distance signaling in floral induction of Arabidopsis. Science 316:1030–1033

Doi K, Izawa T, Fuse T et al (2004) *Ehd1*, a B-type response regulator in rice, confers short-day promotion of flowering and controls *FT*-like gene expression independently of *Hd1*. Genes Dev 18:926–936

Dubcovsky J, Loukoianov A, Fu D et al (2006) Effect of photoperiod on the regulation of wheat vernalization genes *VRN1* and *VRN2*. Plant Mol Biol 60:469–480

Evans L (1960) Inflorescence initiation in *Lolium temulentum* L. II. Evidence for inhibitory and promotive photoperiodic processes involving transmissible products. Aust J Biol Sci 13:429–440

Fornara F, Panigrahi K, Gissot L et al (2009) *Arabidopsis* DOF transcription factors act redundantly to reduce *CONSTANS* expression and are essential for a photoperiodic flowering response. Dev Cell 17:75–86

Garner W, Allard H (1920) Effect of the relative length of day and night and other factors of the environment on growth and reproduction in plants. J Agric Res 18:553–606

Greenham K, McClung C (2015) Integrating circadian dynamics with physiological processes in plants. Nat Rev Genet 16:598–610

Guttridge C (1959) Further evidence for a growth-promoting and flower-inhibiting hormone in strawberry. Ann Bot 23:612–621

Hayama R, Yokoi S, Tamaki S et al (2003) Adaptation of photoperiodic control pathways produces short-day flowering in rice. Nature 422:719–722

Hayama R, Aagashe B, Luley E et al (2007) A circadian rhythm set by dusk determines the expression of *FT* homologs and the short-day photoperiodic flowering response in *Pharbitis*. Plant Cell 19:2988–3000

Heide O, Stavang J, Sønsteby A (2013) Physiology and genetics of flowering in cultivated and wild strawberries. J HortScience Biotechnol 88:1–18

Hepworth J, Dean C (2015) Flowering locus C's lessons: conserved chromatin switches underpinning developmental timing and adaptation. Plant Physiol 168:1237–1245

Higuchi Y, Hisamatsu T (2015) CsTFL1, a constitutive local repressor of flowering, modulates floral initiation by antagonising florigen complex activity in chrysanthemum. Plant Sci 237:1–7

Higuchi Y, Sage-Ono K, Sasaki R et al (2011) Constitutive expression of the *GIGANTEA* ortholog affects circadian rhythms and suppresses one-shot induction of flowering in *Pharbitis nil*, a typical short-day plant. Plant Cell Physiol 52:638–650

Higuchi Y, Sumitomo K, Oda A et al (2012) Day light quality affects the night-break response in the short-day plant chrysanthemum, suggesting differential phytochrome-mediated regulation of flowering. J Plant Physiol 169:1789–1796

Higuchi Y, Narumi T, Oda A et al (2013) The gated induction system of a systemic floral inhibitor, antiflorigen, determines obligate short-day flowering in chrysanthemums. Proc Natl Acad Sci U S A 110:17137–17142

Imaizumi T, Schultz T, Harmon F et al (2005) FKF1 F-box protein mediates cyclic degradation of a repressor of *CONSTANS* in *Arabidopsis*. Science 309:293–297

Imamura S (1967) Physiology of flowering in *Pharbitis nil*. Japanese Society of Plant Physiologist, Tokyo

Ishikawa R, Tamaki S, Yokoi S et al (2005) Suppression of the floral activator *Hd3a* is the principal cause of the night break effect in rice. Plant Cell 17:3326–3336

Itoh H, Nonoue Y, Yano M et al (2010) A pair of floral regulators sets critical day length for Hd3a florigen expression in rice. Nat Genet 42:635–638

Iwata H, Gaston A, Remay A et al (2012) The *TFL1* homologue *KSN* is a regulator of continuous flowering in rose and strawberry. Plant J 69:116–125

Izawa T, Oikawa T, Tokutomi S et al (2000) phytochromes confer the photoperiodic control of flowering in rice (a short-day plant). Plant J 22:391–399

Izawa T, Oikawa T, Sugiyama N et al (2002) Phytochrome mediates the external light signal to repress FT orthologs in photoperiodic flowering of rice. Genes Dev 16:2006–2020

Jaeger K, Pullen N, Lamzin S et al (2013) Interlocking feedback loops govern the dynamic behavior of the floral transition in Arabidopsis. Plant Cell 25:820–833

Kardailsky I, Shukla V, Ahn J et al (1999) Activation tagging of the floral inducer *FT*. Science 286:1962–1965

Kobayashi Y, Kaya H, Goto K et al (1999) A pair of related genes with antagonistic roles in mediating flowering signals. Science 286:1960–1962

Komiya R, Yokoi S, Shimamoto K (2009) A gene network for long-day flowering activates *RFT1* encoding a mobile flowering signal in rice. Development 136:3443–3450

Koskela EA, Mouhu K, Albani M et al (2012) Mutation in *TERMINAL FLOWER 1* reverses the photoperiodic requirement for flowering in the wild strawberry *Fragaria vesca*. Plant Physiol 159:1043–1054

Lang A (1957) The effect of gibberellin upon flower formation. Proc Natl Acad Sci U S A 43:709–717

Lang A, Melchers G (1943) Die photoperiodische reaktion von *Hyoscyamus niger* (in German). Planta 33:653–702

Lang A, Chailakhyan M, Frolova I (1977) Promotion and inhibition of flower formation in a day neutral plant in grafts with a short-day plant and a long-day plant. Proc Natl Acad Sci U S A 74:2412–2416

Lees A (1973) Photoperiodic time measurement in the aphid Megoura viciae. J Insect Physiol 19:2279–2316

Li C, Dubcovsky J (2008) Wheat FT protein regulates VRN1 transcription through interactions with FDL2. Plant J 55:543–554

Lifschitz E, Eviatar T, Rozman A et al (2006) The tomato FT ortholog triggers systemic signals that regulate growth and flowering and substitute for diverse environmental stimuli. Proc Natl Acad Sci U S A 103:6398–6403

Lin M, Belanger H, Lee Y et al (2007) FLOWERING LOCUS T protein may act as the long-distance florigenic signal in the cucurbits. Plant Cell 19:1488–1506

Liu J, Yu J, McIntosh L et al (2001) Isolation of a *CONSTANS* ortholog from *Pharbitis nil* and its role in flowering. Plant Physiol 125:1821–1830

MacMillan C, Blundell C, King R (2005) Flowering of the grass Lolium perenne L.: effects of vernalization and long days on gibberellin biosynthesis and signalling. Plant Physiol 138:1794–1806

Matsoukas I (2015) Florigens and antiflorigens: a molecular genetic understanding. Essays Biochem 58:133–149

Mouhu K, Kurokura T, Koskela E et al (2013) The *Fragaria vesca* homolog of *SUPPRESSOR OF OVEREXPRESSION OF CONSTANS 1* represses flowering and promotes vegetative growth. Plant Cell 25:3296–3310

Nakano Y, Higuchi Y, Sumitomo K et al (2013) Flowering retardation by high temperature in chrysanthemums: involvement of *FLOWERING LOCUS T-like 3* gene expression. J Exp Bot 64:909–920

Nakano Y, Higuchi Y, Yoshida Y et al (2015) Environmental responses of the FT/TFL1 gene family and their involvement in flower induction in Fragaria × ananassa. J Plant Physiol 177:60–66

Nelson R, Denlinger D, Somers D (2010) Photoperiodism: the biological calendar. Oxford University Press, New York

Oda A, Narumi T, Li T et al (2012) *CsFTL3*, a chrysanthemum *FLOWERING LOCUS T-like* gene, is a key regulator of photoperiodic flowering in chrysanthemums. J Exp Bot 63:1461–1477

Ogawa Y, King R (1990) The inhibition of flowering by non-induced cotyledons of *Pharbitis nil*. Plant Cell Physiol 31:129–135

Ogiso-Tanaka E, Matsubara K, Yamamoto S et al (2013) Natural variation of the *RICE FLOWERING LOCUS T 1* contributes to flowering time divergence in rice. PLoS One 8: e75959

Oliver S, Finnegan J, Dennis E et al (2009) Vernalization-induced flowering in cereals is associated with changes in histone methylation at the VERNALIZATION1 gene. Proc Natl Acad Sci U S A 106(20):8386–8391

Park S, Jiang K, Tal L et al (2014) Optimization of crop productivity in tomato using induced mutations in the florigen pathway. Nat Genet 46:1337–1342

Pin P, Benlloch R, Bonnet D et al (2010) An antagonistic pair of FT homologs mediates the control of flowering time in sugar beet. Science 330:1397–1400

Pittendrigh C (1972) Circadian surfaces and the diversity of possible roles of circadian organization in photoperiodic induction. Proc Natl Acad Sci U S A 69:2734–2737

Pittendrigh C, Minis D (1964) The entrainment of circadian oscillations by light and their role as photoperiodic clocks. Am Nat 108:261–295

Rantanen M, Kurokura T, Mouhu K et al (2014) Light quality regulates flowering in *FvFT1/FvTFL1* dependent manner in the woodland strawberry *Fragaria vesca*. Front Plant Sci 5:271

Rantanen M, Kurokura T, Jiang P et al (2015) Strawberry homologue of TERMINAL FLOWER 1 integrate photoperiod and temperature signals to inhibit flowering. Plant J 82:163–173

Ratcliffe O, Bradley D, Coen E (1999) Separation of shoot and floral identity in Arabidopsis. Development 126:1109–1120

Sawa M, Nusinow D, Kay S et al (2007) FKF1 and GIGANTEA complex formation is required for day-length measurement in *Arabidopsis*. Science 318:261–265

Shalit A, Rozman A, Goldshmidt A et al (2009) The flowering hormone florigen functions as a general systemic regulator of growth and termination. Proc Natl Acad Sci U S A 106:8392–8397

Song Y, Shim J, Kinmouth-Schltz H et al (2015) Photoperiodic flowering: time measurement mechanisms in leaves. Annu Rev Plant Biol 66:441–464

Sumitomo K, Higuchi Y, Aoki K et al (2012) Spectral sensitivity of flowering and *FT-like* gene expression in response to a night break treatment in the chrysanthemum cultivar 'Reagan'. J HortScience Biotechnol 87:461–469

Sung S, Amasino R (2004) Vernalization in *Arabidopsis thaliana* is mediated by the PHD finger protein VIN3. Nature 427:159–164

Takano M, Inagaki N, Xie X et al (2009) Phytochromes are the sole photoreceptors for perceiving red/far-red light in rice. Proc Natl Acad Sci U S A 106:14705–14710

Tamaki S, Matsuo S, Wong H et al (2007) Hd3a protein is a mobile flowering signal in rice. Science 316:1033–1036

Tanaka T (1967) Studies on the regulation of Chrysanthemum flowering with special reference to plant regulators I. The inhibiting action of non-induced leaves on floral stimulus. J Jpn Soc HortScience 36:77–85

Taoka K, Ohki I, Tsuji H et al (2011) 14-3-3 proteins act as intracellular receptors for rice Hd3a florigen. Nature 476:332–335

Thomas B, Vince-Prue B (1997) Photoperiodism in plants, 2nd edn. Academic, London

Trevaskis B, Hemming M, Peacock J et al (2006) HvVRN2 responds to daylength, whereas HvVRN1 is regulated by vernalization and developmental status. Plant Physiol 140:1397–1405

Trevaskis B, Hemming M, Dennis E et al (2007) The molecular basis of vernalization-induced flowering in cereals. Trends Plant Sci 12:352–357

Valverde F, Mouradov A, Soppe W et al (2004) Photoreceptor regulation of CONSTANS protein in photoperiodic flowering. Science 303:965–966

Wickland D, Hanzawa Y (2015) *The FLOWERING LOCUS T/TERMINAL FLOWER 1* gene family: functional evolution and molecular mechanisms. Mol Plant 8:983–997

Wigge PA, Kim M, Jaeger K et al (2005) Integration of spatial and temporal information during floral induction in *Arabidopsis*. Science 309:1056–1059

Xue W, Xing Y, Weng X et al (2008) Natural variation in *Ghd7* is an important regulator of heading date and yield potential in rice. Nat Genet 40:761–767

Yan L, Loukoianov A, Blechl A et al (2004) The wheat VRN2 gene is a flowering repressor down-regulated by vernalization. Science 303:1640–1644

Yanovsky M, Kay S (2002) Molecular basis of seasonal time measurement in *Arabidopsis*. Nature 419:308–312

Zeevaart J (2008) Leaf-produced floral signals. Curr Opin Plant Biol 11:541–547

Chapter 7
Light Environment in the Cultivation Space of Plant Factory with LEDs

Takuji Akiyama and Toyoki Kozai

Abstract Spatial distributions of photosynthetic photon flux density (PPFD) in the cultivation space of a plant factory with light-emitting diodes (LEDs) are simulated using the free software package DIALux under different design conditions including (1) reflectance of cultivation panel surfaces, (2) width of side reflectors, (3) layout of LED tubes on the ceiling, (4) angular light distribution of LED lamp, and (5) height of the plant canopy. The simulation shows that the average and uniformity of PPFD above the cultivation panels and percent loss of photosynthetic photons to the outside of the cultivation space are affected not only by the characteristics of LED tubes but also by those of the cultivation space.

Keywords PPFD (photosynthetic photon flux density) distribution • Optimal light environment • Reflectance of culture panels

7.1 Introduction

Light is one of the most important environmental factors affecting plant growth and development. In plant factories with artificial lighting (PFALs), electricity consumption for lighting is a major component of production cost. As described in Chap. 4, the light environment in the cultivation space of a PFAL is affected by the optical and geometric characteristics of the light source, cultivation space, and plant canopy.

Computer software for simulating the light environment is a powerful tool to analyze and design the light source, cultivation space, and plant canopy architecture. This chapter presents the effects of these factors on the light environment in the cultivation space of PFALs.

T. Akiyama
PlantX Corporation, Wakashiba, Kashiwa, Chiba 277-0882, Japan

T. Kozai (✉)
Japan Plant Factory Association (NPO), Kashiwa-no-ha, Kashiwa, Chiba 277-0882, Japan
e-mail: kozai@faculty.chiba-u.jp

7.2 Materials and Methods

7.2.1 Software

DIALux (version 4.12) (DIAL GmbH), a free software package, was used in the present simulation of photosynthetic photon flux density (PPFD) distribution in the cultivation space, which can be downloaded from https://www.dial.de/en/dialux/. A similar simulation can be conducted using Relux, another free software package downloaded from http://www.relux.biz/. In this chapter, the simulated results for tube-type LEDs alone are presented, although similar simulations can be conducted for surface-type and point-source-type LEDs.

7.2.2 Variables and Their Values Assumed as Unique Input Data

Table 7.1 gives variable names and their assumed values as unique input data for the simulations. The assumed spectral light distribution of white LEDs (tube-type lamps) containing fluorescent substances as additives is shown in Fig. 7.1. Since white LEDs are basically blue LEDs, fluorescent substances are added to convert substantial portion of the blue light to red/green/yellow light.

The cross section of the cultivation space (0.3 m high, 1.2 m wide) is assumed to be as shown in Fig. 7.2a. The reflectance (r) of the ceiling is assumed to be 0.9. Averaged reflectance (r) and transmittance of the plant canopy cover are assumed to be 0.08 and 0.04, respectively, for the light source with spectral light distribution of LEDs shown in Fig. 7.1.

7.2.3 Factors Examined to Show Their Effects on PPFD Distribution

Table 7.2 gives factors (variables) examined in the simulation to determine their effects on PPFD distribution in the cultivation space: (1) r (reflectance, ratio of photosynthetic photons reflected to those received) of cultivation panel surfaces, (2) width of light reflectors at the upper sides of the cultivation space, (3) even and uneven spacing between LED tubes for the parallel layout (Fig. 7.3C), (4) layout of LED tubes parallel and perpendicular to the longitudinal cultivation space (Fig. 7.3A, B), (5) wide- and narrow-angle light distributions of LED tubes (Fig. 7.4), and (6) plant canopy height (h), in the cultivation space.

7 Light Environment in the Cultivation Space of Plant Factory with LEDs

Table 7.1 Parameters and their values assumed as unique input data in the simulation

No.	Category	Variable name	Symbol	Unit	Values
1	LED (tube-type LED lamp)	Electric energy consumption per LED tube	E	W $(=\text{J s}^{-1})$	20
2		Photosynthetic photon number efficacy	h	mmol J^{-1}	2
3		Number of LEDs /(1.2 m wide × 1.2 m long)	n	–	6
4		Photosynthetic photon flux of LEDs	$F = E \times h \times n$	mmol s^{-1}	240
5		Length, width, and thickness of tube-type LED lamp		m	1.2, 0.03, and 0.03 m
6		Spectral distribution of LED light		–	Fig. 7.1
7	Cultivation space	Height, width, and length of cultivation space (0.3 high, 1.2 wide, and 10 m long)		m	Fig. 7.2
8		Reflectance of inner surfaces of ceiling and vertical reflectors at both upper sides	r	–	0.9
9		Spectral distribution of reflectance (r) and transmittance (t) of leaves		–	$r = 0.08$ $t = 0.04$

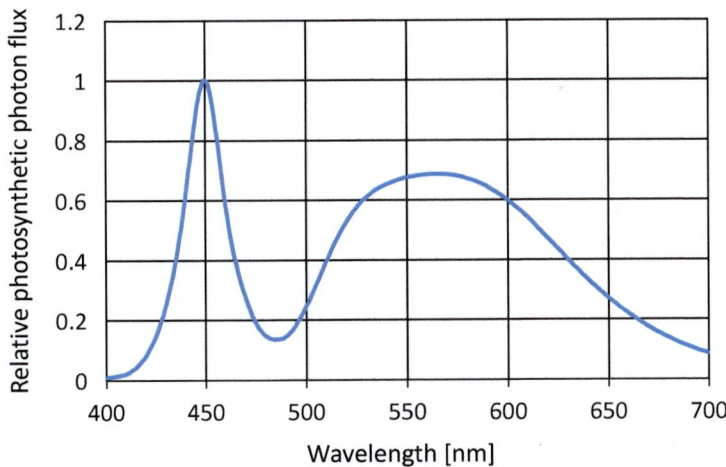

Fig. 7.1 Spectral light distribution of white LEDs (tube-type lamps) with fluorescent substances as additives to modify the spectrum, assumed in the simulation

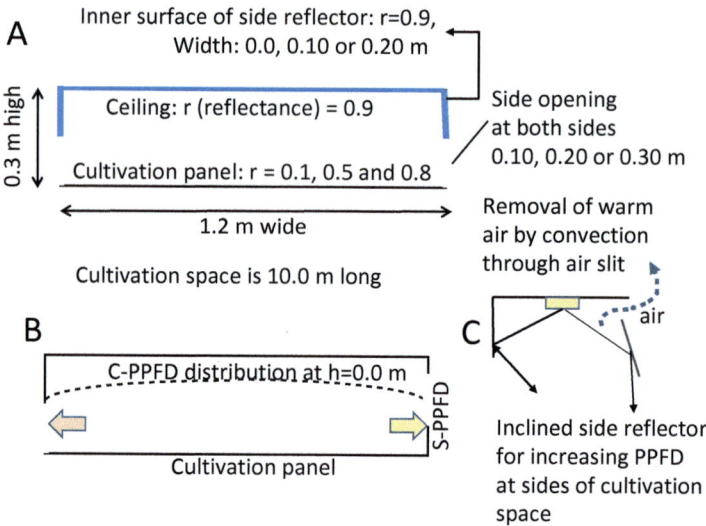

Fig. 7.2 (a) Cross section of the cultivation space (0.3 m high and 1.2 m wide) assumed in the simulation. Reflectance of inner surfaces of ceiling and upper side reflectors: 0.9, Reflectance of cultivation panel surface: 0.1, 0.5 and 0.8. (b) Definitions of C-PPFD and S-PPFD (see also, Table 7.2). (c) Inclined side reflectors for increasing C-PPFD at sides, and air gap for enhancing the air exchange between inside and outside the cultivation space

Table 7.2 Factors examined in the simulation to show their effects on C-PPFD and S-PPFD distributions in the cultivation space

No.	Category	Factors (variables) examined in the simulation	Variable name	Values
1	LED (tube-type LED lamp)	Layout of LED tubes on ceiling (perpendicular or parallel to longitudinal cultivation space)	Parallel or perpendicular	Fig. 7.4a, b
2		Angular light distribution curves of wide- and narrow-angle LED tubes	Wide or narrow	Fig. 7.5
3		Layout of LED tubes on ceiling (even or uneven spacing between LED tubes placed parallel to longitudinal cultivation space)	Even or uneven	Fig. 7.4a, c
4	Cultivation space	Reflectance (r) of culture panel surface	r	0.1, 0.5, and 0.8
5		Width of side vertical reflectors at both upper sides	W	0.0, 0.1, and 0.20 m
6		Height of plant canopy. "$h=0.0$ m" means that the cultivation space is empty (no plant canopy)	h	0.0, 0.15, and 0.20 m

7 Light Environment in the Cultivation Space of Plant Factory with LEDs

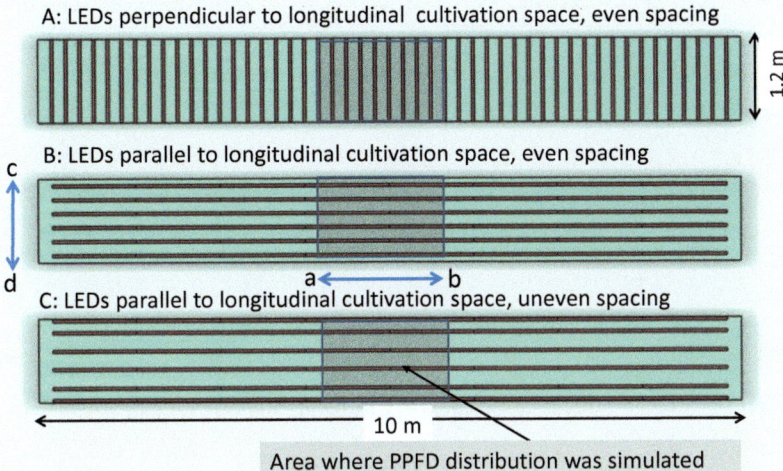

Fig. 7.3 Three types of LED tube layout on the ceiling in the cultivation space. A: Perpendicular to the longitudinal cultivation space and even distance between LED tubes. B: Parallel to the longitudinal cultivation space and even distance between LED tubes. C: Parallel to the longitudinal cultivation space and uneven distance between LED tubes

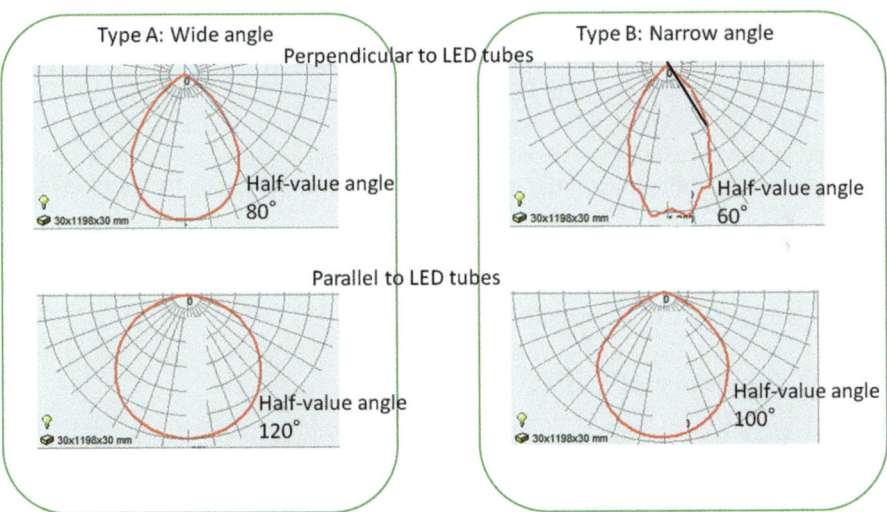

Fig. 7.4 Angular light distribution curves of type A (*wide angle*) LED and type B (*narrow angle*) LED. Half value angle means the angle at which the photosynthetic photon flux is 50 % of its maximum value. Upper: cross-sectional angular distribution (perpendicular to LED tubes), *Lower*: longitudinal angular distribution (parallel to LED tubes)

Table 7.3 Parameters characterizing C-PPFD ($\mu mol\ m^{-2}\ s^{-1}$) distribution in the cultivation space

A

No.	Variables characterizing PPFD distribution in the cultivation space	Symbol
1	Average	AVE
2	Maximum	Max
3	Minimum	Min
4	Standard deviation	SD
5	Percentage of photosynthetic photons lost to outside the cultivation space	%L

B

Variable name	Types of PPFD simulated
C-PPFD	Horizontal PPFD across the cultivation space. Average PPFD over the longitudinal direction shown as line a–b in Fig. 7.4b
S-PPFD	PPFD facing perpendicular to inside of the cultivation space at side openings (Fig. 7.2b)

7.2.3.1 Variables Characterizing Horizontal and Vertical PPFD Distributions

Table 7.3A gives parameters (average [AVE], minimum [Min], maximum [Max], and standard deviation [SD]) characterizing C-PPFD ($\mu mol\ m^{-2}\ s^{-1}$) distribution at $h = 0.00$ m (on the cultivation panel) across the cultivation space and the PPFD at side openings facing inside the cultivation space (S-PPFD). Table 7.3B and Fig. 7.2B give definitions of C-PPFD and S-PPFD.

Percent loss (%L) of photosynthetic photons emitted by LEDs (Table 7.4) is defined as the percentage of photosynthetic photons emitted by LED tubes but lost through the side openings to the outside of the cultivation space (F in Table 7.1).

7.3 Results and Discussion

7.3.1 Summary of C-PPFD and %L

In Table 7.4, simulated results of C-PPFD, its average (AVE), minimum (Min), maximum (Max), standard deviation (SD), and percent loss (%L) as affected by the variables shown in Table 7.2 are summarized.

7.3.2 Summary of S-PPFD

Table 7.5 summarizes the simulated results of S-PPFD and its AVE, Min, Max, and SD values as affected by the variables shown in Table 7.2.

7 Light Environment in the Cultivation Space of Plant Factory with LEDs

Table 7.4 Summary of simulated results of C-PPFD ($\mu mol\ m^{-2}\ s^{-1}$) distribution in the cultivation space. "%L" denotes as the percentage of photosynthetic photons lost through the openings at sides to the outside of cultivation stage to the total photosynthetic photons emitted by LEDs

Category	Variables affecting C-PPFD (Table 7.2b) distribution	Case	Value of variables	C-PPFD AVE	Min	Max	SD	%L
Cultivation space	Effect of reflectance of culture panel surface (r) (Fig. 7.6) Variable values fixed: wide angle, parallel orientation, even spacing $W = 0.1\ m, h = 0.0\ m$ (see Table 7.2 for variable names)	1-1	$r = 0.1$	158	101	176	22	13
		1-2	$r = 0.5$	219	137	247	33	26
		1-3	$r = 0.8$	309	192	353	49	47
	Effect of width (W) of vertical reflectors at both upper sides of cultivation space (Fig. 7.8) Variable values fixed: wide angle, parallel orientation, even spacing, $r = 0.8, h = 0.0\ m$ (see case 1-3 for $W = 0.1\ m$)	2-1	$W = 0.00\ m$	273	166	319	48	55
		2-2	$W = 0.15\ m$	341	214	383	50	39
LED tubes and their layout	Effect of uneven spacing between LED tubes (Fig. 7.9) Parameter values fixed: wide, parallel orientation, $r = 0.8, W = 0.1\ m, h = 0.0\ m$ (see case 1-3 for even spacing)	3-1	Uneven	290	240	305	16	46
	Effects of LED tube orientation (perpendicular) and r (Fig. 7.10) Variable values fixed: wide angle, perpendicular orientation, even spacing, $W = 0.1\ m, h = 0.0\ m$ (see cases 1-1 and 1-3 for parallel orientation)	4-1	$r = 0.1$	153	99	172	22	15
		4-2	$r = 0.8$	299	187	345	49	48
	Effects of narrow-angle distribution of LED tubes and r (Fig. 7.11) Variable values fixed: narrow angle, parallel orientation, even spacing, $W = 0.1\ m, h = 0.0\ m$ (see cases 1-1 and 1-3 for wide angle)	5-1	$r = 0.1$	160	100	178	21	11
		5-2	$r = 0.8$	313	192	357	48	46
Plant canopy height	Effect of plant canopy height (h) (Fig. 7.12) Variable values fixed: wide angle, parallel orientation, even spacing, $W = 0.1\ m$; reflectance (r) and transmittance of plant canopy cover are 0.08 and 0.4, respectively	6-1	$h = 0.00\ m$	156	99	173	22	12
		6-2	$h = 0.15\ m$	162	97	204	41	10
		6-3	$h = 0.20\ m$	164	44	290	118	10

It is noted that significant amounts of photosynthetic photons are lost through the side openings to the outside. Percent loss (%L) is higher at higher r and lower W. Wide side reflectors are beneficial to reduce %L for perpendicular layout of LED tubes (case 3) and uneven distance layout (case 4).

7.3.3 Case 1: Reflectance (r) of Culture Panel Surface

Figure 7.5a shows the C-PPFD curve at $h=0.0$ m across the cultivation space for r values of 0.1, 0.5, and 0.8. The AVE of C-PPFD at $h=0.0$ m and of S-PPFD is about twofold greater when $r=0.8$ (case 1-3) and 1.4-fold greater when $r=0.5$ (case 1-2) than when $r=0.1$ (case 1-1) due to multiple light reflections of photosynthetic photons, mostly between the ceiling and cultivation panel and/or plant canopy (Table 7.4 and Fig. 7.5b).

%L is three fold and twofold greater, respectively, when $r=0.8$ and $r=0.5$ than that when $r=0.1$ due to the increased loss of reflected photosynthetic photons to the outside (Table 7.4), while the number of photosynthetic photons emitted by LED tubes and lost directly to the outside (without reflection) remains the same regardless of the r value.

Figure 7.6a shows a schematic diagram of multiple reflections between the cultivation panel surface ($r=0.8$) and ceiling ($r=0.9$) for a photosynthetic photon flux at LED tubes (F) of 240 mol s^{-1}. As shown in Fig. 7.6b, C-PPFD increases exponentially with increasing r or with decreasing leaf area index (LAI; total leaf area divided by cultivation area) of the plant canopy. Therefore, in order to maintain the C-PPFD at the plant canopy at a fixed level throughout the culture period, the F of LED tubes must be increased as plants grow (as total leaf area increases). For example, when C-PPFD is 325 (relative value) at $r=0.8$, it will decrease to 200 at $r=0.6$, 145 at $r=0.4$, 106 at $r=0.2$, and 100 at $r=0.1$. Thus, to maintain C-PPFD at 325 at $r=0.6$, 0.4, 0.2, and 0.1, F needs to be increased by 1.6-, 2.2-, 3.1-, and 3.3-fold, respectively, compared with F at $r=0.8$.

7.3.4 Case 2: Width (W) of Vertical Side Reflectors

As shown in Table 7.4 and Fig. 7.7a, the AVE C-PPFD for the side reflector width (W) of 0.15 m is 341 µmol m^{-2} s^{-1}, which is 10 % greater than that for side reflector W of 0.1 m and 25 % greater than that for W of 0.0 m (with no side reflectors). Thus, side reflectors are beneficial in increasing the AVE C-PPFD.

The nonuniformity or SD of C-PPFD can be improved by inclining the side reflectors inward at angle of approximately 30° (Fig. 7.2c). On the other hand, wide side reflectors restrict the air exchange between the inside and outside of the cultivation space, resulting in less air movement and higher air temperatures in the cultivation space during photoperiod. Heat removal from LED tubes and/or the

7 Light Environment in the Cultivation Space of Plant Factory with LEDs

Table 7.5 Summary of simulated results of S-PPFD ($\mu mol\ m^{-2}\ s^{-1}$) at the side openings facing to the inside of cultivation space

Category	Variables affecting C-PPFD (Table 7.2b) distribution	Case	Value of variables	S-PPFD AVE	Min	Max	SD
Cultivation space	Effect of reflectance of culture panel surface (r) (Fig. 7.6)	1-1	$r=0.1$	63	57	68	4
	Variable values fixed: wide, parallel orientation, even spacing, $W=0.1$ m, $h=0.0$ m (see Table 7.2 for variable names)	1-2	$r=0.5$	132	121	140	6
		1-3	$r=0.8$	233	215	245	11
	Effect of width (W) of vertical reflectors at both upper sides of cultivation space (Fig. 7.11)	2-1	$W=0.00$ m	184	139	211	25
	Variable values fixed: Wide, parallel orientation, even spacing, $r=0.8$, h 0.0 m (see case 1-3 for $W=0.10$ m)	2-2	$W=0.15$ m	260	242	275	13
LED tubes and their layout	Effect of uneven spacing between LED tubes (Fig. 7.10)	3-1	Uneven	232	220	245	9
	Parameter values fixed: wide, parallel orientation, $r=0.8$, $W=0.1$ m, $h=0.0$ m (see case 1-3 for even spacing)						
	Effects of LED tube orientation (perpendicular) and r (Fig. 7.8)	4-1	$r=0.1$	74	64	95	9
	Variable values fixed: wide, perpendicular orientation, even spacing, $W=0.1$ m, $h=0.0$ m (see cases 1-1 and 1-3 for parallel orientation)	4-2	$r=0.8$	239	223	266	13
	Effects of angular distribution (narrow) of LED tubes and r (Fig. 7.9)	5-1	$r=0.1$	57	48	65	4
	Variable values fixed: narrow, parallel orientation, even spacing, $W=0.1$ m, $h=0.0$ m (see cases 1-1 and 1-, for wide-angle LED tubes)	5-2	$r=0.8$	230	212	244	10
Plant canopy height	Effect of plant canopy height (h) (Fig. 7.12)	6-1	$h=0.00$ m	60	55	66	4
	Variable values fixed: wide, parallel orientation, even spacing, $W=0.1$ m; reflectance (r) and transmittance of plant canopy cover are 0.08 and 0.4, respectively	6-2	$h=0.15$ m	52	49	57	3
		6-3	$h=0.20$ m	52	49	57	3

Fig. 7.5 (Case 1) (**a**) Effect of reflectance ($r = 0.1, 0.5,$ and 0.8) of cultivation panel surface on C-PPFD distribution across the cultivation space. (**b**) S-PPFD for $r = 0.1, 0.5,$ and 0.8. $W = 0.1$ m, $h = 0.0$ m, wide angle, even distance, and parallel orientation

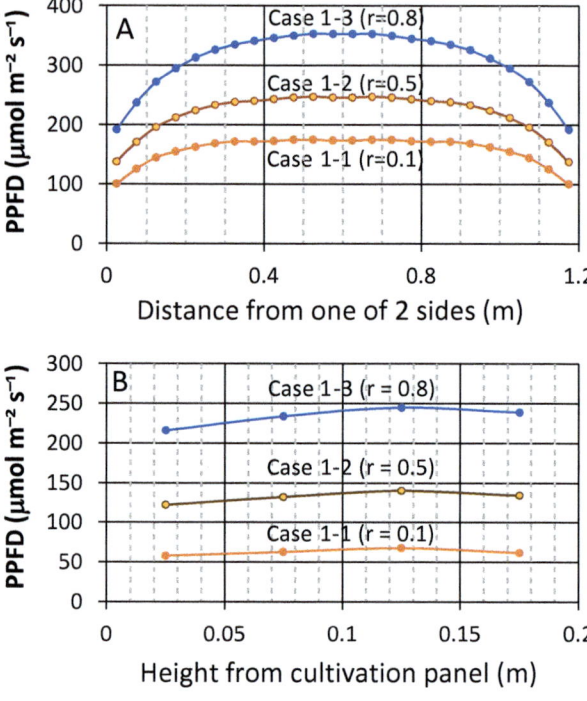

Fig. 7.6 (**a**) Schematic diagram of PPFD under multiple reflection between two infinite flat surfaces ($r;$ $= 0.8$ and 0.9, respectively) placed in parallel under photosynthetic photon flux at LEDs (F) of 240 µmol s^{-1}. (**a**) The PPFD is roughly expressed by

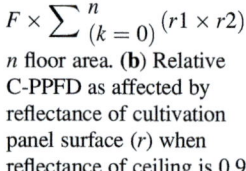

$$F \times \sum_{(k=0)}^{n} (r1 \times r2)$$

n floor area. (**b**) Relative C-PPFD as affected by reflectance of cultivation panel surface (r) when reflectance of ceiling is 0.9

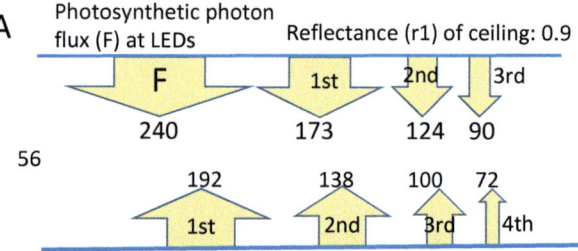

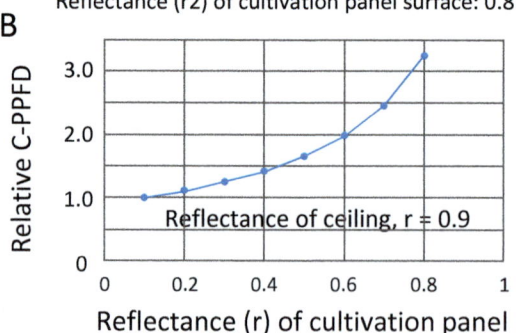

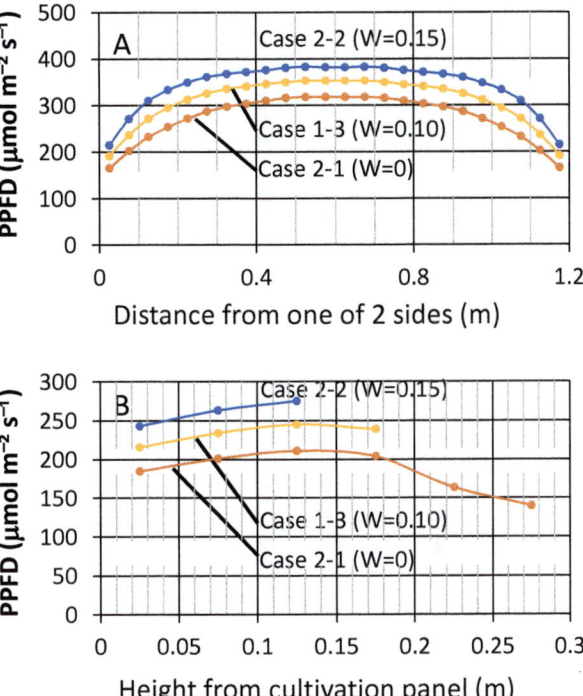

Fig. 7.7 (Case 2) (**a**) Effect of side reflectors width ($W = 0.0$, 0.10 and 0.15 m) at upper sides on C-PPFD distribution. (**b**) Vertical profile of S-PPFD at the side openings. $r = 0.8$, $W = 0.1$ m, $h = 0.0$ m, wide angle, even distance, and parallel orientation

cultivation space with minimum loss of photosynthetic photons can be enhanced, for example, by promoting air exchange through air slits near the ceiling (Fig. 7.2c).

7.3.5 Case 3: Uneven Distance Between LED Tubes

In Fig. 7.8a, C-PPFD distribution with uneven spacing between LED tubes is significantly flatter across the cultivation space compared with that with even spacing between LED tubes, although the AVE distribution for the uneven layout is 7 % less than that for the even layout. This reduction in C-PPFD at both sides can also be improved by the use of inclined reflectors instead of vertical reflectors (Fig. 7.2c).

7.3.6 Case 4: Perpendicular Layout (Fig. 7.4a)

There are no significant differences in AVE, SD, and %L between the perpendicular layout and parallel layout of LED tubes to the longitudinal cultivation space (Table 7.4, Figs. 7.5a and 7.9a).

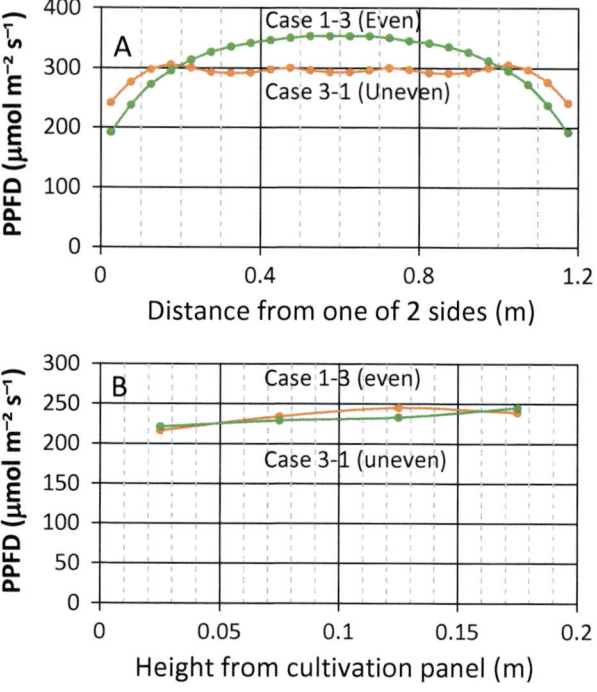

Fig. 7.8 (Case 3) (a) C-PPFD distribution for layout of uneven distance between LED tubes. (b) Vertical profile of S-PPFD at the side openings. $r = 0.8$, $W = 0.1$ m, $h = 0.0$ m, wide angle, and parallel orientation

In actuality, however, a reduction in C-PPFD should occur just below the joint of LED tubes in the parallel layout, because there is a metal socket (about 2 cm long) at each end of the tubes, which is not considered in the present simulation. If the presence of sockets were considered, the reduction in C-PPFD would be about 4 % for the AVE and about 10 % immediately below the joints. To compensate for the reduction in C-PPFD in the parallel layout, another LED tube can be placed perpendicular to the longitudinal LED tubes at the joints in the parallel layout.

Installation, maintenance and exchange of LED tubes may be easier in the perpendicular layout than in the parallel layout. In addition, the distance between LED tubes and number of LED tubes per unit of cultivation area can be changed after initial installation more easily in the perpendicular layout.

7.3.7 Case 5: Narrow Angular Light Distribution

There are no significant differences in C-PPFD between wide- and narrow-angle light distribution LED tubes examined in the present simulation (Fig. 7.10). It

Fig. 7.9 (Case 4) (**a**) C-PPFD distribution for perpendicular LED tube orientation for $r = 0.1$ and 0.8. (**b**) Vertical profile of S-PPFD at the side openings. $W = 0.1$ m, $h = 0.0$ m, wide angle, and even distance

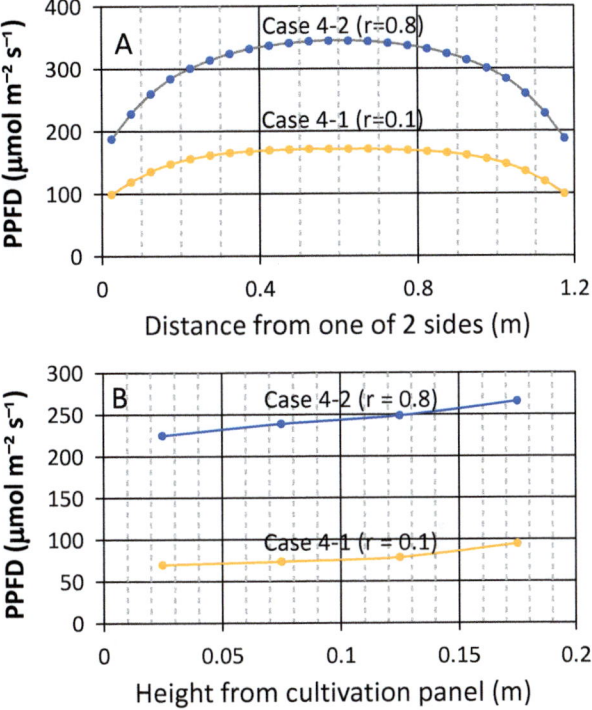

should be noted, however, that LED tubes with a narrower angle of light distribution than shown by type B in Table 7.5 are commercially available, which might be useful to improve the reduction in C-PPFD at the sides of the cultivation space. A mix of wide- and narrow-angle LED tubes may give a better C-PPFD distribution than that shown in Fig. 7.10.

7.3.8 Case 6: Height of Plant Canopy (h)

As the plant canopy grows and covers the cultivation panel, the average r for both the canopy and cultivation panel decreases. At the same time, the plant canopy cover approaches the ceiling with growth.

As shown in Table 7.4, the AVE and %L are not significantly affected by h. However, the SD is 22, 41, and 118 at an h of 0.00, 0.15, and 0.10 m, respectively. This is because the C-PPFD at the plant canopy cover becomes greater immediately below the LED tubes as well as lower between the LED tubes (Figs. 7.11a and 7.12). This large difference in C-PPFD may cause significant variation in plant growth rates.

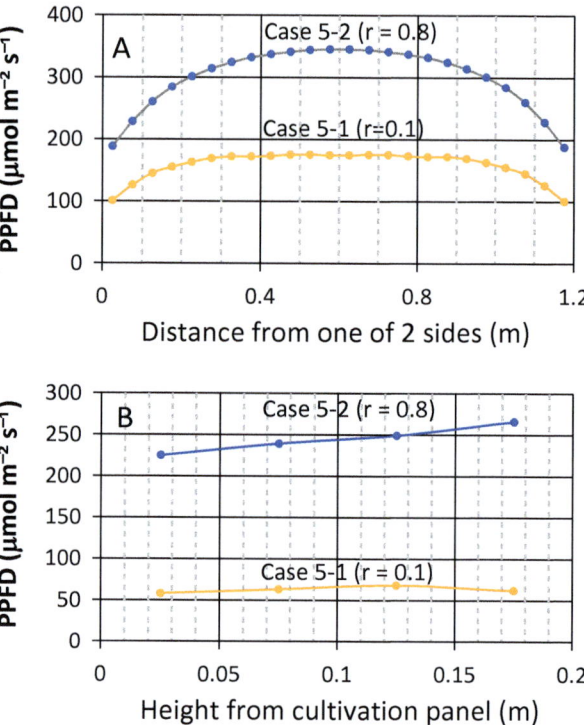

Fig. 7.10 (Case 5) (**a**) C-PPFD distribution for narrow and wide angular light distributions of LED tubes and $r = 0.1$ and 0.8. (**b**) Vertical profile of S-PPFD at the side openings. $W = 0.1$ m. $h = 0.0$ m., even distance, and parallel orientation

7.4 Some Consideration on Optimal Light Environment

This section discusses general questions on the optimal light environment in PFALs in terms of PPFD, light quality, photoperiod, and daily light integral (DLI; = PPFD × photoperiod). Similar questions are often asked about the optimal temperature, CO_2 concentration, and other environmental factors. Answers to these questions about PFALs differ from those about greenhouses, because PPFD, light quality, and photoperiod as well as temperature and CO_2 concentration can be controlled in PFALs more precisely than in greenhouses, regardless of the weather. On the other hand, the costs for lighting and air-conditioning in PFALs are significant.

7.4.1 Optimal PPFD?

The PPFD in a PFAL can be controlled at its optimal value, if known, to maximize plant growth or profitability. On the other hand, the optimal PPFD can change significantly with the cultivar, plant species, planting density (plants/m^2), LAI, light

Fig. 7.11 (Case 6) (**a**) Effects of plant canopy height ($h = 0.0$, 0.15 and 0.20 m) on C-PPFD distribution. (**b**) Vertical profile of S-PPFD at the side openings. $W = 0.1$ m, wide angle, even distance, and parallel orientation. r of plant canopy = 0.08 and transmittance of plant canopy = 0.04

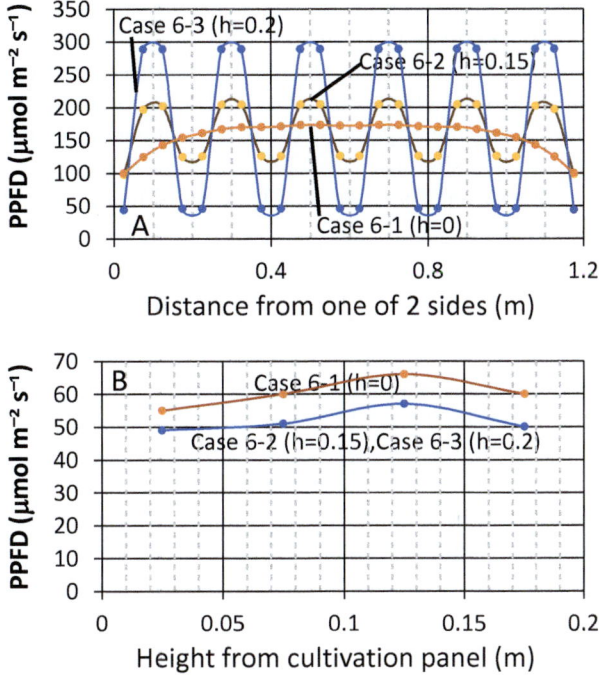

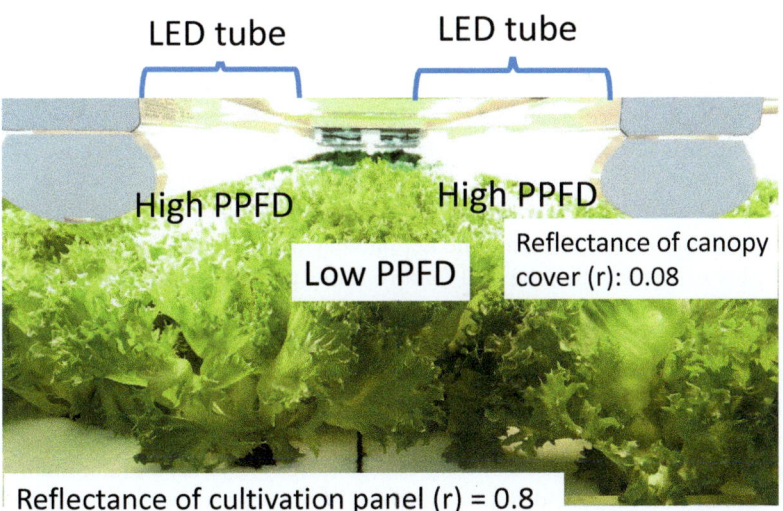

Fig. 7.12 Photograph showing high C-PPFD just under LED tube and low C-PPFD between two LED tubes in the cross section of cultivation space

quality, photoperiod or lighting schedule, and lighting direction (downward, horizontal, and upward lighting).

C-PPFD decreases exponentially with the depth of the plant canopy from the top, since the lower leaves receive much fewer photosynthetic photons than the upper ones. Thus, the optimal PPFD of a plant canopy is generally higher than that of a single leaf. Similarly, the optimal PPFD increases with increasing LAI and CO_2 concentration, for example. In short, the optimal PPFD is affected by many factors.

7.4.1.1 Optimal Lighting Direction

Theoretically, to maximize the photosynthesis of a plant canopy, photosynthetic photons emitted from LEDs should be distributed equally over all leaves, providing the same PPFD perpendicular to each point of the leaf surface. In this case, the light environment in the plant canopy would be better expressed by spherical PPFD rather than horizontal PPFD. A novel LED lighting system can be developed to provide photosynthetic photons more evenly over all leaves by a combination of downward, horizontal, and upward lighting (Chap. 4).

7.4.2 Optimal Photo- and Dark Periods?

Questions are often asked regarding the light/dark periods or lighting cycle in PFALs. Is plant growth at PPFD of 200 $\mu mol\ m^{-2}\ s^{-1}$ with a 16-h light/8-h dark period (24-h cycle) the same as growth with 2 cycles/day of 8-h light and 4-h dark periods or with 4 cycles/day of 4-h light and 2-h dark? The DLI under these 3 conditions is the same ($3.2 = 16 \times 200/10^6$) $mol\ m^{-2}\ d^{-1}$ at a PPFD of 200 $\mu mol\ m^{-2}\ s^{-1}$), so that the daily net photosynthetic rate of the plant canopy would be similar. However, plant height or stem internode and other growth parameters may differ, affecting the net photosynthesis of the canopy.

Another example is that the DLI for a 16-h/day photoperiod at 200 $\mu mol\ m^{-2}\ s^{-1}$ is the same as for a 12-h/day photoperiod at 267 $\mu mol\ m^{-2}\ s^{-1}$ and for a 20-h/day photoperiod at 160 $\mu mol\ m^{-2}\ s^{-1}$ if the PPFD of 267 $\mu mol\ m^{-2}\ s^{-1}$ is not too high. Therefore, the net photosynthetic rate and thus the plant canopy growth rate should be similar under these conditions. In actuality, plant growth differs in some cases due to the interactions among photosynthesis, leaf area expansion, and stem elongation. In case that the electricity charge (price) depends on maximum power consumption rate (kW) and time of day (night time reduction), the cost for electricity varies under the same DLI. The initial cost for LED installation also varies.

The lighting schedule and DLI must be decided by taking into account the yield and quality of produce, electricity costs for lighting, etc. (Kubota et al. 2016). Since the lighting schedule can be set relatively arbitrarily in PFALs, this is important to maximize profitability. On the other hand, the PPFD over the plant canopy can be changed relatively arbitrarily if the DLI can be maintained at a specific level. This is important when variable natural energy sources such as solar energy, wind power, and biomass are used for generation of electricity in PFALs.

7.4.3 Optimal Light Quality?

The spectral light distribution of LEDs used in PFALs (see Chap. 1 for definition) is roughly divided into 5 wavelength bands: ultraviolet (UV), blue, green, red, and far-red (some LEDs do not emit ultraviolet and/or far-red). For growing high-quality plants, photon flux ratios of the red/far-red and blue/red bands are often critical.

To determine the quasi-optimal light quality, experiments with 243 ($= 3^5$) conditions must be performed for three levels of photon flux density for each wavelength band. Since the optimal light quality is often affected by PPFD (400–700 nm), 729 ($= 243 \times 3$) conditions must be examined to determine the optimal light quality under three PPFD levels. In addition, the optimal combinations of light quality and PPFD depend on other environmental factors and plant growth stage. Then, we need simulation models and muti-variable data analyses, in addtion to experiments (See Chap. 32).

7.4.3.1 Light Source for Far-Red and Ultraviolet (UV)

In addition to photosynthetically active radiation (PAR; 400–700 nm), far-red and/or UV radiation is often necessary at specific growth stages to improve the morphology and functional components of plants. Required flux densities for far-red and UV are much lower than those for PAR. Thus, it may be beneficial to install far-red/UV lamps separately from PAR lamps or to turn far-red/UV lamps on and off independent of PAR lamps.

7.4.4 Interactions Among Environmental Factors

When light quality changes, leaf thickness (Shibuya et al. 2015), leaf inclination angle, internode (stem) length, leaf area, light reflectance/transmittance, etc. also change. Those ecophysiological characteristics of plants affect the subsequent growth of the plant canopy.

Changes in plant growth in turn affect the microenvironment within the plant canopy, again affecting plant growth. Plant growth is thus a result of multiple interactions among light quality, changes in ecophysiological characteristics, and microenvironments. All of these must be kept in mind when interpreting the results of experiments on light quality effects.

7.5 Future Work

In the present simulations, r of the LAI of the plant canopy is considered as a variable, but its three-dimensional (3D) structure of the canopy is not. In the near future, a 3D plant canopy model needs to be introduced, which is not too difficult technically and theoretically, as described in Parts 3 and 4 of this volume. That will allow light environment simulation models to be integrated with models for plant growth and development, heat and mass balance, spatial distributions of environmental factors, cost and benefit analysis, etc. (Fig. 7.13).

7.5.1 Challenges

To determine the optimal light environment for PFALs, a combination of the following methodologies may be useful, which will be challenges in the development of next-generation PFALs (See also Chap. 32):

1. Using self-learning systems with artificial intelligence (AI) including big data mining and image-processing technologies
2. Using models for simulating the growth, development, and functional components of the plant canopy under different environmental conditions

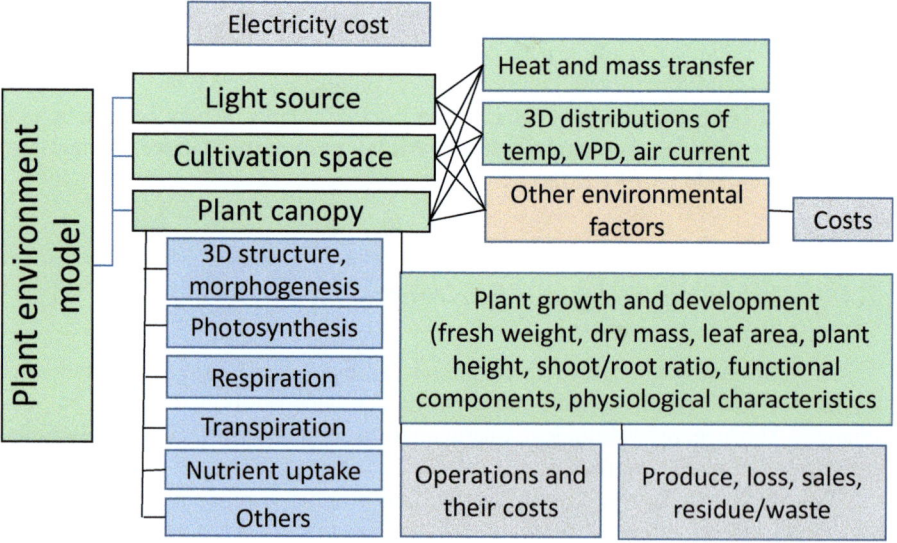

Fig. 7.13 Scheme showing the plant environment model and its components for plant production process in PFAL

3. Developing a flexible, adaptive LED lighting system using the outputs from self-learning and simulation systems
4. Noninvasive capture and processing of 3D camera images of plant canopy architecture for estimating its 3D structure (leaf inclination angle distribution, LAI, etc.) and biochemical characteristics of plants (chlorophyll fluorescence, functional components, etc.)
5. Developing an intelligent, distributed autonomous PFAL system that can generate electricity using natural energy sources such as solar energy, wind power, biomass, and geothermal energy

References

Kubota, C, Kroggel M, Both AJ, Burr JF, Whalen M (2016) Does supplemental lighting make sense for my crop? – empirical evaluations. Acta Hortic 1134 (Light in Horticulture) 403–411

Shibuya T, Endo R, Yuba T, Kitaya Y (2015) The photosynthetic parameters of cucumber as affected by irradiances with different red: far-red ratios. Biol Plant 59:198–200

Part III
Optical and Physiological Characteristics of a Plant Leaf and a Canopy

Chapter 8
Optical and Physiological Properties of a Leaf

Keach Murakami and Ryo Matsuda

Abstract The most important role of a leaf is capturing light energy and fixing CO_2 into carbohydrates (i.e., photosynthesis). Fundamental knowledge on the optical and physiological properties of an individual leaf of a C_3 plant is summarized below. A leaf adjusts light absorption, at the scales of both whole-leaf and intra-leaf, in order to efficiently capture light energy and to avoid photodamage caused by excessive light energy. Several interacting factors involved in orchestrating these optical properties, such as leaf orientation, mesophyll structure, chloroplast movement, and the absorption properties of phytopigments, are outlined. Photosynthesis consists of two reactions that are spatially separated within the chloroplast. Light energy is converted into reducing power and chemical energy via the electron transport chain. These are then consumed during CO_2 fixation in the carbon assimilation process. The electron transport reaction is affected significantly by the spectral distribution of light due to the optical properties of the leaf. Photosynthesis is closely related to other physiological processes. CO_2 uptake accompanies water vapor release (transpiration). Produced photosynthates are transported to the other plant organs (translocation). Brief information about the significance and the machinery of these photosynthesis-related processes is provided.

Keywords Carbon assimilation • Electron transport • Light absorption • Photosynthetic quantum yield • Phytopigment • Transpiration • Translocation

8.1 Introduction

The most important role of a leaf is capturing light energy and fixing CO_2 into carbohydrates—i.e., photosynthesis—and serving them as the energy source to the plant. In greenhouses and plant factories, LEDs are used to enhance the photosynthetic rate of plants, as well as to regulate development, such as floral differentiation

---

K. Murakami (✉) • R. Matsuda
Graduate School of Agricultural and Life Sciences, The University of Tokyo, Bunkyo-Ku, Tokyo, Japan
e-mail: keach.murakami@gmail.com

© Springer Science+Business Media Singapore 2016
T. Kozai et al. (eds.), *LED Lighting for Urban Agriculture*,
DOI 10.1007/978-981-10-1848-0_8

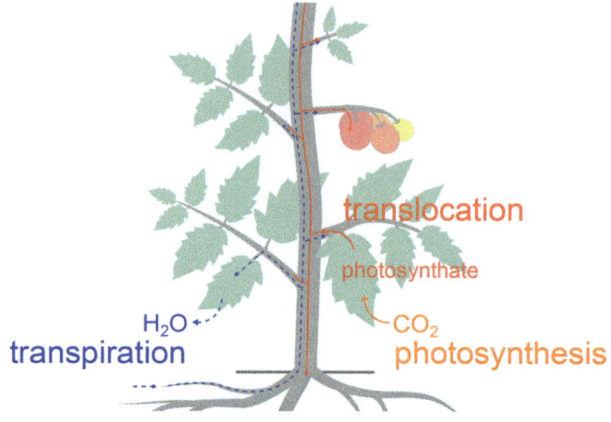

Fig. 8.1 Schematic diagram of leaf functions and flows of photosynthate (*solid lines*) and water (*dashed lines*) in plants

and morphogenesis. The aim of this chapter is to outline the optical properties and photosynthetic physiology of a leaf in a C_3 plant. A leaf is a site not only for photosynthesis but also for other physiological processes, such as transpiration and translocation, which are essential for maintaining efficient leaf photosynthesis (Fig. 8.1). Brief information about the significance and the machinery of transpiration and translocation will be provided in connection to photosynthesis. These physiological functions of a leaf are variable in response to changes in environmental factors, such as photosynthetic photon flux density (PPFD), air temperature, relative humidity, CO_2 concentration, and air current. The close relationships between leaf physiology and environmental factors will be reviewed in the following chapters. Some other physiological responses, such as respiration and senescence, occur commonly in plant cells (not uniquely in leaf cells) and are beyond the scope of this work.

8.2 Optical Properties of a Leaf

An individual leaf has several mechanisms that function to capture light efficiently. A leaf also has several mechanisms to dissipate any excess energy from incident light, in order to avoid photodamage. In this section, we will introduce how a leaf regulates the absorption of incident light and the vertical profiles of PPFD and spectral distribution of light within the leaf. Then we will describe the general functions of phytopigments and their light absorption properties, which originate the leaf absorption spectrum. The vertical light profiles within a leaf, as outlined in this section, are analogous to that observed in plant canopies and communities (see Chap. 9 for more details).

8.2.1 Leaf Orientation and the Vertical Light Profiles Within a Leaf

A leaf changes its orientation in response to the incident light angle from a light source. This light-tracking movement, namely, heliotropism, contributes to maximizing light capture while avoiding excess light energy absorption (Muraoka et al. 1998). These leaf movements have been proven to be mediated by blue light (BL) photoreceptors in *Arabidopsis thaliana*, a model plant species (e.g., Inoue et al. 2008).

Light reflection properties on a leaf surface also change in association with the change in leaf angle (e.g., Brodersen and Vogelmann 2010). A large proportion of photons, those not reflected by the leaf surface, enter the leaf. Note that a portion of the light that enters the leaf is reflected by the numerous air/cell interfaces and exits from the leaf without being absorbed (Vogelmann 1993). A dorsiventral leaf consists of vertically differentiated two-layer mesophyll tissues (palisade and spongy tissues) sandwiched between epidermal cells (Fig. 8.2). The epidermis is covered with a waxy covering called the cuticle, which protects against water evaporation and pathogen infection. The palisade cells are cylindrically shaped, vertically elongated, and closely arranged along the upper epidermis. Chloroplasts play a central role in photon absorption as they contain the photosynthetic apparatus and are the site of photosynthesis (see Sect. 8.3.1) in the mesophyll cells. The PPFD at a given paradermal plane decreases exponentially according to the depth within the mesophyll layers, and a large proportion of the photons is absorbed by the palisade chloroplasts (e.g., Brodersen and Vogelmann 2010). Light that penetrates the palisade layers enters the spongy layers. The spongy cells are loosely packed below the palisade tissues. Owing to the complex shape of spongy cells, the light is well scattered and the optical path is lengthened (Vogelmann 1993). The longer path length contributes to an increased opportunity for light absorption by the pigments.

The chloroplasts accumulate at the upper and lower sides of palisade cells at low PPFDs and redistribute on the side walls of palisade cells at high PPFDs (Fig. 8.2), as mediated by the BL photoreceptor phototropin 2 (Wada 2013). Under high-PPFD conditions, the chloroplasts in the upper cell layers absorb enough photons to become light saturated. Therefore, increasing photon absorption by these chloroplasts does not enhance whole-leaf photosynthesis. In this case, distributing light toward the chloroplasts in the lower cell layers enhances the whole-leaf photosynthesis, because these chloroplasts are not yet photosynthetically light saturated. In addition to the vertical distribution of the light, the photosynthetic characteristics of the chloroplasts also exhibit a vertical profile within a leaf (e.g., Terashima et al. 2009). For better comprehension of the vertical light profiles within a leaf and photosynthetic properties of a leaf, see Terashima et al. (2009) and the cited articles therein.

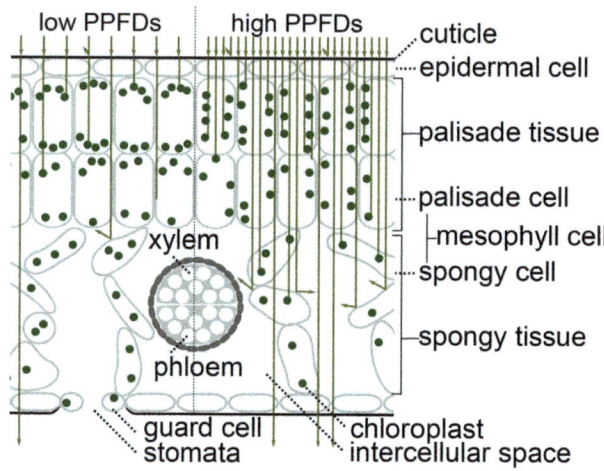

Fig. 8.2 Vertical cross section of a leaf and the distribution of chloroplasts at high and low PPFDs

8.2.2 Pigments and Spectral Absorption of a Leaf

As they travel through the leaf, photons are absorbed by phytopigments. The representative pigment of photosynthetic organisms is chlorophyll (Chl). In higher plants, two types of Chls (Chl a and Chl b) play a dominant role in photosynthetic light absorption. The absorption peaks of these pigments are observed in the blue and red wavebands (Fig. 8.3). A portion of the energy of photons absorbed by Chls is consumed by photochemical reactions (see Sect. 8.3.1). Other pigments are also involved in photosynthetic photon absorption in a leaf. Some carotenoids—insoluble pigments in chloroplasts with an absorption band ranging approximately 350–500 nm (Fig. 8.3)—absorb photons and transfer the energy to Chls. Owing to this function, they are called accessory pigments. On the other hand, at high PPFDs, different types of the carotenoids receive a certain amount of energy from Chls and dissipate it as heat. This system (xanthophyll cycle) protects the photosynthetic apparatus from photodamage (Niyogi et al. 1997). Another major phytopigment group is the flavonoids like anthocyanin mainly accumulating in the vacuole of epidermal cells. Flavonoids absorb shorter wavelengths of light, including ultraviolet (UV) radiation, but the energy is not utilized for photosynthesis. The main functions of flavonoids are cell protection by the screening of UV radiation from incident light and the deactivation of reactive oxygen species (Agati and Tattini 2010). The BL photoreceptor cryptochrome 1 is reported to be involved in the expression of genes related to flavonoids biosynthesis in *Arabidopsis* (Jackson and Jenkins 1995).

As a result of the spectral absorption properties of the pigments, light absorptance of a leaf depends on the wavelength of the light (Fig. 8.4). Note that the term absorptance (α) indicates the fraction of light absorbed, while absorbance (A) indicates the attenuation of the light in the logarithmic scale. These parameters can be defined at a given wavelength as

Fig. 8.3 Absorbance spectra of chlorophyll *a* and *b* and α-carotene (Adapted from Gates et al. 1965)

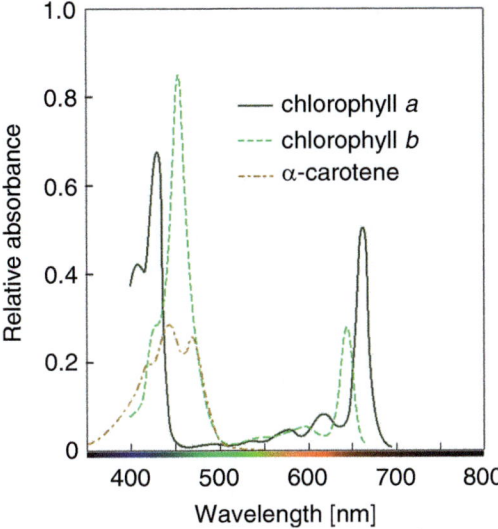

Fig. 8.4 Absorptance, reflectance, and transmittance spectra of a cucumber leaf (Adapted from Murakami et al. 2016)

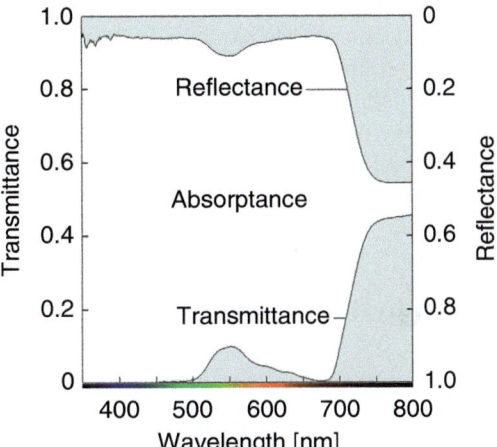

$$\alpha = \frac{I - (I_R + I_T)}{I}, \quad \text{and} \quad A = -\log_{10}\left(\frac{I_T}{I}\right)$$

where I, I_R, and I_T are spectral photon flux densities (PFDs) or irradiance of incident, reflected, and transmitted light, respectively. Because of the fundamental principle of light absorption (the Beer-Lambert law), the absorbance is in proportion to the optical path length and the concentration of the pigments. As shown in Fig. 8.3, the in vitro absorbances of the photosynthetic pigments in the green waveband are quite low. The low absorbances result in a higher proportion of PFD in the green waveband in reflected light; thus, a leaf generally looks green.

However, although the absorbances are very low, a considerable proportion of the photons in the green waveband is still absorbed by a leaf (Fig. 8.4). This is because the remarkably elongated optical path length, caused by the scattering effect, increases the opportunity of photon absorption in the green waveband (Terashima et al. 2009). Considering these data, the extent to which green light (GL) actually drives photosynthesis of a leaf is greater than what is inferred from the pigment absorbance spectrum and the leaf color. Moreover, the lower absorbance in the green waveband indicates that GL can reach chloroplasts located in cells deeper within the leaf, more so than BL and red light (RL), as shown by Brodersen and Vogelmann (2010). GL might enhance the leaf photosynthetic rate by increasing the photosynthetic rate of the chloroplasts located in the lower cells, i.e., those that are not yet photosynthetically light saturated, when the upper mesophyll chloroplasts are light saturated (Terashima et al. 2009).

Even with the relatively planar leaf absorptance spectrum resulting from the scattering effect, the transmitted incident light on the lower leaves still contains more GL and far-red light (FRL) than does untransmitted light. The relative spectral PFD distribution, especially in the far-red waveband, seems to be used by plants as the key stimulus for inducing changes in photosynthetic characteristics (i.e., light acclimation) of the irradiated leaves (e.g., Murakami et al. 2016) and in whole-plant growth (Demotes-Mainard et al. 2015). The light environment of a leaf not only affects the photosynthetic characteristics of the irradiated leaf but also those of unirradiated, newly developing leaves (e.g., Murakami et al. 2014). In higher plants, environmental changes around an organ sometimes induce morphogenetic and physiological reactions not only to the organ but also to other parts of organs. Such phenomena are called systemic regulation.

8.3 Physiological Properties of a Leaf

Considering the above mentioned light absorption properties of a leaf, we will now outline the machinery of photosynthesis. We will pay particular attention to the light-driven process of photosynthesis and its dependency on spectral PFD distribution, due to the importance in the horticultural application of LEDs. We also briefly summarize the importance and the machinery of transpiration and translocation in this section, as representative photosynthesis-related physiological processes.

8.3.1 Photosynthesis

Photosynthesis consists of two reactions that are spatially separated within the chloroplast (Fig. 8.5): (1) light-driven electron transport on the thylakoid membrane (the lipid bilayer within the chloroplast) and (2) enzymatic reactions for carbon

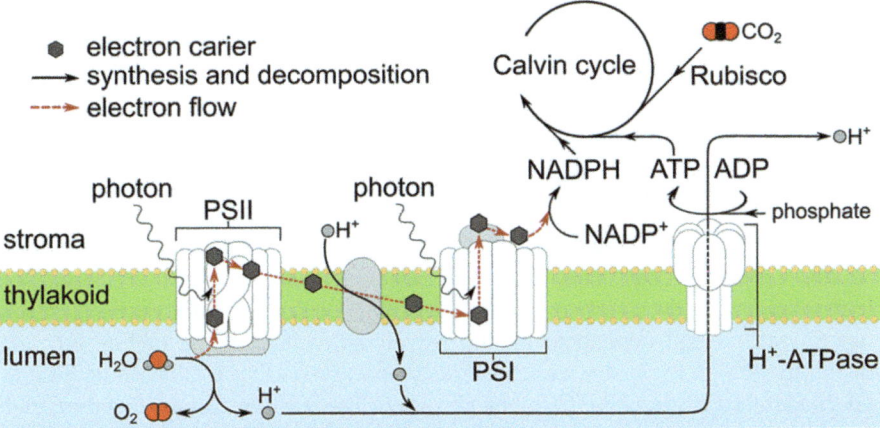

Fig. 8.5 Schematic diagram of the photosynthetic electron transport reactions on the thylakoid membrane and carbon assimilation in the stroma of a C_3 plant

assimilation in the stroma (the fluid surrounding the thylakoid membranes within the chloroplast). The reducing power and chemical energy produced by electron transport are consumed during the carbohydrate synthesis of the carbon assimilation. An appropriate balance between these two reactions is essential for efficient photosynthesis. These have traditionally been called the "light reaction" and the "dark reaction", respectively. However, further research has clarified that several processes in the light reaction progress without light and that a part of the dark reaction needs light in order to activate some of the necessary enzymes. Thus, these terms are not currently accepted. Though the terminology seems to be under discussion among the specialists, we will use the terms "electron transport" and "carbon assimilation" to avoid confusion in this chapter.

Electron Transport The light energy captured by Chls and accessory pigments is converted into reducing power as the reduced form of nicotinamide adenine dinucleotide phosphate (NADPH) and chemical energy as adenosine triphosphate (ATP), through the electron transport chain (ETC). The ETC is anchored by the two photochemical reactions, occurring in series, at the two photosystem complexes, PSII and PSI, within the thylakoid membrane (Fig. 8.5). The electron transports are powered by photon energy transferred to the respective photosystems, thus enabling electron transfer against the redox potential (called the Z scheme). The series of the two reactions drive (1) O_2 evolution, proton (H^+) production in the lumen (the fluid inside of the thylakoid membrane), and electron (e^-) subtraction, by water decomposition, (2) transfer of the electron via carriers and accompanying H^+ movement from stroma to lumen, and (3) $NADP^+$ reduction into NADPH by consumption of the electrons. The accumulated protons in the lumen form a pH-gradient across the membrane. This electrochemical potential drives ATP synthesis from adenosine diphosphate (ADP) and phosphate, via H^+-ATPase in the membrane. As the results of ETC, NADPH and ATP are generated in the stroma. These end products are used widely as a reductant and an energy source, respectively, in plant metabolism.

Because of the optical properties of a leaf (see Sect. 8.2), the gross photosynthetic rate of a leaf (P_g) also depends on the spectral PFD distribution of light (e.g., McCree 1972, Inada 1976). When a leaf surface is irradiated with light at the same PPFD incident, the P_g is highest in the red waveband, followed by the blue and then the green wave bands (solid line in Fig. 8.6). Note that FRL and UV, which deviate from the definition of photosynthetically active radiation (PAR), also drive photosynthesis slightly. Even when P_g is represented per unit of absorbed photons by a leaf (photosynthetic quantum yield) to compensate for the differences in absorptance, yield is still dependent on the wavelength (dashed line in Fig. 8.6). The quantum yield in the green waveband is comparable to that in blue, indicating that GL moderately drives the ETC when absorbed. The relatively lower quantum yields in the blue and green wavebands, compared to the red wave band, should be attributable to the absorption by the non-photosynthetic pigments (flavonoids) and to the heat dissipation by the carotenoids.

As for the longer wavelengths (>680–690 nm), the quantum yield decreases sharply. This sharp decrease can be accounted for by the difference in the light absorption properties between PSII and PSI. PSI contains much Chl *a* and thereby preferentially absorbs longer wavelength of light than PSII absorbs (e.g., Evans 1986). Because of this preferential absorption by PSI, electron transport is limited by the lower electron transport reaction rate of PSII. Under this condition, photon absorption by PSI exceeds supply, and the light-use efficiency of the reaction in PSI is hence lowered. In contrast, BL and RL have been reported to be preferentially absorbed by PSII (e.g., Evans 1986). Therefore, a measured quantum yield under the mixture of FRL+RL or FRL+BL can be higher than that estimated from the separately obtained yields (Emerson effect).

Carbon Assimilation Dissolved CO_2 in the stroma is fixed into sugar phosphates. This carbon assimilation process, called the Calvin cycle, consists of over a dozen enzymatic reactions. The most important and abundant enzyme in the Calvin cycle is ribulose-1,5-bisphosphate carboxylase/oxygenase (Rubisco). This enzyme catalyzes the entry-point reaction of CO_2 into the cycle. Triose phosphate (TP), the end product of the Calvin cycle, is synthesized from the fixed CO_2, consuming the NADPH and ATP produced by the ETC. A portion of the TP created is exported to the cytosol and converted into sucrose. TP is also required for starch synthesis in the chloroplasts. At the same time, Rubisco catalyzes the oxygenation reaction, consuming O_2. This reaction triggers a series of enzymatic reactions wherein CO_2 is released and NADPH and ATP are consumed (the glycolate cycle). Although the physiological role of this process is still debated, these reactions decrease the photosynthetic carbon assimilation rate and photosynthetic light-use efficiency (see also Chap. 12).

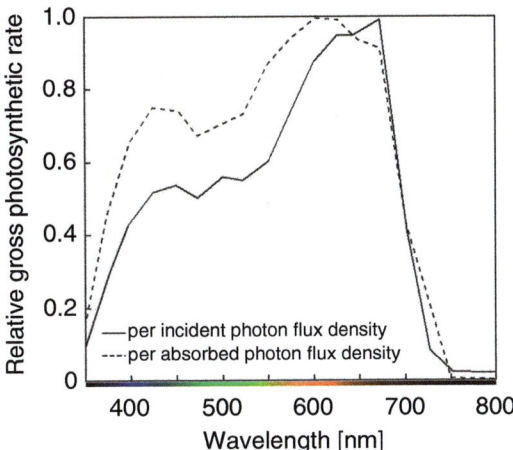

Fig. 8.6 Relative gross photosynthetic rates per incident (*solid line*) and absorbed (*dashed line*) photons. The mean values of 22 plant species are shown (Adapted from McCree 1972)

8.3.2 Transpiration

Transpiration is the process of water vapor movement from plants to the atmosphere. Although transpiration can occur at any aerial part of the plant, such as stems and flowers, leaves are the predominant organ for transpiration. CO_2 uptake occurs at the expense of inevitable water vapor release in leaves. The amount of H_2O molecules released from plant tissues is approximately 500 times as many as that of CO_2 molecules taken in during this gas exchange. Through transpiration, water vapor diffuses along the concentration gradient between the transpiring surface of the plant and the atmosphere. The water vapor moves through stomatal pores or across the cuticle layer (called stomatal or cuticular transpiration, respectively). In stomatal transpiration, water vapor evaporates at the surface of mesophyll cells, diffuses in the intercellular air spaces, and moves to the outside of the leaf through stomata located within the leaf epidermis. The flux of water vapor in this pathway is largely regulated by the extent of the stomatal openings. In cuticular transpiration, water vapor from the epidermal cell surface diffuses across the cuticular layer to the atmosphere. In either pathway, water vapor present immediately outside of the leaf surface further diffuses through the leaf boundary layer—a layer of boundary air flow formed on the leaf surface—to the "free" atmosphere (see Chaps. 12 and 13 for more details).

Too much transpiration can result in growth inhibition due to water shortage. Although there is the risk of wilting, this water movement process plays essential roles in plant growth. First, it serves as a driving force of the transpiration stream in the xylem, contributing to the absorption of water and inorganic ions from the root zone and to the transportation of water, ions, and organic compounds, such as phytohormones, throughout the plant. Thus, too little transpiration can lead to nutrient deficiency disorders.

Some of the major problems observed in greenhouses and plant factories are blossom-end rot in tomato fruits (e.g., Ho 1999) and tipburn in the leaves of lettuce (e.g., Collier and Tibbitts 1982) and strawberry (e.g., Bradfield and Guttridge 1979); these are caused by low concentrations of calcium in the cells. Promotion of transpiration by controlling vapor pressure deficit (VPD) and air current might be a key solution for these disorders (e.g., Goto and Takakura 1992). Second, transpiration contributes to leaf cooling by facilitating evaporative heat transfer. The leaf dissipates a part of the absorbed radiant energy so as not to overheat. The water vapor released by transpiration removes the energy as latent heat and suppresses the temperature increase of a leaf. Owing to the smaller thermal radiant intensity of LEDs in general lighting applications, leaf and plant temperatures can be kept lower than when using traditional light sources such as high-intensity discharge lamps (HIDs) and fluorescent lamps (FLs).

8.3.3 Translocation

The photosynthates manufactured in the leaf are transported to other organs as sucrose dissolved in phloem sap. This transportation through the phloem, called translocation, is usually explained by the pressure flow model (Münch 1930). The term translocation also indicates inorganic nutrition transport and/or cell-to-cell short-distance transport of the substrates in the broad sense, but we only refer to the long-distance sugar transport here. The organs that demand the substrates, such as the newly growing stems, leaves, flowers, fruits, and tubers, are referred to as "sinks", while those supplying them (i.e., mature leaves) are "sources". The translocated sucrose is converted into monosaccharides, and these are consumed as respiratory substrates to generate ATP and carbon skeletons. In fruits and storage organs, a portion of the sucrose is converted into starch and stored.

When the rate of photosynthate production exceeds the translocation rate, the produced photosynthates accumulate in the source leaves. Such a carbohydrate accumulation can be observed in particular during the day with a high PPFD. The accumulated photosynthates are reported to suppress instantaneous leaf photosynthesis and the synthesis of photosynthetic proteins such as Rubisco (e.g., Araya et al. 2006). The appropriate management of the balance between the source and sink has been believed to be important for maintaining a higher photosynthetic rate and enhancing the yields and quality of the harvests. A review article on this balance, dealing with tomato, is available (Ho 1988).

References

Agati G, Tattini M (2010) Multiple functional roles of flavonoids in photoprotection. New Phytol 186:786–793

Araya T, Noguchi K, Terashima I (2006) Effects of carbohydrate accumulation on photosynthesis differ between sink and source leaves of *Phaseolus vulgaris* L. Plant Cell Physiol 47:644–652

Bradfield EG, Guttridge CG (1979) The dependence of calcium transport and leaf tipburn in strawberry on relative humidity and nutrient solution concentration. Ann Bot 43:363–372

Brodersen CR, Vogelmann TC (2010) Do changes in light direction affect absorption profiles in leaves? Funct Plant Biol 37:403–412

Collier GF, Tibbitts TW (1982) Tipburn in lettuce. In: Janick J (ed) Horticultural reviews, vol 4. Wiley, Hoboken, pp 49–65

Demotes-Mainard S, Péron T, Corot A et al (2015) Plant responses to red and far-red lights, applications in horticulture. Environ Exp Bot 121:4–21

Evans JR (1986) A quantitative analysis of light distribution between the two photosystems, considering variation in both the relative amounts of the chlorophyll-protein complexes and the spectral quality of light. Photobiochem Photobiophys 10:135–147

Gates DM, Keegan HJ, Schleter JC et al (1965) Spectral properties of plants. Appl Opt 4:11–20

Goto E, Takakura T (1992) Prevention of lettuce tipburn by supplying air to inner leaves. Trans ASAE 35:641–645

Ho LC (1988) Metabolism and compartmentation of imported sugars in sink organs in relation to sink strength. Annu Rev Plant Physiol Plant Mol Biol 39:355–378

Ho LC (1999) The physiological basis for improving tomato fruit quality. Acta Hortic 487:33–40

Inada K (1976) Action spectra for photosynthesis in higher plants. Plant Cell Physiol 17:355–365

Inoue S, Kinoshita T, Takemiya A et al (2008) Leaf positioning of *Arabidopsis* in response to blue light. Mol Plant 1:15–26

Jackson JA, Jenkins GI (1995) Extension-growth responses and expression of flavonoid biosynthesis genes in the *Arabidopsis hy4* mutant. Planta 197:233–239

McCree KJ (1972) The action spectrum, absorptance and quantum yield of photosynthesis in crop plants. Agric Meteorol 9:191–216

Münch E (1930) Die Stoffbewegungen in der Pflanze. Gustav Fischer, Jena

Murakami K, Matsuda R, Fujiwara K (2014) Light-induced systemic regulation of photosynthesis in primary and trifoliate leaves of *Phaseolus vulgaris*: effects of photosynthetic photon flux density (PPFD) *versus* spectrum. Plant Biol 16:16–21

Murakami K, Matsuda R, Fujiwara K (2016) Interaction between the spectral photon flux density distributions of light during growth and for measurements in net photosynthetic rates of cucumber leaves. Physiol Plant 158(2):213–224. http://onlinelibrary.wiley.com/journal/10.1111/(ISSN)1399-3054/earlyview

Muraoka H, Takenaka A, Tang Y et al (1998) Flexible leaf orientations of *Arisaema heterophyllum* maximize light capture in a forest understorey and avoid excess irradiance at a deforested site. Ann Bot 82:297–307

Niyogi KK, Björkman O, Grossman AR (1997) The roles of specific xanthophylls in photoprotection. Proc Natl Sci USA 94:14162–14167

Terashima I, Fujita T, Inoue T et al (2009) Green light drives leaf photosynthesis more efficiently than red light in strong white light: revisiting the enigmatic question of why leaves are green. Plant Cell Physiol 50:684–697

Vogelmann TC (1993) Plant tissue optics. Annu Rev Plant Physiol Plant Mol Biol 44:231–251

Wada M (2013) Chloroplast movement. Plant Sci 210:177–182

Chapter 9
Optical and Physiological Properties of a Plant Canopy

Yasuomi Ibaraki

Abstract The characteristics of masses of leaves should be considered while controlling the light environment for plant production. The term "canopy" is used to express the mass of leaves of multiple plants. The optical characteristics and photosynthetic properties of a canopy are sometimes different from those of a single leaf. One of the most important characteristics of a canopy is that there exists a photosynthetic photon flux density (PPFD) distribution on and within the canopy according to its leaf distribution pattern. In addition, the light spectral distribution within the canopy is different from that outside the canopy, owing to light absorption and reflection by leaves. Moreover, the light-photosynthetic curve of a canopy differs from that of a single leaf, showing higher or no light-saturated points. A basic method for estimating canopy photosynthesis is to sum up the photosynthetic rates of small parts of canopy that are assumed to be under uniform environmental conditions. An extinction coefficient is used to express light attenuation within a canopy as a function of LAI and to estimate average PPFD at each horizontal layer.

Keywords Canopy photosynthesis • Extinction coefficient • Growth analysis • Leaf area • Light attenuation • Net assimilation rate • PPFD distribution • RGR

9.1 Introduction

As described in the previous chapter (Chap. 8), the optical and physiological properties of a plant leaf are important to understand the light environment needed for plant production. Note that plants have multiple leaves and only single plants are rarely cultivated in agricultural practice. Therefore, the characteristics of masses of leaves should be considered while controlling the light environment for plant production. The term "canopy" is used to express the mass of leaves of multiple plants. According to Moffett (2000), the term "canopy" is defined as the aboveground plant organs within a community. As plants grow, their biomass increases due to photosynthesis. Photosynthesis of a canopy/individual plant is

Y. Ibaraki (✉)
Faculty of Agriculture, Yamaguchi University, 1677-1 Yoshida, Yamaguchi 753-8515, Japan
e-mail: ibaraki@yamaguchi-u.ac.jp

the sum total of the photosynthesis of each leaf composing the canopy or plant. However, light environments differ among leaves owing to differences in the spatial positions in the canopy. In addition, a leaf's photosynthetic properties depend on the leaf's physiological states, including leaf age, water content, and environmental history. Thus, the process of summing the photosynthetic amounts of individual leaves over a canopy is complicated. In addition, note that the optical characteristics of a canopy are sometimes different from those of a single leaf. For example, the reflectance of a plant canopy tends to be rather lower than that of its components, i.e., single leaves, because multiple reflections between adjacent leaves and between leaves and stems lead to trapping of radiation (Jones 1992).

One of the most important characteristics of a canopy that should be considered in plant cultivation is that there exists a photosynthetic photon flux density (PPFD; μmol m^{-2} s^{-1}) distribution on and within the canopy according to its leaf distribution pattern. This PPFD distribution is critical for photosynthesis of the plant canopy and consequently affects growth. Proper understanding of PPFD distribution within/on a plant canopy is likely to be required in the application of artificial lighting for plant production.

Normally, the total leaf area of a plant canopy that has grown well is greater than the ground area below it. The ratio of the total leaf area to the ground area (which is referred to as the leaf area index (LAI); see Chap. 10 for details) depends on plant species, the growth stage, and cultivation density. The concept of LAI is useful when considering light distribution and canopy photosynthesis.

In this chapter, the optical and physiological properties of a plant canopy are introduced with the concept of extinction coefficients to express light attenuation as a function of LAI in order to understand the PPFD distribution in a canopy. Methods for estimating canopy photosynthesis will also be introduced.

9.2 Light Attenuation Through Plant Canopy

Light falling on a canopy surface enters the canopy and attenuates through absorption and reflection by leaves. Under the assumption of a horizontally uniform structure for the canopy, the vertical light distribution within canopy can be expressed by the following equations:

$$I = I_0 e^{-k \cdot F} \quad \text{or} \quad \ln\left(\frac{I}{I_0}\right) = -k \cdot F, \tag{9.1}$$

where I is the irradiance (W m^{-2}) or photon flux density (μmol m^{-2} s^{-1}) at vertical position z from the top surface of the canopy, I_0 is the irradiance/photon flux density at the top of the canopy, F is the leaf area index summed from the top to z (which is sometimes referred to as cumulative LAI), and k is the extinction coefficient. This equation is analogous to Beer's law expressing light attenuation in a homogeneous

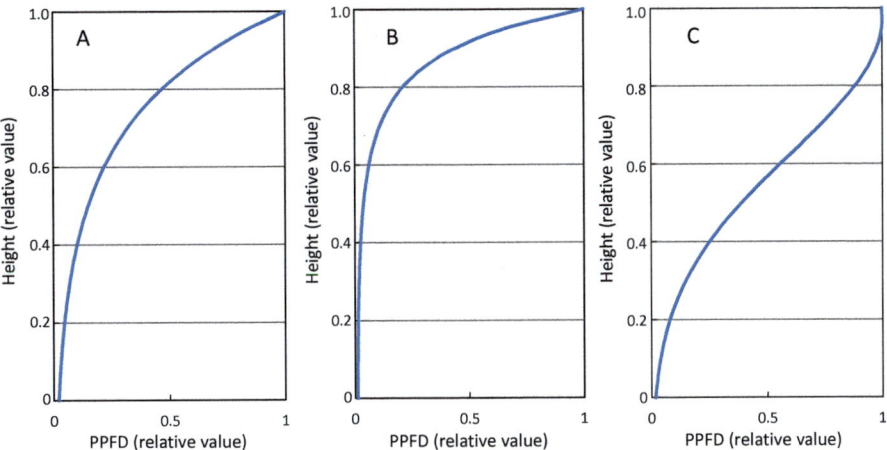

Fig. 9.1 Typical vertical PPFD distribution patterns within plant canopy depicted using Eq. 9.1 for canopies in which leaf area is distributed evenly (**a**), decreases linearly toward the *bottom* (**b**), and increases linearly toward the *bottom* (**c**)

medium and expresses the exponential attenuation of light as a function of vertically accumulated LAI within the canopy. The extinction coefficient depends not only on the leaf optical properties but also on the geometrical structure of the canopy, including the size, shape, and inclination of leaves, as well as on the azimuth and altitude of the sun. Figure 9.1 shows a typical PPFD distribution pattern within a plant canopy depicted using an Eq. 9.1 for a case in which light enters from the top of canopy.

The concept of considering the light distribution within a canopy as a function of LAI using the extinction coefficient has been commonly used for various types of plant canopies, in particular for forests and crops in which the plants have considerable height and there is a vertical leaf distribution, such as in orchards or cereal crops. In the case of leaf lettuce, which is a dominant species for cultivation in plant factories, the concept for vertical light distribution may be difficult to apply because it is difficult to define the horizontal multilayer due to its lower height or the leaf distribution pattern. However, note that there still is a variation in irradiance/photon flux density within the canopy or on the canopy surface even in such lower canopies.

9.3 Extinction Coefficients in Plant Canopy

Examples of extinction coefficients in various types of canopies are listed in Table 9.1. The extinction coefficients depend on the canopy structure and direction of incident light as described above. Although the differences among plant species is mainly due to differences in their canopy structures, note that for a given species,

Table 9.1 Examples of extinction coefficient k for PAR of various types of canopies

Type of canopy/species	Extinction coefficient k	Reference
Mainly horizontal leaves, such as sunflower or cotton	0.7–1.0	Fukai (1999)
Mainly erect leaves, such as barley and sugar cane	0.3–0.6	Fukai (1999)
Lettuce	*0.80, **0.66	*Javanovic et al. (1999)
		** Tei et al. (1996)
Onion	*1.06, **0.47	*Javanovic et al. (1999)
		**Tei et al. (1996)
Cabbage	1.17	Javanovic et al. (1999)

the structure may change with growth. For a given leaf area profile, large k values generally imply that photon irradiance (photon flux density) decreases rapidly with depth, whereas a canopy with a small k would allow solar radiation to penetrate deeply (Fukai 1999).

Under simple assumptions, k can be calculated. For example, in a canopy comprising only randomly distributed horizontal opaque leaves that can be divided into horizontal layers with the same leaf area without leaves overlapping, the extinction coefficient will be 1. As individual leaves are opaque in this case, the change in irradiance/photon flux density (dI) when light passes through a layer with LAI of dF is equal to $-I\,dF$:

$$dI = -I\,dF \qquad (9.2)$$

where I is the irradiance/photon flux density at the top of the layer.

Integrating this relationship downward through total LAI gives the average irradiance/photon flux density on a horizontal surface below the position with that LAI (F):

$$I = I_0 e^{-F} \qquad (9.3)$$

where I_0 is I at top layer (i.e., I when $F = 0$). Thus, $k = 1$ in this case.

For a canopy comprising leaves at other angles under the same assumptions, the extinction coefficient k is assumed to be the ratio of the shadow area to the actual leaf area. Thus, if the solar elevation angle is β, k equals $\cot \beta$ for vertical leaves and $\cos \alpha$ for leaves with an inclination angle of α (Fig. 9.2).

In real canopies, the distribution pattern of leaves is complicated and varies with time. Therefore, calculation of k is difficult without simple assumptions of leaf distribution. Examples of calculating k using simple leaf distribution models are described by Jones (1992).

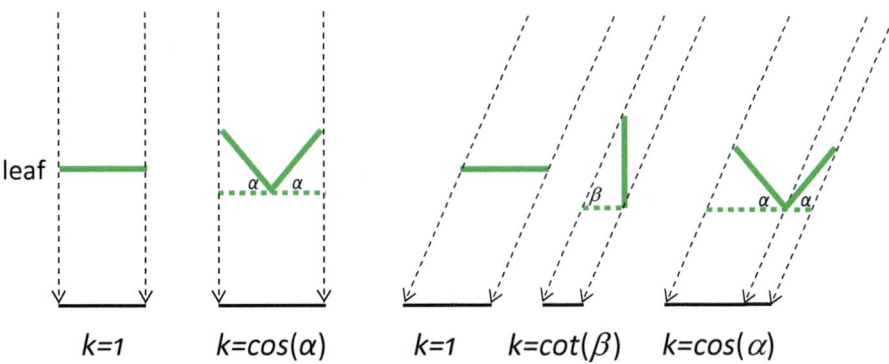

Fig. 9.2 Extinction coefficient k as the ratio of the shadow area to the actual leaf area for several cases under sun light. α is leaf inclination angle and β is the solar elevation angle

In contrast, k can be estimated for a real canopy based on light distribution measurements (Fig. 9.3).

As a more simple method, using measurements of irradiance/photon flux density I_0 above the canopy surface and I_n on the ground, k can also be estimated by the following equation:

$$k = -\frac{\ln\left(\frac{I_n}{I_0}\right)}{\text{LAI}}. \tag{9.4}$$

In this case, both the direction of incident light (solar elevation angle) and clumping index (a nonrandom spatial distribution parameter) are implicitly included in k, which implies that k may vary both temporally and spatially (Zhang et al. 2014). In actual canopies, the value of k calculated from Eq. 9.4 is sometimes greater than 1. In a canopy in which horizontal leaves are gathered at upper layers, light attenuates mainly by passing through only one thin layer with horizontally nonuniform distribution. For example, if k is estimated based on the irradiance/photon flux density measured just below one large leaf such as a lotus, it will be proportional to the negative logarithm of the transmittance of the leaf, being greater than 1 when the transmittance is less than $1/e$ (=0.37).

Note that the extinction coefficient k can be defined both in terms of total solar radiation (including all ranges of wavelength) and PAR (400–700 nm); the k value for solar radiation is likely to be smaller than that for PAR because of the higher absorbance and lower transmittance of leaves for PAR than for total solar radiation. In addition, k for artificial lighting may be different from that for solar radiation due to the differences in the light spectral distribution and the direction of incident light depending on the properties of the light source.

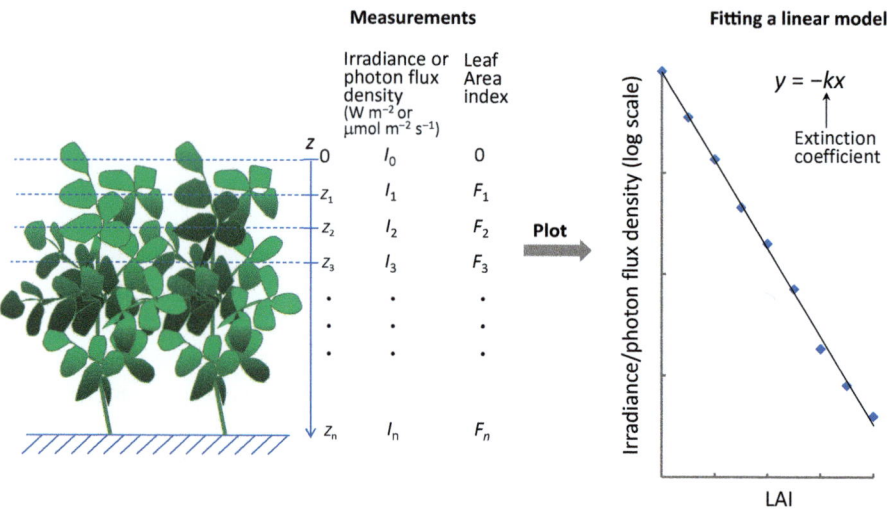

Fig. 9.3 Schematic diagram of estimation of extinction coefficient k based on the light distribution measurements within the canopy

9.4 Consideration of Spectral Properties Within the Canopy

As previously described, the light spectral distribution within the canopy is different from that outside the canopy, owing to light absorption and reflection by leaves. Normally, the proportion of green and/or far-red light is increased within the canopy (Kasperbauer 1971). Figure 9.4 shows examples of light spectra inside and outside the lettuce canopy under artificial lighting (with white LEDs). The spectral distribution of light within a canopy depends not only on the optical properties of the leaves but also on the spectral properties of lamps that are used for artificial lighting. In particular, for light from narrowband LED lamps, which contain no green and far-red light, light spectral distribution within a canopy greatly differs from those under sunlight. Green light may offer benefits for canopy photosynthesis, given that it can better penetrate the plant canopy and potentially increase plant growth by increasing photosynthesis from the leaves in the lower canopy (Kim et al. 2004). Far-red light may affect the morphogenesis of plants.

It should be noted that spectral properties of lamps also affect the extinction coefficient k and, thereby, the PPFD distribution within a canopy. Under light comprising only blue and red light, the extinction coefficient k is high and PPFD within a canopy is low.

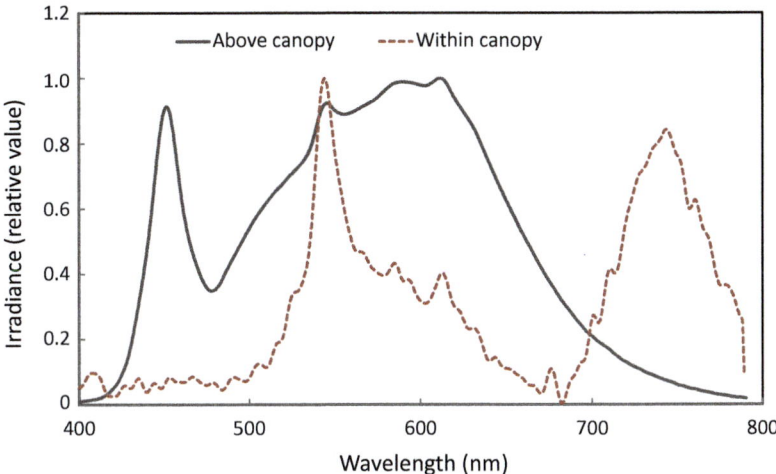

Fig. 9.4 Examples of light spectra measured above and within the lettuce canopy under white LED lighting

9.5 Canopy Photosynthesis

9.5.1 Characteristics of Canopy Photosynthesis

A plant accumulates biomass through photosynthesis and normally has numerous leaves as organs for photosynthesis. In plant production, including plant factories, control of photosynthesis is critical for cultivation management, especially for cases in which there is vegetative growth. Although the photosynthetic properties of a single leaf is a baseline for considering the photosynthetic amount or rate of an individual plant or plant canopy, the photosynthetic properties of individual leaves vary over a plant or canopy depending on leaf age and position; for example, young leaves may be sink leaves with a low photosynthetic capacity. Furthermore, as described in the previous chapter, the spectral distribution of light will change when passing through a leaf.

Normally the light-photosynthetic curve of a canopy differs from that of a single leaf, showing higher or no light-saturated points in extreme cases (Fig. 9.5). This is caused by the existence of a PPFD distribution within the canopy. Increments of light irradiating a canopy can contribute to increasing the photosynthetic rates at the lower layers even if the PPFD at the top layer has reached the saturation point.

To estimate the photosynthetic rate of a canopy properly, the spatial distribution of environmental factors that affect photosynthesis, such as PPFD, CO_2 concentration, air temperature, humidity, and air current, should be identified.

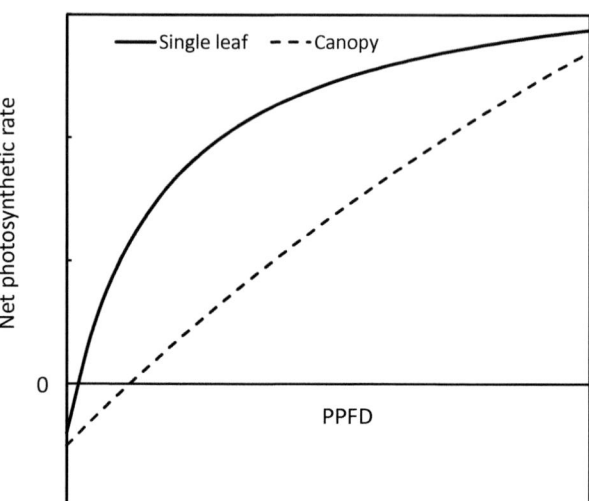

Fig. 9.5 Typical patterns of light-photosynthetic *curves* of a single leaf and canopy. The *curves* were depicted using a rectangular hyperbolic function (Eq. 9.7) changing parameter values

9.5.2 Simple Method for Estimating Canopy Photosynthetic Rate

A basic method for estimating canopy photosynthesis is to sum up the photosynthetic rates of small parts of canopy that are assumed to be under uniform environmental conditions, in particular under uniform PPFDs. After the average PPFDs for these portions of canopy are estimated, their average photosynthetic rates are calculated. The photosynthetic rate of a single leaf depends on PPFD (see Chap. 8 for details), and the light-photosynthetic curve is not linear. Therefore, the average PPFDs are calculated for portions with similar PPFDs and used to estimate the average photosynthetic rate of these portions.

Monsi and Saeki (2005) proposed a simple estimation method for canopy photosynthetic rates using the extinction coefficient. The absorbed PPFD per unit leaf area at a given horizontal layer (in which LAI is dF) is expressed as $-dI/dF$; correspondingly, the PPFD on leaves at that layer is given by the following equation:

$$I' = -\frac{dI}{dF}\frac{1}{(1-\tau)}, \quad (9.5)$$

where τ is the leaf transmittance.

Because dI/dF can be calculated by differentiating I in Eq. 9.1 with respect to F, I' can be calculated using the following equation:

$$I' = I_0 k e^{-kF} \frac{1}{(1-\tau)}, \qquad (9.6)$$

where I' represents the average PPFD on leaves at a given horizontal layer and can be used as an input to a model for the relationship between net photosynthetic rate and PPFD. Several functions have been proposed to express this relationship, with a common characteristic that they represent convex curves with saturated photosynthetic rates. Monsi and Saeki (2005) used a rectangular hyperbolic curve expressed by the following equations:

$$p = \frac{aI'}{(1+bI')} - r, \qquad (9.7)$$

where a and b are constant and r is dark respiration. By inputting I' from Eq. 9.6 into this equation and integrating p with respect to F, canopy photosynthesis P can be expressed as:

$$P = \frac{a}{kb} \ln \frac{(1-\tau) + kbI_0}{(1-\tau) + kbI_0 e^{-kF}} - rF. \qquad (9.8)$$

From this equation, we can estimate the optimal LAI that gives the maximum canopy photosynthesis (Hirose 2004).

In this calculation, photosynthetic properties of an individual leaf are assumed to be constant over the entire canopy; however, this will actually depend on the position within the canopy. For example, leaves at lower layers where PPFD is low may show the photosynthetic properties of shade leaves, i.e., high photosynthetic rates at low PPFD and low saturated rates. Models considering not only light distribution but also distribution of the leaf photosynthetic properties have also been proposed (e.g., Hirose and Werger 1987).

Note that simpler models can be adequate in several cases. Jones (1992) listed two limiting cases in which simpler models can be applied to estimate canopy photosynthesis:

(a) All leaves are at an acute angle to any direct radiation.

- In this case, the total canopy photosynthesis is proportional to light interception with a proportionality constant ε_p. At high LAIs, where all radiation is intercepted, canopy photosynthesis is simply $\varepsilon_p I_0$.

(b) Very low LAI canopies.

- In this case, single leaf photosynthetic models can be applied.

9.5.3 Growth Analysis

We can also estimate canopy photosynthesis using growth analysis, which evaluates how much biomass can be acquired by photosynthesis during a given period. The relative growth rate (RGR) defined as:

$$\text{RGR} = \frac{1}{W}\frac{dW}{dt} \quad (d^{-1}) \tag{9.9}$$

is commonly used for vegetative growth analysis, where W (g) is the dry weight of a plant and t is time (in days). In addition, the net assimilation rate (NAR) as growth rate per unit leaf area can be derived from the following equation and is related to the photosynthetic rate:

$$\text{NAR} = \frac{1}{L}\frac{dW}{dt} \quad (g\ m^{-2}\ d^{-1}), \tag{9.10}$$

where L is leaf area (m^2).

Under the assumption that RGR is constant during a given period, it can be calculated from the dry weights (W_1, W_2) and leaf areas (L_1, L_2) measured at different times (t_1, t_2) by the following equation:

$$\text{RGR} = \frac{\ln(W_2/W_1)}{t_2 - t_1}, \tag{9.11}$$

$$\text{NAR} = \frac{(W_2 - W_1)(\ln L_2 - \ln L_1)}{(L_2 - L_1)(t_2 - t_1)}, \tag{9.12}$$

RGR and NAR can be connected using the leaf area ratio (LAR):

$$\text{RGR} = \text{NAR} \times \text{LAR} \tag{9.13}$$

$$\text{LAR} = \frac{L}{W} \tag{9.14}$$

These calculations are based on invasive (destructive) measurement in which sampling is required for determining the dry weight or leaf area at different times. By contrast, relative leaf growth rate (RLGR) can be estimated nondestructively from images:

$$\text{RLGR} = \frac{\ln(L_2/L_1)}{t_2 - t_1} \tag{9.15}$$

For example, the RLGR of *Arabidopsis thaliana* was estimated automatically and used for phenotyping (Arvidsson et al. 2011). Normally, a projected area can be measured from an image. If a linear relationship is observed between the projected area and the actual leaf area, RLGR can be estimated simply through image analysis (Ibaraki and Dutta Gupta 2014).

References

Arvidsson S, Pérez-Rodríguez P, Mueller-Roeber B (2011) A growth phenotyping pipeline for *Arabidopsis thaliana* integrating image analysis and rosette area modeling for robust quantification of genotype effects. New Phytol 191:895–907

Fukai S (1999) Leaf area index and canopy light climate. In: Atwell BJ et al (eds) Plants in action. Macmillan Education Australia Pty Ltd, Melbourne, Available via DIALOG. http://plantsinaction.science.uq.edu.au/edition1/of subordinate document. Accessed 15 Dec 2015

Hirose T (2004) Development of the Monsi-Saeki theory on canopy structure and function. Ann Bot 95:483–494

Hirose T, Werger MJA (1987) Maximizing daily canopy photosynthesis with respect to the leaf nitrogen allocation pattern in the canopy. Oecologia 72:520–526

Ibaraki Y, Dutta Gupta S (2014) Image analysis for plants: basic procedures and techniques. In: Dutta Gupta S, Ibaraki Y (eds) Plant image analysis. CRC Press, Boca Raton, pp 25–40

Javanovic NZ, Annandale JG, Mhlauli NC (1999) Field water balance and SWB parameter determination of six winter vegetation species. Water S A 25:191–196

Jones HG (1992) Plants and microclimate: a quantitative approach to environmental plant physiology. Cambridge University Press, London

Kasperbauer MJ (1971) Spectral distribution of light in a tobacco canopy and effects of end-of-day light quality on growth and development. Plant Physiol 47:775–778

Kim HH, Goins GD, Wheeler RM et al (2004) Green-light supplementation for enhanced lettuce growth under red- and blue-light-emitting diodes. HortScience 39:1617–1622

Moffett WM (2000) What's "Up"? A critical look at the basic terms of canopy biology. Biotropica 32:569–596

Monsi M, Saeki T (2005) On the factor light in plant communities and its importance for matter production. Ann Bot 95:549–567

Tei F, Scaife A, Aikman DP (1996) Growth of lettuce, onion, and red beet. 1. Growth analysis, light interception and radiation use efficiency. Ann Bot 78:633–643

Zhang D, Hu Z, Fan J et al (2014) A meta-analysis of the canopy light extinction coefficient in terrestrial ecosystems. Front Earth Sci 8:599–609

Chapter 10
Evaluation of Spatial Light Environment and Plant Canopy Structure

Yasuomi Ibaraki

Abstract It is important to properly understand the light environment for improving plant production efficiency in a plant factory. Plant canopy structure affects the photosynthetic photon flux density (PPFD) distribution, thereby strongly affecting plant growth. Therefore, it is important to understand the canopy structure and control it in some cases for plant production. Indices for evaluating the canopy structure include extinction coefficient, leaf area index (LAI) or leaf area density (LAD) distribution, and leaf angle distribution. LAI is an important index not only for plant growth but also for evaluating the light environment of the canopy. Several indirect but nondestructive ways for LAI estimation have been proposed, including methods using plant canopy analyzer (PCA), hemispherical photography, terrestrial laser scanner (TLS), spectral reflectance, and image analysis. Moreover, the canopy surface has a PPFD distribution because of its variation in leaf inclination and orientation. A simple method for evaluating PPFD distribution on plant canopy surface using the reflection image of canopy was proposed and successfully constructed PPFD histograms of the canopy surface of several plant species under both natural and artificial lights.

Keywords Canopy surface • Image analysis • Leaf angle • LAI • PPFD • Quantum sensor • Reflection image

10.1 Introduction

Light environment, particularly photosynthetic photon flux density (PPFD; μmol $m^{-2} s^{-1}$) distribution, critically affects biomass productivity. Therefore, it is very important to properly understand the distribution for improving plant production efficiency in a plant factory. As described in the previous chapter (Chap. 9), light is absorbed by leaves, and a vertical PPFD distribution is present within a plant canopy. Moreover, the canopy surface has a PPFD distribution because of its variation in leaf inclination and orientation. Determining PPFD on the leaf surface

---

Y. Ibaraki (✉)
Faculty of Agriculture, Yamaguchi University, 1677-1 Yoshida, Yamaguchi 753-8515, Japan
e-mail: ibaraki@yamaguchi-u.ac.jp

is essential for accurately evaluating plant photosynthetic status, particularly in methods using remotely acquired images (Ibaraki and Dutta Gupta 2014a).

In this chapter, methods for evaluating the PPFD distribution within and on the plant canopy will be introduced. Evaluation methods for plant canopy structure and leaf area, which are critical factors for determining the PPFD distribution in a canopy, will also be discussed.

10.2 Measurement of PPFD Distribution in a Plant Canopy

In plant production, "light intensity" on leaf surface is often evaluated as PPFD using a quantum sensor. Furthermore, "light intensity" on leaf surface can be measured as irradiance (W m^{-2}), which is a measure of energy per unit area per unit time. PPFD is used for evaluating "light intensity" in plant production because it offers direct information on number of photons to be used for photosynthesis under the assumption that photons ranging from 400 to 700 nm can be equally used for photosynthesis. Although this assumption is not strictly true, it is acceptable for practical use. The relationship between irradiance and PPFD is constant under lighting conditions using the same light source, and they can be accordingly converted to each other. The metrics for expressing "light intensity" are described in detail in Chap. 25.

PPFD per se can be measured by several types of quantum sensor, including point-, line-, and globe-type sensors. However, determining the PPFD distribution on leaves that are located at the canopy surface is difficult even with these sensors because of dynamic variation in solar radiation and large variations in leaf angle and orientation in a plant canopy (Ibaraki et al. 2012b). For a plant factory using artificial lighting, dynamic variation in PPFD is unusual. In this case, spatial distribution can be potentially measured with a point sensor, although it is time-consuming and labor intensive. In addition, the size of the sensor may be too large to determine PPFD on a leaf for plant species with small leaves.

Line sensors can be used for estimating average PPFD over a given area. For example, a commercially available line quantum sensor (LI-COR, LI-191R) measures PPFD that is integrated over its 1-m length. It is used to measure sunlight under a plant canopy, where the light field is nonuniform, and is suitable for rapid and reproducible measurement of light within the canopy. This type of sensor is useful for determining the average PPFD in a horizontal layer as an input for calculating canopy photosynthesis.

The globe-type sensor is designed for measuring light from all directions. In a commercially available spherical quantum sensor (LI-COR, LI-193), photosynthetic photon flux fluence rate (PPFFR; μmol m^{-2} s^{-1}), which is the integral of photon flux radiance (μmol m^{-2} s^{-1} sr^{-1}) at a point over all directions, is measured. The ideal PPFFR sensor has a spherical collecting surface and exhibits the properties of a cosine receiver at every point on its surface (William 2015). This type of sensor may be useful for evaluating "light intensity" in plant cultivation under combination of sideward and downward lighting, which is a possible occurrence in plant factories using artificial lights.

10.3 Evaluation of Plant Canopy Structure

Plant canopy structure affects the PPFD distribution, thereby strongly affecting plant growth. Therefore, it is important to understand the canopy structure and control it in some cases for plant production. However, evaluating the canopy structure is not simple because of the complexity of the leaf spatial distribution patterns and their changes with growth. As described in Chap. 9, determining the extinction coefficient k provides information regarding the canopy structure. k can be estimated by measuring the vertical distribution of PPFD (by light interception measurements) and leaf area index (LAI). However, this approach, requiring the measurement of PPFD distribution, may not be reasonable if the main objective of evaluating the canopy structure is to characterize the PPFD distribution in the canopy. Instead, k can be estimated simply by Eq. 9.5 (in Chap. 9) under the assumption of vertical uniformity of k, requiring only photosynthetically active radiation (PAR) measurements above and below the canopy and estimation of canopy LAI.

Furthermore, the leaf angle distribution expresses the canopy structure. The leaf angle distribution can be determined by measuring the leaf angle one by one with a clinometer although this method requires time-consuming and labor intensive work and may be difficult to apply for some cases. In real canopies, leaves assume a range of orientations, with some (planophile) canopies having predominantly horizontal leaves and others (erectophile) having predominantly vertically oriented leaves; however, many other distributions are also found (Jones 1992). The ellipsoidal leaf angle density function (Campbell 1999) is often used as a close approximation of real plant canopies (Flerchinger and Yu 2007). The function expresses the probability density of leaf angle, namely, the fraction of leaf area per unit leaf inclination angle, based on the assumption that the angular distribution of leaves in a canopy is similar to the distribution of area on the surface of an ellipsoid (Wang et al. 2007). To compute the fraction of leaves between leaf inclination angles α_1 and α_2, leaf angle distribution function can be integrated from α_1 to α_2 (Wang et al. 2007). In the function, a single parameter, x (the ratio of the horizontal to the vertical semiaxis of the ellipsoid), is used for determining the shape of the distribution (Jones 1992). Examples of leaf angle distribution by the ellipsoidal leaf angle density function are shown in Fig. 10.1.

Other important indices for evaluating the canopy structure are LAI and leaf area density (LAD). Although LAI is a measure of the amount of leaves in the plant canopy, it provides information regarding the canopy structure. In contrast, LAD (m^{-1}) is the total one-sided leaf surface area (m^2) per unit volume (m^3) in the canopy, and the integral of LAD up to the canopy height is LAI (Fig. 10.2). The vertical distribution of LAD is a direct measure of the canopy structure, and

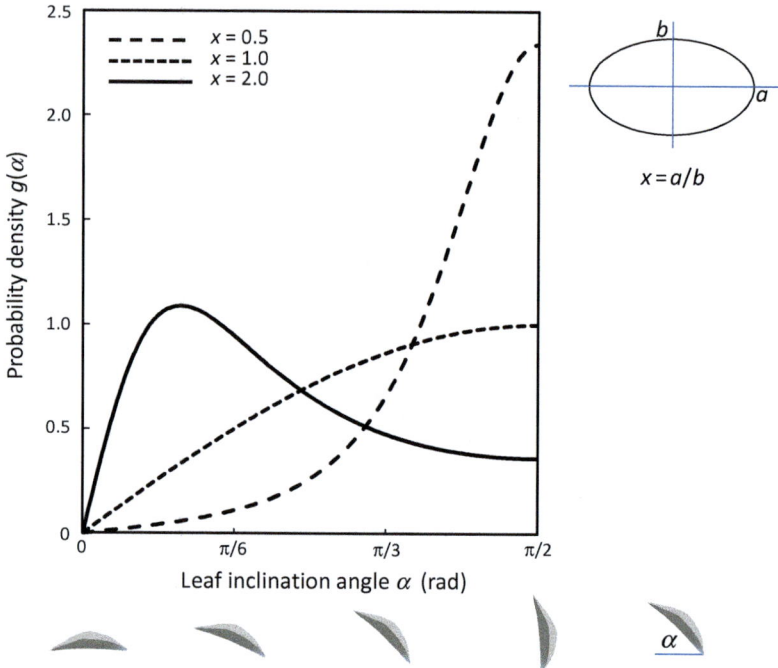

Fig. 10.1 Examples of leaf inclination angle distribution determined by the ellipsoidal leaf angle density function. Calculations were made by the following equation for $x = 0.5$, $x = 1$, and $x = 2$. $g(\alpha) = 2x^3 \sin \alpha / (A(\cos^2 \alpha + x^2 \sin^2 \alpha)^2)$, where $g(\alpha)$ is a probability density of inclination angle α, x is the ratio of the horizontal semiaxis length to the vertical semiaxis length of an ellipsoid, and A is a parameter determined by x. See Campbell (1999) or Wang et al. (2007) in detail for calculation

knowing it is helpful in analyzing the PPFD distribution. The vertical distribution patterns of LAD for crops may be roughly classified into several types: increasing toward the bottom, decreasing toward the bottom, increasing to the center and then decreasing, and being distributed uniformly.

Methods for estimating LAI or LAD will be introduced in Sect. 10.4.

10.4 LAI Estimation

10.4.1 Direct and Indirect Estimation

LAI is defined as the total projected leaf area (one side only) per unit area of ground and is an important index not only for plant growth but also for evaluating the light environment of the canopy, as described in Chap. 9. Here we should note the

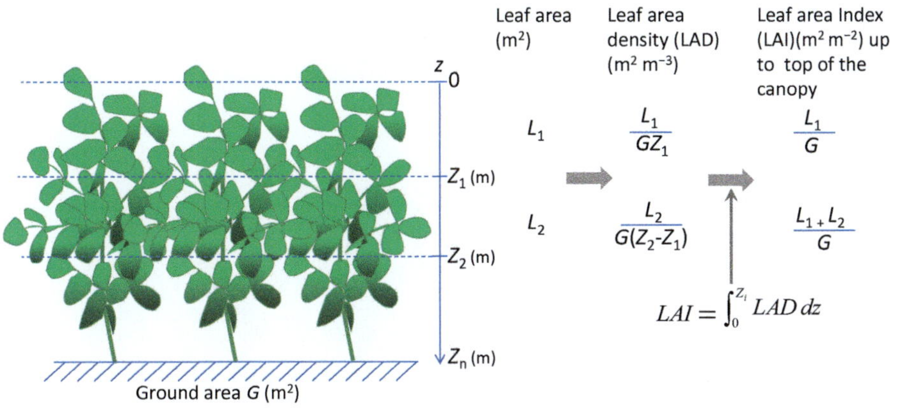

Fig. 10.2 Leaf area density (LAD) and leaf area index (LAI). LAD (m^{-1}) is the total one-sided leaf surface area (m^2) per unit volume (m^3) in the canopy, and the integral of LAD up to the canopy height is LAI

difference in units between LAI and LAD. The integral of LAD (m^{-1}) up to the canopy height (m) is LAI (dimensionless), as previously described.

Normally LAI can be measured by destructive means, such as detaching all leaves within a unit area of ground and measuring their area with a leaf area meter or scanning them with an optical scanner. Similarly, LAD can be measured by sampling at every unit of the horizontal layer. This requirement for destructive sampling is a defect of LAI measurement, precluding time-course analysis. Therefore, it is desirable to develop nondestructive methods. For rough estimation, measuring the number of leaves and area of the representative leaf which is determined from length and width of the leaf measured with a ruler can be used. Several indirect but nondestructive ways have been proposed, including methods using plant canopy analyzer, hemispherical photography, terrestrial laser scanner, spectral reflectance, and image analysis.

10.4.2 Methods Using Gap Fraction

As a nondestructive method for measuring plant leaf area, a plant canopy analyzer was developed and is commercially available (LI-COR, LAI-2200). The device estimates LAI from measurements of light made above and below the canopy and uses them to determine canopy light interception at five angles (Fig. 10.3). These data are used to compute leaf area index, mean tilt angle, and canopy gap fraction.

Canopy gap fraction is defined as the probability of a ray of light passing through the canopy without encountering leaves or other plant elements (Danson et al. 2007). Canopy directional gap fraction indicates the probability that a beam will not be intercepted by canopy elements in a given direction (Danson

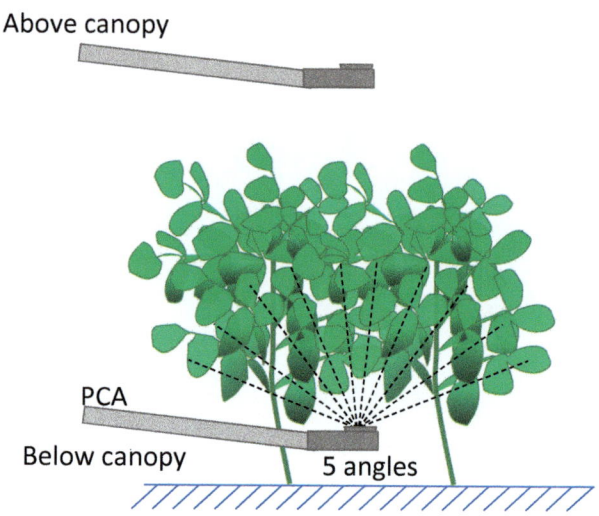

Fig. 10.3 Schematic diagram of LAI measurement using a plant canopy analyzer (PCA). PCA determines canopy light interception at five angles by measurements made *above* and *below* the canopy

et al. 2007). It is an important index for analyzing a forest canopy, which is related to canopy LAI, and can be measured by hemispherical photography or laser scanners. Methods of deriving LAI from gap fraction measurements are reviewed by Weiss et al. (2004).

For estimation of a gap fraction using hemispherical photography, the images are taken with a circular fish-eye lens from below the canopy looking upward (Fig. 10.4). Gap fractions are computed from such an image by determining the fraction of exposed background within rings or bands about the center of the image (Anderson 1964). Whereas hemispherical photography provides a two-dimensional record of the canopy structure, terrestrial laser scanners (TLS) have the potential to provide a 3-D record of canopy structure (Danson et al. 2007). TLS uses range-finding measurement techniques, which measure the distance to objects by emitting a series of laser pulses, to determine the 3-D positions of the objects within the scanner field of view (Danson et al. 2007) (Fig. 10.5). Gap fraction can be estimated from records of return to a laser "shot" in a given direction. In addition, laser scanning may be used for the direct estimation of vertical LAD distribution. Hosoi and Omasa (2009) proposed a method for estimating vertical LAD profiles using 3-D portable lidar (*l*ight *d*etection *a*nd *r*anging), which is referred to as the voxel-based canopy profiling method. The method successfully estimated the LAD profile of an individual tree of *Camellia sasanqua* (Hosoi and Omasa 2009).

10.4.3 The Use of Spectral Reflectance

With respect to the use of spectral reflectance, methods using vegetation index or reflectance in specific wavelength regions have been proposed. As plant leaves

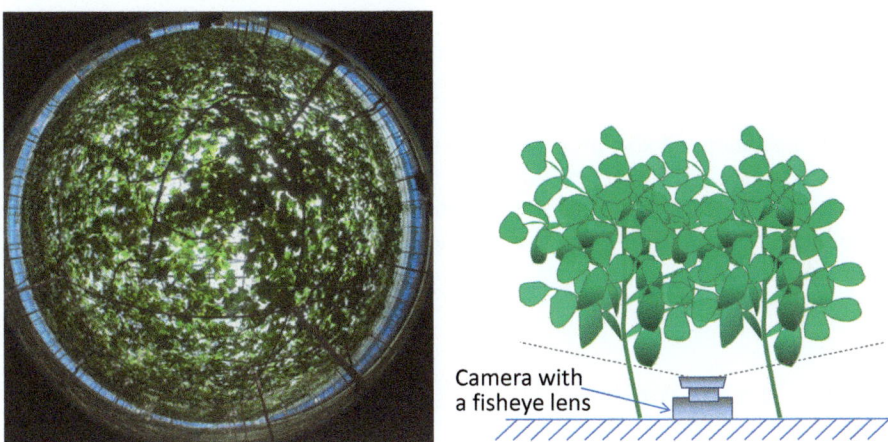

Fig. 10.4 Schematic diagram of hemispherical photography for LAI estimation and an example of the image of a grape canopy. The image was taken with a circular fish-eye lens from below the canopy looking *upward* (The photograph was provided courtesy of Dr. Kiyoshi Iwaya, Yamaguchi University)

Fig. 10.5 Conceptual diagram of analysis of canopy structure by using a terrestrial laser scanner (TLS)

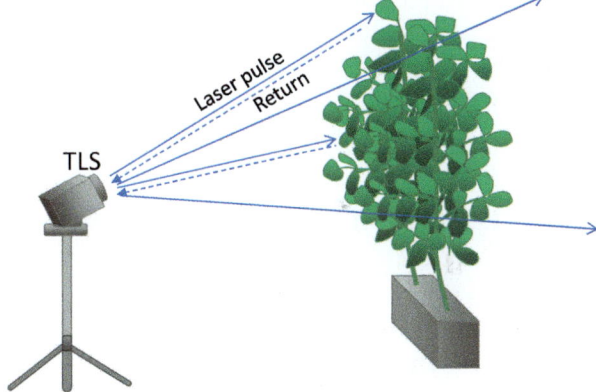

absorb red light and reflect near-infrared (NIR) light, canopies with more leaves demonstrate low reflectance in the red region and high reflectance for NIR. Using this principle, we can estimate LAI from the measurement of reflectance in red and NIR regions. Normalized difference vegetation index (NDVI), which has been widely used for evaluating plant cover in remote sensing, is sometimes used for estimating LAI. NDVI can be derived as follows:

$$\text{NDVI} = \frac{R_{\text{NIR}} - R_{\text{Red}}}{R_{\text{NIR}} + R_{\text{Red}}} \tag{10.1}$$

where R_{NIR} is the reflectance of NIR (e.g., 730–1100 nm) and R_{Red} is the reflectance in the red region (e.g., 620–690 nm). These methods using spectral reflectance are based on an empirical relationship, i.e., the regression model for estimation of LAI should be determined in advance, and the model parameters depend on the canopy type (plant species and growth stage).

As an alternative approach, a method using a ratio of PAR and NIR irradiance within the canopy is also proposed. The LAI sensor (Environmental Measurement Japan, MIJ-15LAI, Japan) estimates LAI of a target canopy, being set within the canopy (on the ground), using the predetermined relationship between the ratio of irradiance at PAR (400–700 nm) and at IR (700–1000 nm) within the canopy and LAI.

10.4.4 Image Analysis

Image analysis using digital cameras potentially permits the acquisition of size information of objects (Ibaraki and Dutta Gupta 2014b). Normally, a projected area can be nondestructively measured from an image. If a linear relationship between a projected area and actual leaf area is observed, LAI can be estimated simply by image analysis. Liu and Pattey (2010) reported the effectiveness of digital photography for LAI estimation of agricultural crops using a rectilinear lens at the top of the canopy. Campillo et al. (2010) also developed a nondestructive method for estimating LAI in vegetable crops using digital images that were acquired from above the crop canopy.

10.5 Estimation of PPFD Distribution on Plant Canopy Surface

10.5.1 Importance of Understanding Light Distribution on the Canopy Surface

Recently, nondestructive methods for evaluating photosynthetic properties of plants using images remotely acquired have been used for environmental control on the basis of the understanding of plant status in plant production. In such methods, proper understanding of the irradiance/photon flux density distribution on the target surface of the canopy is essential (Ibaraki and Dutta Gupta 2014a). For example, parameters that are derived from chlorophyll fluorescence measurement, such as PSII quantum yield, strongly depend on PPFD on leaves and would be meaningless without PPFD information. Moreover, irradiance on leaves affects leaf temperature and may be a limiting factor for evaluating plant stress by leaf temperature measurement. Thus, it is essential to evaluate the irradiance/photon flux density distribution on the canopy surface for analyzing the distribution of photosynthetic properties on the canopy surface using such image-based methods. In addition, the light distribution on the surface may be critical for considering plant productivity (canopy photosynthesis) in some types of canopy, in which leaves are concentrated in the upper layer or leaf layer is difficult to define because of its low height.

Because the reflection image of a leaf includes information regarding light irradiating the leaf, it has the potential to be used for the simple estimation of irradiance/photon flux density. Ibaraki et al. (2012a, b) developed a simple method for evaluating PPFD distribution on plant canopy surface using the relationship between PPFD on the leaf surface and pixel value in the reflection image. In this section, the method is introduced with applications.

10.5.2 Reflection Image-Based Estimation Method of PPFD on Canopy Surface

Reflection images of plant canopies were acquired with a monochrome 14-bit CCD camera (Bitran, BU-41L, Japan), digital camera (Canon, SX130, Japan), or Android tablet (Sony, Tablet-S, Japan) through a blue–green band-pass filter (Suruga, S76-BG28 or S76-BG7, Japan). The filter had a peak wavelength of approximately 450–500 nm and was used for improving the correlation between PPFD on the leaf surface and pixel value, which is a digital value assigned to each pixel to express brightness or color, in the reflection image. To minimize the effect of specular reflection, images were acquired from several directions. Cameras were horizontally moved at a fixed distance from the plant material and at a fixed angle of depression to acquire images from different directions (Fig. 10.6). In this method, the linearity of output (pixel value in an image) and input (PPFD entering the camera) should be confirmed for each camera before imaging. In other words, if possible, the gamma value should be set to 1.0, and if it cannot be changed, it is determined in advance and used for gamma correction in analyses (Ibaraki and Dutta Gupta 2014a). An example of gamma correction is shown in Fig. 10.7.

Ibaraki et al. (2012b) compared actual PPFD with a pixel value in the reflection images for several plant canopies. Linear relationships between these values were observed in both outdoor (under natural sunlight) and indoor (under artificial light) conditions for several plant species. However, they indicated that the slope and intercept of the linear regression model between the PPFD and pixel values depended on the canopy type (plant species), and thus, a regression model should be fitted for each measurement.

On the basis of these results, a simple method was proposed to estimate the PPFD distribution on a plant canopy surface in a greenhouse (Ibaraki et al. 2012b). In this method, the actual PPFD was measured with a quantum sensor at one point on the canopy surface and was used for constructing a linear regression model for calculating PPFD from the pixel value at any point in the reflection image. A small quantum sensor was placed adjacent to a leaf on which PPFD appeared to be high in the target canopy. After imaging, slope a in the linear regression model ($y = ax$)

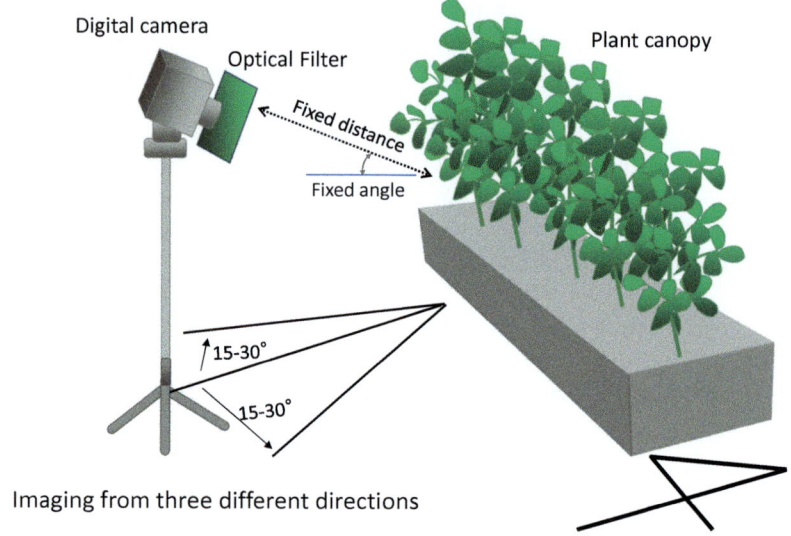

Fig. 10.6 Conceptual diagram of estimation of PPFD distribution on canopy surface using reflection images of the canopy. The images are acquired from three different directions

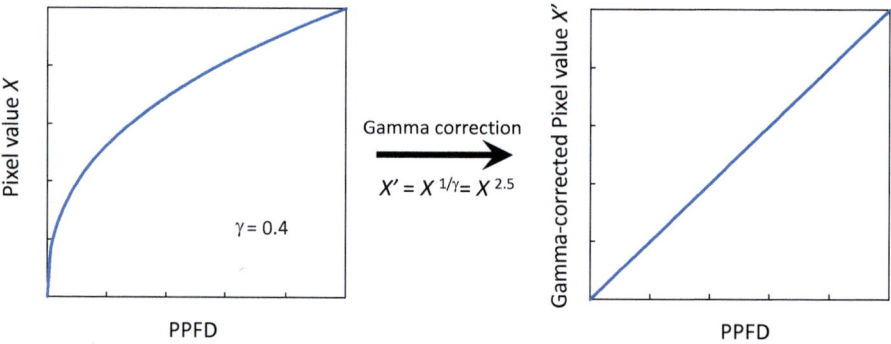

Fig. 10.7 An example of gamma correction of an image to assure the linearity of pixel value and PPFD

was calculated as the ratio of measured PPFD versus pixel value at the measurement point in the image. PPFD at any point on the leaf was estimated using the linear regression model.

Figure 10.8 shows the concept of the estimation method. The circles in the figure denote the values of the points at which PPFD was measured with the PPFD sensor, and the lines represent the regression models for estimating PPFD. On both clear and cloudy days, the method could roughly estimate PPFD on the leaves of a rose canopy in a greenhouse (Ibaraki et al. 2012b). This method, in which a regression model is fitted for each image on the basis of actual measurement at one point, can

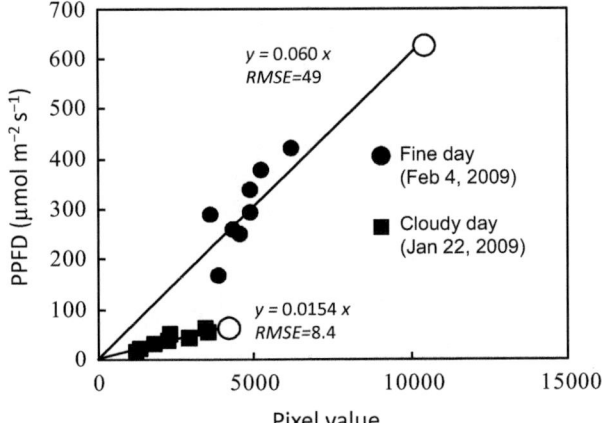

Fig. 10.8 A concept of estimation of PPFD distribution on canopy surface. Examples for a rose canopy in the greenhouse on fine (*closed circle*) and cloudy days (*closed squire*) (Reproduced from Ibaraki et al. 2012b). The slope *a* in the linear model ($y = ax$) was determined as a ratio of measured PPFD (*open circle*) to pixel value at a measuring point in the image and used for estimation of PPFD at other points

minimize the effects of canopy type (plant species) and/or light spectral distribution (Ibaraki and Dutta Gupta 2014a).

10.5.3 Applications

Ibaraki et al. (2012a) successfully constructed a PPFD histogram of the canopy surface of tomato plants in a greenhouse from reflection images that were acquired between rows. The histogram of pixel values after gamma correction was converted into a PPFD histogram using a linear model on the basis of the relationship between the observed PPFD and pixel value at one point in each image. They demonstrated that the pattern of PPFD histograms on the tomato canopy surface changed over time during the day.

An Android application for easy, real-time estimation of the PPFD distribution on a plant canopy surface has also been developed (Miyoshi et al. 2016). The application was designed to semiautomatically analyze PPFD distribution on the canopy from a reflection image that was acquired by a tablet (Fig. 10.9). An image preview frame was captured and used for analysis. Plant canopy areas were extracted from the image using a combination of the threshold values of "luminance" and "g" ["G"/"(R + G + B)"], which users could input with Android SeekBars. When histogram data from three different directional images were stored, these data were averaged and analyzed. Figure 10.10 shows an example of PPFD distribution on a lettuce canopy under artificial light.

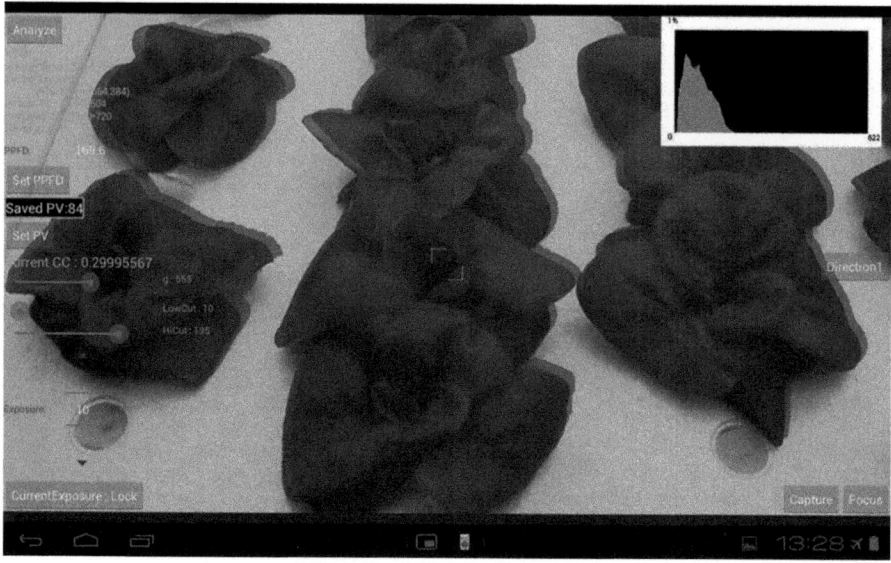

Fig. 10.9 A screenshot of the Android tablet-based system (Miyoshi et al. 2016), when being applied to lettuce cultivated under artificial lighting. The PPFD histogram is shown in *upper right*

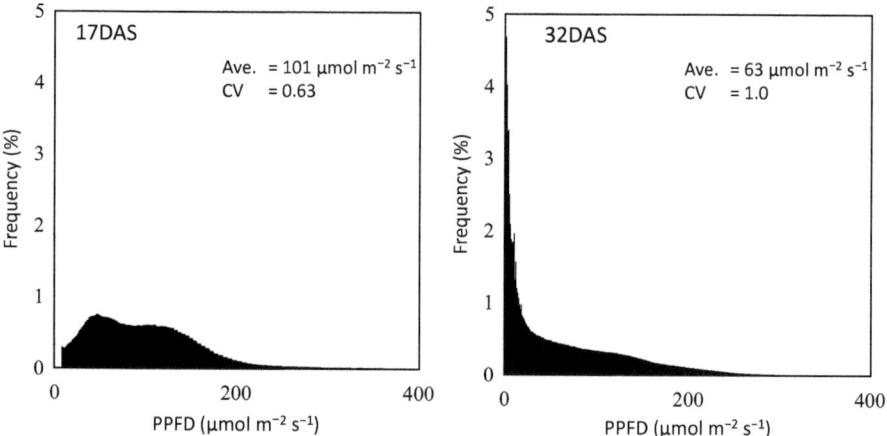

Fig. 10.10 Examples of PPFD distribution (PPFD histogram) on the surface of lettuce canopies estimated with the Android tablet-based system (Miyoshi et al. 2016) under white fluorescent lamps. The distribution pattern changed with days after sowing (DAS). On 32 DAS, the frequencies in low PPFD increased possibly due to overlapping leaves

References

Anderson MC (1964) Studies of the woodland light climate. 1. The photographic computation of light conditions. J Ecol 52:27–41

Campbell GS (1999) Derivation of an angle density function for canopies with ellipsoidal leaf angle distributions. Agric For Meteorol 49:173–176

Campillo C, Garcia MI, Daza C et al (2010) Study of non-destructive method for estimating the leaf area index in vegetation crops using digital images. HortScience 45:1459–1463

Danson FM, Hetherington D, Morsdorf F et al (2007) Forest Canopy gap fraction from terrestrial laser scanning. IEEE Geol Remote Sens Lett 4:157–160

Flerchinger GN, Yu Q (2007) Simplified expressions for radiation scattering in canopies with ellipsoidal leaf angle distributions. Agric For Meteorol 144:230–235

Hosoi F, Omasa K (2009) Estimating vertical leaf area density profiles of tree canopies using three-dimensional portable lidar imaging. In: Bretar F et al (eds) Laser scanning. Vol. XXXVIII, Part 3/W8. IAPRS, Paris, pp 152–157

Ibaraki Y, Dutta Gupta S (2014a) PRI imaging and image-based estimation of light intensity distribution on plant canopy surface. In: Dutta Gupta S, Ibaraki Y (eds) Plant image analysis. CRC Press, Boca Raton, pp 229–244

Ibaraki Y, Dutta Gupta S (2014b) Image analysis for plants: basic procedures and techniques. In: Dutta Gupta S, Ibaraki Y (eds) Plant image analysis. CRC Press, Boca Raton, pp 25–40

Ibaraki Y, Kishida T, Shigemoto C (2012a) Image-based estimation of PPFD distribution on the canopy surface in a greenhouse. Acta Hortic 956:577–582

Ibaraki Y, Yano Y, Okuhara H et al (2012b) Estimation of light intensity distribution on a canopy surface from reflection images. Environ Control Biol 50:117–126

Jones HG (1992) Plants and microclimate: a quantitative approach to environmental plant physiology. Cambridge University Press, London

Liu J, Pattey E (2010) Retrieval of leaf area index from top-of-canopy digital photography over agricultural crops. Agric For Meteorol 150:1485–1490

Miyoshi T, Ibaraki Y, Sago Y (2016) Development of an android-tablet-based system for analyzing light intensity distribution on a plant canopy surface. Comput Electron Agric 122:211–217

Wang WM, Li ZL, Sub HB (2007) Comparison of leaf angle distribution functions: effects on extinction coefficient and fraction of sunlit foliage. Agric For Meteorol 143:106–122

Weiss M, Baret F, Smith GJ et al (2004) Review of methods for in situ leaf area index (LAI) determination: part II. Estimation of LAI, errors and sampling. Agric For Meteorol 121:37–53

William WB (2015) Principle of radiation measurement. LI-COR Biosciences. https://www.licor.com/env/pdf/light/Rad_Meas.pdf. Accessed 20 Dec 2015

Chapter 11
Lighting Efficiency in Plant Production Under Artificial Lighting and Plant Growth Modeling for Evaluating the Lighting Efficiency

Yasuomi Ibaraki

Abstract As it is critical that plant growers improve the efficiency of their lighting when it uses artificial lighting, the lighting efficiency should be evaluated properly. One possible way to evaluate the lighting efficiency is to compare the amount of biomass produced per unit of energy used to irradiate the plants. A simpler index uses the fraction of the light energy or photons received by plants. Lighting efficiency can also be evaluated from the viewpoint of how much the irradiance/photon flux density on leaf surfaces can be improved. It is useful to obtain information of canopy structure or leaf spatial distribution in addition to determining plant mass (dry weight, fresh weight, or LAI) increments for evaluating the lighting efficiency. Modeling leaf growth and development can be used for this purpose.

Keywords Electrical energy use efficiency • Energy consumption • Functional–structural plant model • Light use efficiency • L-system • PPFD distribution • Radiation use efficiency • Reflection image

11.1 Introduction

Although the use of artificial lighting in plant production has been increasing, little attention has been paid to the efficiency of the lighting (Ibaraki and Shigemoto 2013). As artificial lighting consumes energy, thereby increasing the cost of production, it is critical that plant growers improve the efficiency of their lighting. One possible solution is to use lamps with high luminous efficacy. However, lighting efficiency also depends on the arrangement of the lamps and/or the plant canopy structure being irradiated. The total luminous flux emitted by a lamp may not always irradiate the plant body, and unnecessary irradiation is often produced by

---

Y. Ibaraki (✉)
Faculty of Agriculture, Yamaguchi University, 1677-1 Yoshida, Yamaguchi 753-8515, Japan
e-mail: ibaraki@yamaguchi-u.ac.jp

artificial lighting, particularly when the plant canopy has a low leaf area index (LAI). The plant canopy structure changes as the plants grow during cultivation. Accordingly, the light environment also changes in line with the change in the canopy structure, even if the light source, lighting direction, and distance from plants remain constant. The lighting efficiency may therefore change with plant growth, and the dynamics of this process should be evaluated properly.

Lighting efficiency can be evaluated from several viewpoints. First, the efficiency can be evaluated in terms of energy conversion efficiency, comparing biomass production per unit of energy used for the irradiating light. A simpler index uses the fraction of the light energy or photons received by plants. It is also important to understand the extent to which the irradiance (W m^{-2}) or photon flux density (mol m^{-2} s^{-1}) is improved by the artificial lighting because the objective of artificial lighting is to irradiate the leaves and increase the irradiance/photon flux density on them for photosynthesis or other light-induced biological processes.

In this chapter, the evaluation methods for lighting efficiency are introduced, focusing on the energy use efficiency and the photosynthetic photon flux density (PPFD) distribution on the canopy surface provided by artificial lighting. The use of plant growth modeling for estimating the lighting efficiency will also be discussed.

11.2 Light Energy Received by Leaves

11.2.1 Light Use Efficiency

Plants absorb light energy and convert it into chemical energy stored as organic matter (biomass). The lighting efficiency can therefore be assessed by the energy conversion efficiency. One possible way to evaluate the efficiency is to compare the amount of biomass produced per unit of energy used to irradiate the plants or those absorbed by the plants.

A ratio between accumulated biomass and the photosynthetically active radiation (PAR) absorbed by plants is sometimes referred to as light use efficiency (LUE) or radiation use efficiency (RUE), having units of µg J^{-1}, and has been used as an index for assessing canopy productivity (Gitelson and Gamon 2015) for natural ecosystems or field crops. This approach is based on Monteith's observation (1972) that the net primary productivity of the plant canopy is proportional to the intercepted solar radiation (Rosati and Dejong 2003). However, the lack of a universally agreed definition of LUE may cause difficulties in comparison of the results from different studies (McCallum et al. 2009). The denominators of LUE range from simple incident PAR (or PPFD), through total PAR absorbed (intercepted), to total PAR absorbed by green vegetation (photosynthetically active leaves) (Gitelson and Gamon 2015). The numerator is also variable and may be net primary production (NPP) (g C), gross primary production (GPP) (g C), weight of biomass (g), or weight of aboveground biomass (g). In botanical studies, LUE is

Table 11.1 LUE values for several crops reported in the literature

Species	LUE value ($\mu g\ J^{-1}$)	Description of the term in the literature	Reference
Tomato	2.8–4.0	Light use efficiency	Dorais (2003)
Sweet pepper	2.1	Light use efficiency	Dorais (2003)
Lettuce	1.44–2.43	Conversion efficiency of absorbed PAR	Tei et al. (1996)
	1.26	Radiation conversion efficiency	Javanovic et al. (1999)
Onion	0.99–5.08	Conversion efficiency of absorbed PAR	Tei et al. (1996)
	1.08	Radiation conversion efficiency	Javanovic et al. (1999)
Rice	4.15	Efficiency of light utilization for DM production	Sands (1999)
Maize	3.4	Efficiency of light utilization for DM production	Sands (1999)
Soybean	1.29	Efficiency of light utilization for DM production	Sands (1999)

often evaluated as the slope of a light photosynthetic curve or a quantum yield of oxygen evolution, having units of mol mol^{-1}. When referring to LUEs or RUEs reported in the literature, the definition and method of measurement must be specified.

LUE varies between crops, depending on the plant physiological status, including the nitrogen status (Rosati and Dejong 2003), as well as environmental conditions such as temperature or CO_2 concentration. Table 11.1 shows the LUEs for several crops are expressed in terms of dry mass formed (μg) per unit of PAR absorbed (J).

11.2.2 Ratio of Light Energy Received by the Plants

An alternative index uses the fraction of light energy or photons received by the plants to evaluate the lighting efficiency. The ratio of the PAR (PAR_P) received at the plant canopy surface to that (PAR_L) emitted from the lamps, often referred to as the "utilization factor" in illumination engineering (Kozai 2013), can be used for this purpose. The ratio of PAR_P to PAR_L depends not only on the lamp properties (i.e., spatial distribution of the light intensity emitted from the lamp) but also on the canopy structure. The ratio thus changes over time. Improving the ratio of PAR_P to PAR_L is a way to minimize the unnecessary irradiation, reduce energy consumption, and consequently lower the cost of production.

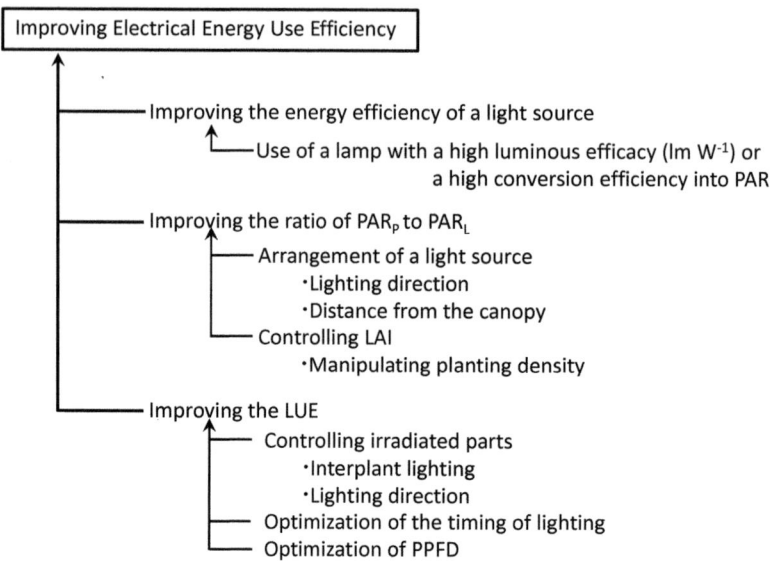

Fig. 11.1 Factors affecting electrical energy use efficiency

11.2.3 Improving Electrical Energy Use Efficiency

For crop production under artificial lighting, electrical energy use efficiency estimated based on the power consumption is also an important index to evaluate the lighting efficiency. The methods of estimation of the electrical energy use efficiency are described in detail in Chap. 29.

Various methods can be considered to improve the electrical energy use efficiency (Kozai 2013). These methods can be divided into the following approaches: improving the energy efficiency of a light source, improving the ratio of PAR_P to PAR_L, and improving the LUE based on the plant physiological (photosynthetic) properties (Fig. 11.1).

A direct method for improving the energy efficiency is to use a light source with a high luminous efficacy (lm W^{-1}) as described before. The energy efficiency of LEDs and LED lighting systems was described in detail in Chap. 29.

To improve the ratio of PAR_P to PAR_L, it is important to minimize unnecessary irradiation. The ratio can be improved by well-designed light reflectors or by a reduction in the vertical distance between lamps and plants (Massa et al. 2008). Reflectors may be placed behind (above) the lamps to direct the backward light to the forward (downward) or on the side of the cultivation tray to minimize the amount of light irradiated outside the tray. The reduction of distance between lamps and plants also leads to minimizing the amount of light irradiated outside the plant canopy. Moreover, controlling the lighting direction may also be effective, depending on the canopy structure and spatial distribution of the lamps. Plant density also affects the ratio of PAR_P to PAR_L (Kozai 2013; Yokoi et al. 2003; Massa et al. 2008).

The electrical energy use efficiency can be improved from the aspects of both irradiation time and position, based on the physiological properties of the plants, i.e., when plants are irradiated and which parts of plants are irradiated affect the efficiency. For example, the net photosynthetic rate of the upper leaves that have already received light at a high level (near the light saturation level) may not be increased by further increasing PPFD by supplemental lighting. On the other hand, the net photosynthetic rate of the lower leaves, which is often negative or nearly zero, will become positive by increasing PPFD. From this point of view, the interplant lighting provides more light energy to the lower leaves than downward lighting only, potentially improving the light energy use efficiency (Kozai 2013; Massa et al. 2008).

The timing of lighting is also important for supplemental lighting. It has been reported that lighting during the night period is effective for promoting the growth of lettuce (Fukuda et al. 2004), and end-of-day lighting is effective in controlling plant morphological events (e.g., Yang et al. 2012). Furthermore, diurnal variation of LUE has been reported (e.g., Mukherjee et al. 2014).

11.3 Lighting Efficiency Based on PPFD Distribution on a Canopy Surface

PPFD on a leaf surface is critical for plant production. Lighting efficiency can also be evaluated from the viewpoint of how much the PPFD on leaf surfaces can be improved.

A method of evaluating the efficiency of supplemental lighting based on PPFD distribution on a canopy surface under artificial lighting conditions was developed (Ibaraki and Shigemoto 2013), and several indices for lighting efficiency derived from the PPFD distribution histogram estimated by using a reflection image of the canopy surface were proposed. In this method, the reflection images of plant canopy surfaces were acquired from three directions with a digital camera, and PPFD on leaf surfaces was estimated from the pixel values of the image by a regression model determined from PPFD measured at one point on the canopy simultaneously with imaging (see Chap. 10 for details). Then, the histogram of the pixel values after gamma correction was converted to a PPFD histogram (Fig. 11.2). To characterize the PPFD distribution, an average PPFD, a median PPFD, and the coefficient of variances (CV) of PPFD over the illuminated canopy surface were calculated from the PPFD histogram. Integrated PPFD over all illuminated leaves per unit power consumption (IPPC) was then proposed as a criterion for evaluating the efficiency of supplemental lighting. IPPC was calculated by the following equation:

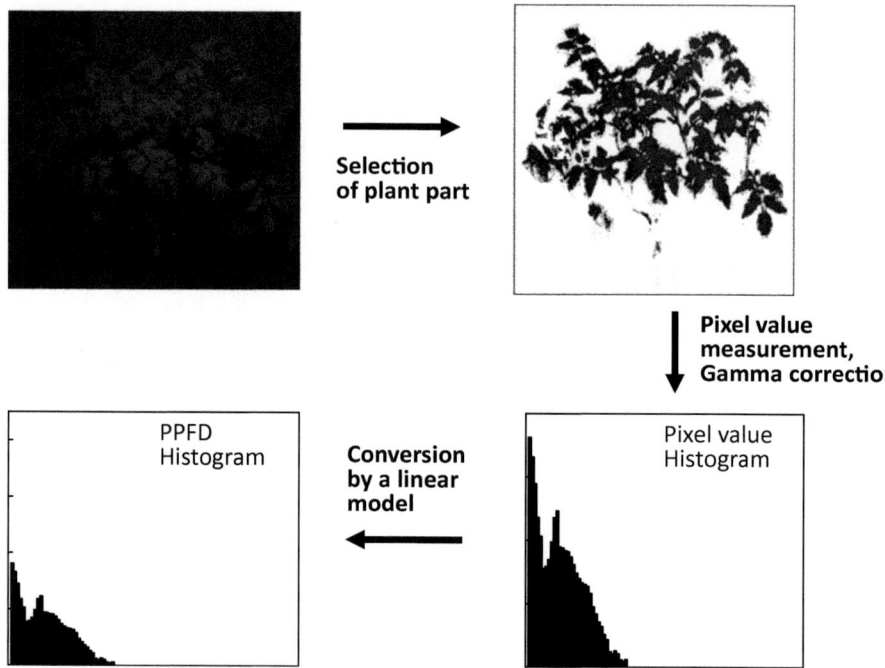

Fig. 11.2 Flow diagram of PPFD histogram construction from reflection images (Reproduced from Ibaraki and Shigemoto (2013))

$$\text{IPPC} \left(\mu\text{mol s}^{-1} \text{W}^{-1} \text{ or } \mu\text{mol J}^{-1} \right) = \frac{\text{Averaged PPFD } (\mu\text{mol m}^{-2} \text{ s}^{-1}) \times \text{Projected leaf area } (\text{m}^{-2})}{\text{Power consumption of light source (W)}} \quad (11.1)$$

The projected leaf area was estimated from the image of the canopy surface by selecting pixels corresponding to leaves. Ibaraki and Shigemoto (2013) reported that the histogram pattern of PPFD on a tomato plant canopy surface under supplemental lighting depended on the light source and canopy structure. Histograms estimated from images could depict the differences, showing average values and CVs close to the measured values. The IPPC also depended on the types of light sources, canopy structures, and the distance between lamps and the canopy surfaces.

Bornwaβer and Tantau (2012) calculated a similar index, the energy efficiency with PPFD (μmol s^{-1} W^{-1}), to evaluate the lighting efficiency of the LED lighting system in in vitro culture. They calculated the index for both average PPFD and PPFD at the center of the irradiated surface to represent the PPFD distribution.

When artificial lights are used, it is easy to convert PPFD into total photon flux density or irradiance because, for the same light source, the light spectrum is constant. Therefore, these PPFD-based methods can be applied for supplemental lighting, which should be evaluated by total photon flux density or irradiance rather than by PPFD. If irradiance is used instead of PPFD, the integrated irradiance per unit power consumption is dimensionless (W m^{-2} × m^2/W).

It is important to know the actual irradiance/photon flux density on the plant canopy surface not only to evaluate the lighting efficiency but also to improve stability and repeatability in controlling the environmental conditions when supplemental lighting is used. The image-based PPFD histogram estimation method is also expected to be used for this purpose (Ibaraki and Shigemoto 2013).

11.4 Plant Growth Modeling for Evaluating Lighting Efficiency

11.4.1 Simple Growth Model

Plant growth modeling is an effective tool for understanding light distribution and estimating the lighting efficiency. For vegetative growth, an exponential model is often used. Assuming that the relative growth rate (RGR) or the relative leaf area growth rate (RLGR) is constant during a given period, growth (in terms of dry weight, W, or leaf area, L) can be expressed as exponential growth (an exponential function of time t, see Fig. 11.3a) by the following equations:

$$W = W_0 e^{RGR t} \qquad (11.2)$$

$$L = L_0 e^{RLGR t} \qquad (11.3)$$

where W_0 and L_0 are initial values of W and L, respectively.

For leafy vegetables and seedlings that are dominant crops in a plant factory with artificial lights, such exponential models are often used to estimate the vegetative growth. For example, Yokoi et al. (2003) used an exponential model to fit the

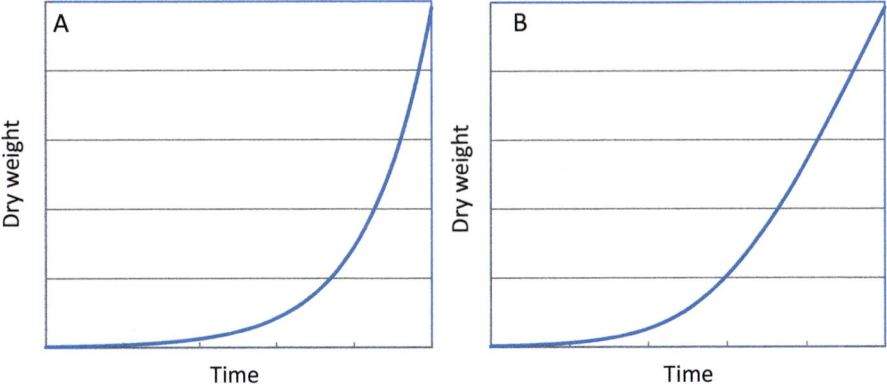

Fig. 11.3 Examples of an exponential growth curve (**a**) and an expolinear growth curve (**b**)

increments in LAI and dry weight and to calculate the electrical energy use efficiency in the production of tomato seedlings under artificial lighting.

For individual plants, such as seedlings growing without competition between neighbors, RGR is assumed to be constant (Monteith 2000) under constant environmental conditions. However, RGR may decline if there is competition for resources (Monteith 2000). In addition, RGR depends on both environmental conditions and plant physiological state, such as leaf nitrogen content. Models for changing RGR include an expolinear model (Goudriaan and Monteith 1990; Dennett and Ishag 1998; Monteith 2000) available for longer period of growth (Fig. 11.3b), a model expressing RGR as a function of temperature and PAR (Aikman and Scaife 1993), and a model using a Gompertz function (Shimizu et al. 2008).

11.4.2 2D and 3D Modeling for Vegetative Growth

It is useful to obtain (simulate) information of canopy structure or leaf spatial distribution in addition to determining plant mass (dry weight, fresh weight, or LAI) increments for evaluating lighting efficiency. Therefore, modeling leaf growth and development is effective. Leaves of vascular plants are arranged in an orderly, often spectacular pattern (Lubkin 1995). Normally, the leaf arrangement pattern, i.e., phyllotaxis, depends on plant species or cultivar and includes alternate, opposite, whorled, and rosulate patterns (Fig. 11.4). It is useful to know the leaf arrangement pattern of the target plant for modeling leaf growth and development. Considering both this pattern and the spectral distribution of light, we may estimate the PAR_P/PAR_L ratio.

Recently, 3D measurements, including lidar (Hosoi and Omasa 2009) and stereo imaging (Biskup et al. 2007; Müller-Linow et al. 2015), have been used to analyze plant canopy structure. From 3D data of plant architecture, leaf angle distribution

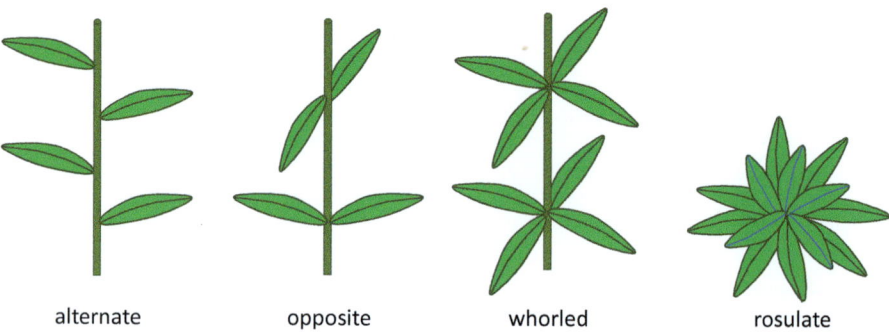

Fig. 11.4 Leaf arrangement patterns

and leaf area density distribution, which are important parameters, can be estimated.

Models for simulating the 3D architecture of plants have been developed based mainly on L-systems or similar approaches (Fournier and Andrieu 1998). An L-system, developed by Lindenmayer (1968), is a string rewriting system and is a powerful tool to model the growth of plants (Fournier and Andrieu 1998). In general, rewriting is a technique for defining complex objects by successively replacing parts of a simple initial object using a set of rewriting rules or productions (Prusinkiewicz and Lindenmayer 1990). In an L-system, plant architecture is represented by a string symbol, each symbol representing a plant component such as a leaf or internode (Kaitaniemi et al. 2000). A simple example of L-systems is shown in Fig. 11.5. Plant growth and development can be simulated by the symbols changing according to the production rules. A comprehensive overview of the simulation of plant development using L-systems is reviewed by Prusinkiewicz and Lindenmayer (1990).

Recently, new computer models of plant functioning and growth, called functional–structural plant models (FSPMs), have been developed (Godin and Sinoquet 2005). FSPMs combine the representation of 3D plant structure with selected physiological functions, consisting of an architectural part (plant structure) and a process part (plant functioning) (Vos et al. 2010). In FSPMs, L-systems are often adopted as a paradigm to model plant development (Godin and Sinoquet 2005). FSPMs were used to compare lamp positioning scenarios to identify the most efficient lighting strategy in greenhouse production of tomatoes, being combined with 3D models of light distribution from the lamps and greenhouse architecture (Visser et al. 2012, 2014).

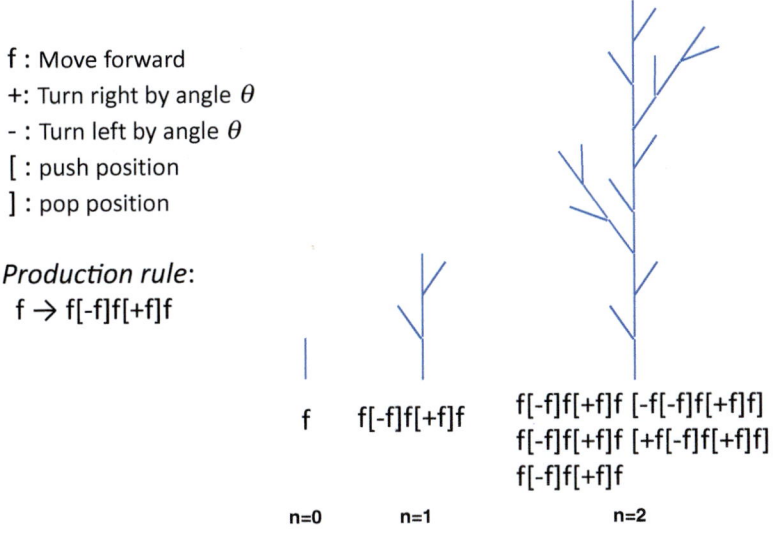

Fig. 11.5 A simple example of L-systems

References

Aikman DP, Scaife A (1993) Modelling plant growth under varying environmental conditions in a uniform canopy. Ann Bot 72:485–492

Biskup B, Scharr H, Schurr U et al (2007) A stereo imaging system for measuring structural parameters of plant canopies. Plant Cell Environ 30:1299–1308

Bornwaβer T, Tantau HJ (2012) Evaluation of LED lighting system in in vitro cultures. Acta Hortic 956:555–560

de Visser PHB, Buck-Sorlin GH, van der Heijden GWAM et al (2012) A 3D model of illumination, light distribution and crop photosynthesis to simulate lighting strategies in greenhouses. Acta Hortic 956:195–200

de Visser PHB, Buck-Sorlin GH, van der Heijden GWAM (2014) Optimizing illumination in the greenhouse using a 3D model of tomato and a ray tracer. Front Plant Sci 5(48):1–7

Dennett MD, Ishag KHM (1998) Use of the expolinear model to analyse the growth of faba bean, peas and lentils at three densities: predictive use of the model. Ann Bot 82:507–512

Dorais M (2003) The use of supplemental lighting for vegetable crop production: light intensity, crop response, nutrition, crop management, cultural practices. Canadian Greenhouse Conference, 9 Oct 2003, Toronto http://www.agrireseau.qc.ca/legumesdeserre/Documents/CGC-Dorais2003fin2.PDF. Accessed on 25 Dec 2015

Fournier C, Andrieu B (1998) A 3D architectural and process-based model of maize development. Ann Bot 81:233–250

Fukuda N, Nishimura S, Fumiki Y (2004) Effect supplemental lighting during the period from middle of night to morning on photosynthesis and leaf thickness of lettuce (*Lactuca sativa* L.) and tsukena (*Brassica campestris* L.). Acta Hortic 633:237–244

Gitelson AA, Gamon JA (2015) The need for a common basis for defining light-use efficiency: implications for productivity estimation. Remote Sens Environ 156:196–201

Godin C, Sinoquet H (2005) Functional-structural plant modeling. New Phytol 166:705–708

Goudriaan J, Monteith JL (1990) A mathematical function for crop growth based on light interception and leaf area expansion. Ann Bot 66:695–701

Hosoi F, Omasa K (2009) Estimating vertical leaf area density profiles of tree canopies using three-dimensional portable lidar imaging. In: Bretar F et al (eds) Laser scanning, vol XXXVIII, Part 3/W8. IAPRS, France, pp 152–157

Ibaraki Y, Shigemoto C (2013) Estimation of supplemental lighting efficiency based on PPFD distribution on the canopy surface. J Agric Meteorol 69:47–54

Javanovic NZ, Annandale JG, Mhlauli NC (1999) Field water balance and SWB parameter determination of six winter vegetation species. Water SA 25:191–196

Kaitaniemi P, Hanan JS, Room PM (2000) Virtual sorghum: visualisation of partitioning and morphogenesis. Comput Electron Agric 28:195–205

Kozai T (2013) Resource use efficiency of closed plant production system with artificial light: concept, estimation and application to plant factory. Proc Jpn Acad Ser B Phys Biol Sci 89:447–461

Lindenmayer A (1968) Mathematical models for cellular interaction in development, I and II. J Theor Biol 18:280–315

Lubkin S (1995) Book review of phyllotaxis: a systemic study in plant morphogenesis (Jean RV, 1994). Bull Math Biol 57:377–379

Massa GD, Kim H-H, Wheeler RM et al (2008) Plant productivity in response to LED lighting. HortScience 43:1951–1956

McCallum I, Wagner W, Schmullius C et al (2009) Satellite-based terrestrial production efficiency modeling. Carbon Balance Manag 4:8. doi:10.1186/1750-0680-4-8

Monteith JL (1972) Solar radiation and productivity in tropical ecosystems. J Appl Ecol 9:744–766

Monteith JL (2000) Fundamental equations for growth in uniform stands of vegetation. Agric For Meteorol 104:5–11

Mukherjee J, Singh G, Bal SK (2014) Radiation use efficiency and instantaneous photosynthesis at different growth stages of wheat (*Triticum aestivum* L.) in semi arid ecosystem of central Punjab, India. J Agrometeorology 16:69–77

Müller-Linow M, Pinto-Espinosa F, Scharr H et al (2015) The leaf angle distribution of natural plant populations: assessing the canopy with a novel software tool. Plant Method 11:11. doi:10.1186/s13007-015-0052-z

Prusinkiewicz P, Lindenmayer A (1990) Graphical modeling using L-systems, in the algorithmic beauty of plants, part of the series the virtual laboratory. Springer, New York, pp 1–50

Rosati A, Dejong TM (2003) Estimating photosynthetic radiation use efficiency using incident light and photosynthesis of individual leaves. Ann Bot 91:869–877

Sands PJ (1999) 6.4.2 light use efficiency. In: Atwell BJ et al (eds) Plants in action. Macmillan Education Australia Pty Ltd, Melbourne, Available via DIALOG. http://plantsinaction.science.uq.edu.au/edition1/ofsubordinatedocument. Accessed 25 Dec 2015

Shimizu H, Kushisa M, Fujinuma W (2008) A growth model for leaf lettuce under greenhouse environments. Environ Control Biol 46:211–219

Tei F, Scaife A, Aikman DP (1996) Growth of lettuce, onion, and red beet. 1. Growth analysis, light interception and radiation use efficiency. Ann Bot 78:633–643

Vos J, Evers JB, Buck-Sorlin GH et al (2010) Functional–structural plant modeling: a new versatile tool in crop science. J Exp Bot 61:2101–2115

Yang Z-C, Kubota C, Chia P-L et al (2012) Effect of end-of-day far-red light from a movable LED fixture on squash rootstock hypocotyl elongation. Sci Hortic 136:81–86

Yokoi S, Kozai T, Ohyama K et al (2003) Effects of leaf area index of tomato seedling population on energy utilization efficiencies in a closed transplant production system. J SHITA 15:231–238 (in Japanese with English abstract)

Chapter 12
Effects of Physical Environment on Photosynthesis, Respiration, and Transpiration

Ryo Matsuda

Abstract Responses of photosynthetic, respiratory, and transpiration rates of plants to the levels of physical environmental factors are outlined. Water vapor movement from a leaf to the atmosphere in transpiration is quantitatively described using a gas diffusion model that incorporates the concentration gradient of water vapor and the conductance for water vapor. Changes in the levels of environmental factors affect transpiration rate directly through changes in the driving force of water vapor diffusion or indirectly through changes in stomatal aperture. In plants, there are two types of respiration, namely, dark respiration and photorespiration, although they are completely different metabolisms. Dark respiratory rate is sensitive to changes in temperature, gas concentrations, and light intensity, and the effects are summarized. Photosynthetic carbon dioxide (CO_2) influx of a leaf is spatially divided into CO_2 diffusion from the atmosphere to the chloroplasts and biochemical CO_2 fixation within the chloroplasts. The former can be described with a model in a similar manner to that for water vapor diffusion. Net photosynthetic rate in C_3 leaves shows a saturating-type increase in response to increases in CO_2 concentration or photosynthetic photon flux density (PPFD), and the response curves are characterized by several parameters. Net photosynthetic rate is low at extremely low and high temperatures and shows an optimum level at intermediate temperatures.

Keywords CO_2 diffusion • Conductance • Environmental factor • Flux • Response curve • Stomata • Water vapor diffusion

---

R. Matsuda (✉)
Department of Biological and Environmental Engineering, Graduate School of Agricultural and Life Sciences, The University of Tokyo, Bunkyo, Tokyo 113-8657, Japan
e-mail: amatsuda@mail.ecc.u-tokyo.ac.jp

12.1 Introduction

Photosynthesis, respiration, and transpiration are fundamental physiological processes of plants and are closely associated with their growth and development. The reaction rates of these processes are largely influenced by the environment of the plant. On the other hand, through these processes, energy and compounds such as water vapor and CO_2 are exchanged between the plant and the surrounding atmosphere, which thereby affects the environment. Understanding this plant–environment interaction is important for controlled environment agriculture in order to attain optimal facility design and operation as well as appropriate plant management practices.

In this chapter, I present an outline of the environmental effects on photosynthesis, respiration, and transpiration of green plants. An environmental factor can be categorized on the basis of its characteristics into the physical, chemical, or biological environment. Here I focus on the physical environment, which includes the light, thermal, humidity, and gas environments. I consider only C_3 photosynthesis and not C_4 and crassulacean acid metabolism (CAM) photosynthesis, because most plant species currently grown in greenhouses and plant factories with artificial lighting (PFAL) are C_3 plants. Refer to Chap. 8 for the physiological basis of photosynthesis and transpiration as well as the light spectral effects on photosynthesis.

12.2 Transpiration

Transpiration of leaves drives the transpiration stream in the xylem and therefore water and nutrient uptake by roots. Regulation of transpiration rate at appropriate levels by environmental control in greenhouses and PFAL is thus important to maintain a favorable water and nutritional status of the plants. Transpiration also contributes to dissipation of energy absorbed by leaves into the atmosphere in the form of latent heat. I introduce this section with a model describing water vapor diffusion from the leaf to the atmosphere and then summarize the effects of environmental factors on transpiration rate, particularly in relation to stomatal responses.

12.2.1 Water Vapor Diffusion Model

Figure 12.1 shows a schematic diagram of water vapor and CO_2 diffusion between a leaf and the atmosphere. Water vapor in the intercellular air spaces, which has evaporated at the mesophyll cell surface, diffuses to the atmosphere immediately external to the leaf through stomatal pores surrounded by guard cells. Water vapor

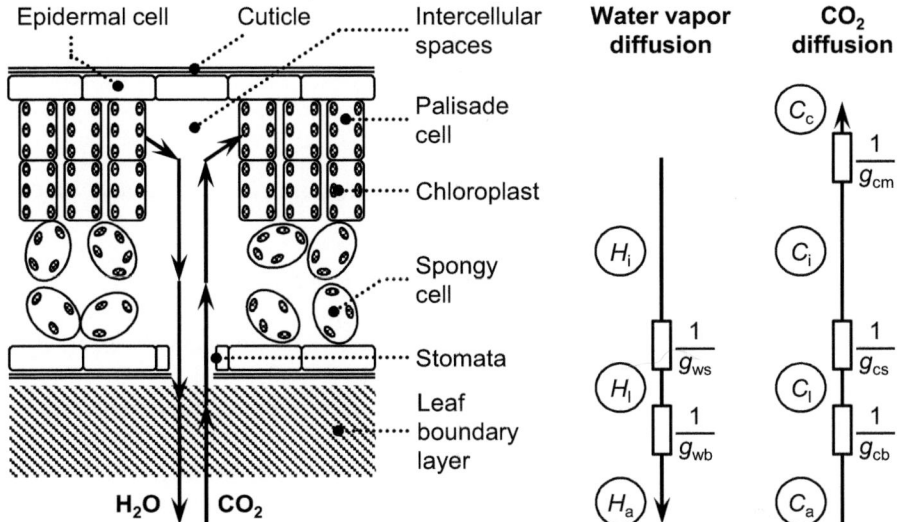

Fig. 12.1 Schematic diagram of cross-sectional view of a dorsiventral leaf (*left*) and models of water vapor and CO_2 diffusion between the leaf and the atmosphere with diffusion resistance (the inverse of conductance) between them (*right*). Cuticular transpiration is omitted. H_i, H_l, and H_a: mole fractions of water vapor in the intercellular air spaces, immediately external to the leaf surface, and in the atmosphere beyond the leaf boundary layer; g_{ws} and g_{wb}: stomatal and leaf boundary layer conductances for water vapor; C_a, C_l, C_i, and C_c: mole fractions of CO_2 in the atmosphere, immediately external to the leaf surface, in the intercellular spaces, and in the hypothetical air equilibrated with the chloroplast stroma; g_{cb}, g_{cs}, and g_{cm}: leaf boundary layer, stomatal, and mesophyll conductances for CO_2

diffuses further to the free atmosphere through the leaf boundary layer. Water vapor or liquid water also moves across the cuticle layer on the leaf epidermis, but the extent is substantially lower than that through stomata and thus is ignorable in most cases. Water vapor diffusion can be quantitatively described on the basis of an analogy of Ohm's law. By this principle, transpiration rate per unit leaf area, T [molH$_2$O m^{-2} s^{-1}], is proportional to the difference in mole fractions of water vapor at given two representative points. Conductance is the inverse of resistance and is often used instead of resistance in the model. T at a steady state is expressed as:

$$T = g_{wt}(H_i - H_a) \quad (12.1)$$

where H_i and H_a [mol mol^{-1}] are the respective mole fractions of water vapor in the intercellular air spaces and in the atmosphere beyond the leaf boundary layer, and g_{wt} [molH$_2$O m^{-2} s^{-1}] is the total water vapor conductance between them. The g_{wt} is divided into stomatal conductance, g_{ws} [molH$_2$O m^{-2} s^{-1}], and leaf boundary layer conductance, g_{wb} [molH$_2$O m^{-2} s^{-1}], for water vapor:

$$\frac{1}{g_{wt}} = \frac{1}{g_{ws}} + \frac{1}{g_{wb}}$$
$$g_{wt} = \frac{g_{ws} g_{wb}}{g_{ws} + g_{wb}} \qquad (12.2)$$

where g_{ws} and g_{wb} are connected in series. Using g_{ws} and g_{wb}, T at a steady state is also written as:

$$T = g_{ws}(H_i - H_l) = g_{wb}(H_l - H_a) \qquad (12.3)$$

where H_l [mol mol^{-1}] is the mole fraction of water vapor immediately external to the leaf surface.

Changes in the levels of environmental factors affect T directly and/or indirectly. The former is the effect on the concentration gradient of water vapor. The latter, on the other hand, is the effect on T through alterations of g_{wb} and/or g_{ws}. For example, g_{wb} decreases with increasing thickness of the leaf boundary layer, and the thickness is influenced by wind speed as well as leaf morphological characteristics (Campbell and Norman 1998; refer also to Chap. 13). Changes in a variety of environmental factors cause a rapid response of g_{ws} within seconds to minutes. Alteration of g_{ws} influences not only T but also photosynthetic CO_2 assimilation, the regulation of which is important for carbon gain by plants (see Sect. 12.4.1).

12.2.2 Effects of Humidity

Humidity directly affects T. Let us consider the effect quantitatively using Eq. 12.1. In general, the mole fraction of water vapor in the intercellular air spaces, H_i, can be assumed to be saturated at the leaf temperature, t_l [°C]:

$$H_i = \frac{e_s(t_l)}{p} \qquad (12.4)$$

where $e_s(t_l)$ [Pa] is the saturated water vapor pressure at t_l, and p is the atmospheric pressure [Pa]. Equation 12.1 is then rewritten using the atmospheric water vapor pressure, e [Pa]:

$$T = g_{wt}(H_i - H_a) = \frac{g_{wt}(e_s(t_l) - e)}{p} \qquad (12.5)$$

where $(e_s(t_l) - e)$ is termed the leaf-to-air water vapor pressure deficit (VPD). It is clear from this equation that, if the total water vapor conductance (g_{wt}) and p are constant, T is proportional to leaf-to-air VPD. Leaf-to-air VPD is thus more appropriate than relative humidity (the percentage of e to e_s at the air temperature) when evaluating the effect of humidity on T.

Humidity also affects T indirectly via a change in stomatal conductance: stomata tend to close when leaf-to-air VPD is high. Stomatal conductance and consequently total water vapor conductance decrease in order to suppress excessive water loss. Such the decrease in stomatal conductance thus counteracts the direct effect of high leaf-to-air VPD on T.

12.2.3 Effects of Rhizosphere Environment

Moisture deficiency and excess salt accumulation in the rhizosphere, which potentially cause water stress in plants, are typical stimuli that trigger stomatal closure. The phytohormone abscisic acid is known to be a chemical inducer of stomatal closure. Abscisic acid is generated in roots in response to water shortage, is transported to leaves via the xylem sap, and induces stomatal closure. The rapid decrease in stomatal conductance triggered by this mechanism, before any negative impacts of water deficiency are experienced, is considered to be a kind of feedforward response of plants, which should prevent subsequent excess water loss (Lambers et al. 2008).

12.2.4 Effects of Light Intensity and Spectrum

Stomatal conductance increases with increasing PPFD (Fig. 12.2a), thus promoting transpiration (Fig. 12.2b). This response facilitates CO_2 uptake for photosynthesis under high PPFD conditions as well as evaporative cooling of the leaf subjected to a high radiative heat load. The light-driven response of stomata occurs via at least two mechanisms (Shimazaki et al. 2007). One is the blue light (BL)-dependent response, mediated by the BL photoreceptor phototropin. The extent of response is considered to be saturated by a relatively low BL intensity. The other mechanism is the photosynthesis-dependent response, which is thought to be driven by the photosynthetic activity of mesophyll cells and/or guard cells, although the details of the mechanism remain unclear.

12.2.5 Effects of CO_2 Concentration

Stomatal conductance tends to increase at low CO_2 concentrations and decrease at high CO_2 concentrations. Such changes in stomatal conductance in response to CO_2 concentration play a role in compensating in part for the altered photosynthetic rate, which is directly affected by CO_2 concentration (see Sect. 12.4.2). The response to high CO_2 concentration also contributes to reduction in water loss relative to CO_2 uptake through stomata. The photosynthetic water-use efficiency, defined as the

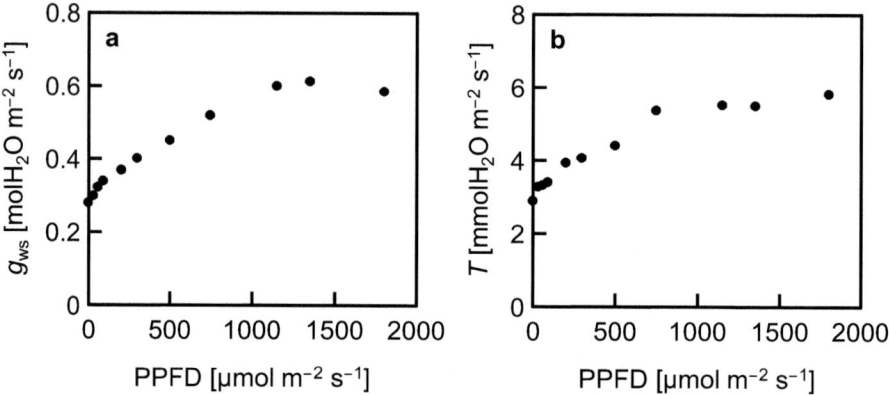

Fig. 12.2 Stomatal conductance for water vapor (g_{ws}, **a**) and transpiration rate (T, **b**) at different incident PPFDs in a spinach leaf (Matsuda et al. unpublished). Measurements were made at an atmospheric CO_2 concentration of 360 μmol mol^{-1}, a leaf temperature of 25 °C, and a leaf-to-air water vapor pressure deficit of 1.1 kPa. Light was provided by a white halogen lamp

ratio of net photosynthetic rate to T, both expressed on a leaf area basis (Lambers et al. 2008), therefore increases at high CO_2 concentrations.

12.2.6 Effects of Temperature

Because leaf-to-air VPD is a function of leaf temperature (Eq. 12.5), leaf temperature affects T. On the other hand, a change in leaf temperature causes alterations in photosynthetic and respiratory rates (see Sects. 12.4.4 and 12.3.2, respectively), which may indirectly affect T through changes in stomatal conductance. Moreover, an increase in T leads to a reduction in leaf temperature via latent heat loss of leaves. Thus, there is a complex interrelationship between leaf temperature and T. At a steady state, leaf temperature and T are determined so that the energy budget of the leaf balances. Refer to Campbell and Norman (1998) and Jones (2014) for detailed description of the leaf energy balance.

12.3 Respiration

12.3.1 Dark Respiration and Photorespiration

In plants, there are two types of the so-called respiration, namely, dark respiration and photorespiration, although they are completely different metabolisms. The pathway of dark respiration consists of the glycolysis and pentose phosphate pathway located in the cytosol and the plastids, followed by the tricarboxylic acid cycle and oxidative phosphorylation in the mitochondria. The role of dark respiration is to provide chemical energy and reducing power by oxidizing respiratory substrates such as carbohydrates, as well as provide intermediate metabolites as materials for a variety of biological compounds. Photorespiration is a metabolism specific to photosynthetic organisms and is localized in the chloroplasts, peroxisomes, and mitochondria. A role of the photorespiratory pathway is to minimize carbon loss that results almost inevitably from O_2 fixation in the Calvin cycle located in the chloroplasts. Photorespiration is observed only in photosynthetic organs, whereas dark respiration basically proceeds in every plant organ.

In spite of the metabolic differences, plants uptake O_2 and release CO_2 through both dark respiration and photorespiration. Dark respiration occurs both in the light and the dark, whereas photorespiration proceeds only in the presence of incident light. Therefore, when the rate of CO_2 efflux is measured without light irradiation, whereby neither photosynthesis nor photorespiration is active, the efflux rate corresponds to dark respiratory rate. It should be noted, however, that dark respiratory rate is reported to be influenced by the light environment (see Sect. 12.3.4).

12.3.2 Effects of Temperature

The dark respiratory rate of plant organs is sensitive to temperature. The rate increases almost exponentially with increasing temperature between approximately 10 and 40 °C (Fig. 12.3), although the extent of increase varies among species and growth conditions. Such temperature dependence of dark respiratory rate is often expressed using the Q_{10} model. According to the model, dark respiratory rate per unit area, R_d [molCO$_2$ m^{-2} s^{-1}], at an organ/plant temperature of t_p [°C] is expressed as:

$$R_d = R_{ref} Q_{10}^{t_{dif}} \tag{12.6}$$

where

$$t_{dif} = \frac{t_p - t_{ref}}{10} \tag{12.7}$$

R_{ref} [molCO$_2$ m^{-2} s^{-1}] is R_d at a reference temperature, t_{ref} [°C]. The Q_{10} value for plant R_d is approximately 2 (Lambers et al. 2008), which means that a 10 °C increase in temperature brings about a twofold increase in R_d.

Suppression of R_d by decreasing temperature is employed for storage of postharvest fruits, vegetables, cut flowers, and sometimes seedlings. On the other hand, over-suppression of R_d of growing plants by lowering temperature at night is not necessarily effective for promoting growth because of the importance of dark respiration in the construction of new tissues and organs.

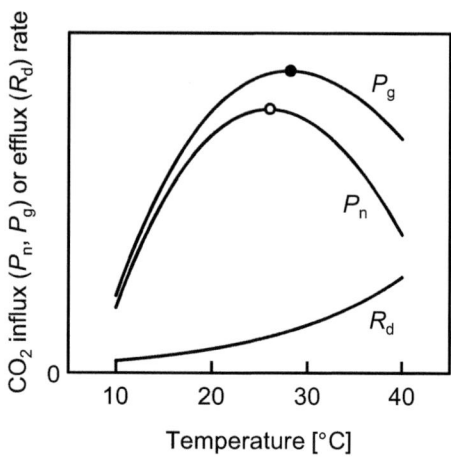

Fig. 12.3 Schematic diagram of responses of net (P_n) and gross (P_g) photosynthetic rates and dark respiratory rate (R_d) to temperature in a C_3 leaf. P_n and P_g: cubic functions; R_d: the Q_{10} model (Sect. 12.3.2). Open and filled circles represent the maximal values for P_n and P_g, respectively

12.3.3 Effects of O_2 and CO_2 Concentrations

Decrease in O_2 concentration and elevation of CO_2 concentration tend to suppress R_d of plants. In greenhouses and PFAL, atmospheric O_2 concentration remains at the ambient level of 21 % and barely fluctuates and therefore has little effect on R_d of the aboveground parts of plants. On the other hand, CO_2 concentration is often enriched during the day for promotion of photosynthesis (see Sect. 12.4.2), but the concentration is generally too low to reduce R_d. In postharvest storage, O_2 and CO_2 concentrations around the harvested products can be decreased and increased, respectively, in order to suppress respiration and prolong storage life. In hydroponics, the dissolved O_2 concentration in the rhizosphere can sometimes decline below the level required for root respiration as a result of insufficient aeration.

12.3.4 Effects of Light Intensity

Several studies have shown that R_d of leaves is influenced by light intensity (Brooks and Farquhar 1985): the estimated rate in the light is significantly lower than the measured rate in the dark for the same leaf. Moreover, the rate drastically decreases as the light intensity increases. For distinction, dark respiration under incident light is sometimes referred to specifically as "day" respiration.

12.4 Photosynthesis

Photosynthesis consists of a series of reactions by which CO_2 is assimilated using light energy for carbohydrate synthesis and is thus an important physiological function that directly influences plant growth and yield. The reactions proceed in the chloroplasts. The net photosynthetic rate per unit leaf area, P_n [molCO_2 m^{-2} s^{-1}], is expressed as:

$$P_n = P_g - R_d \qquad (12.8)$$

where P_g [molCO_2 m^{-2} s^{-1}] is gross photosynthetic rate per unit leaf area. In this chapter, photorespiratory CO_2 efflux is included in P_g.

Photosynthetic CO_2 influx of a leaf can spatially be divided into two processes: CO_2 diffusion from the atmosphere to the chloroplasts and the biochemical CO_2 fixation inside the chloroplasts. In the following sections, the model of CO_2 diffusion from the atmosphere to the chloroplasts is first described. Then, the effects of the physical environment on P_n are summarized. In this chapter, I mainly focus on leaf photosynthesis. Responses of canopy photosynthesis to environmental changes often differ from those of leaf photosynthesis. Several characteristics of canopy photosynthesis are discussed in Chaps. 9 and 13.

12.4.1 CO_2 Diffusion Model

Diffusion of CO_2 from the atmosphere to the chloroplasts can be described in a similar manner to that for water vapor diffusion, although water vapor and CO_2 diffusion occur in opposite directions in a photosynthesizing leaf (Fig. 12.1). Because of resistance against CO_2 diffusion, as observed for water vapor diffusion, the CO_2 concentration in the intercellular air spaces of the leaf may be as low as approximately 70 % of that in the atmosphere (Evans and Loreto 2000). Although water vapor in the intercellular spaces is considered to be equilibrated with liquid water in the mesophyll cells, CO_2 gas in the intercellular spaces is dissolved in the liquid of mesophyll cells and further diffuses to the chloroplast stroma against an additional resistance.

Calculation of P_n at a steady state is written as follows:

$$P_n = g_{ct}(C_a - C_c) \\ = g_{cb}(C_a - C_1) = g_{cs}(C_1 - C_i) = g_{cm}(C_i - C_c) \qquad (12.9)$$

where C_a, C_1, C_i, and C_c [mol mol^{-1}] are the respective mole fractions of CO_2 in the atmosphere, immediately external to the leaf surface, in the intercellular spaces, and in the hypothetical air equilibrated with the chloroplast stroma. The g_{ct} [molCO_2 m^{-2} s^{-1}] is the total CO_2 conductance between the atmosphere and the chloroplast stroma, consisting of three conductances connected in series. Leaf boundary layer conductance, g_{cb} [molCO$_2$ m^{-2} s^{-1}], and stomatal conductance, g_{cs} [molCO$_2$ m^{-2} s^{-1}], for CO$_2$ are analogous to those for water vapor. Mesophyll conductance (or internal conductance) for CO$_2$, g_{cm} [molCO$_2$ m^{-2} s^{-1}], is the additional conductance, present between the mesophyll cell surface and the chloroplast stroma. Several anatomical and biochemical properties of leaves are reportedly involved in g_{cm} (Evans and Loreto 2000).

The conductance for water vapor and CO$_2$ can be converted into each other using the conductance ratio in the still air for stomatal conductance and that in the laminar flow for boundary layer conductance, respectively (Farquhar and Sharkey 1982):

$$g_{ws} = 1.6 g_{cs} \qquad (12.10)$$

$$g_{wb} = 1.37 g_{cb} \qquad (12.11)$$

Changes in stomatal conductance in response to changes in the levels of environmental factors, as seen in Sects. 12.2.2, 12.2.3, 12.2.4, 12.2.5, and 12.2.6, therefore potentially influence P_n via CO$_2$ diffusion. For example, regulation of leaf-to-air VPD at a low level contributes to maintaining a high degree of stomatal opening (Sect. 12.2.2), thereby promoting CO$_2$ diffusion into the leaf.

12.4.2 Effects of CO$_2$ Concentration

In general, P_n in C$_3$ leaves shows a saturating-type increase in response to increasing CO$_2$ concentration (Fig. 12.4). The increment of P_n gradually decreases as CO$_2$ concentration increases, and P_n does not increase further once CO$_2$ concentration increases beyond a certain critical level. The latter CO$_2$ concentration is termed the CO$_2$ saturation point (CSP). The CO$_2$ compensation point (CCP), on the other hand, is the CO$_2$ concentration at which P_n is zero. In Fig. 12.4a, P_n is plotted against atmospheric CO$_2$ concentration (C_a), but a similar trend is also observed when P_n is plotted against intercellular CO$_2$ concentration (C_i) (Fig. 12.4b), which mostly reflects the CO$_2$ fixation characteristics at the chloroplast level. The C_i-response curve of P_n is referred to as the demand function, whereas Eq. 12.9 (dashed line in Fig. 12.4b) is referred to as the supply function (Farquhar and Sharkey 1982). The actual P_n of a leaf at a given external CO$_2$ concentration is determined from the intersection of the two lines.

Enrichment of CO$_2$ is frequently employed in greenhouses and PFAL in order to increase P_n and thereby biomass production. For example, CO$_2$ concentration can be approximately doubled in greenhouses and increased to 1,000–2,000 µmol mol^{-1} in PFAL. Indeed, P_n is increased initially in accordance with the CO$_2$-response curve of P_n. Note, however, that the extent of increase in P_n sometimes gradually decreases when a high CO$_2$ concentration is maintained and prolonged (the downregulation of photosynthesis).

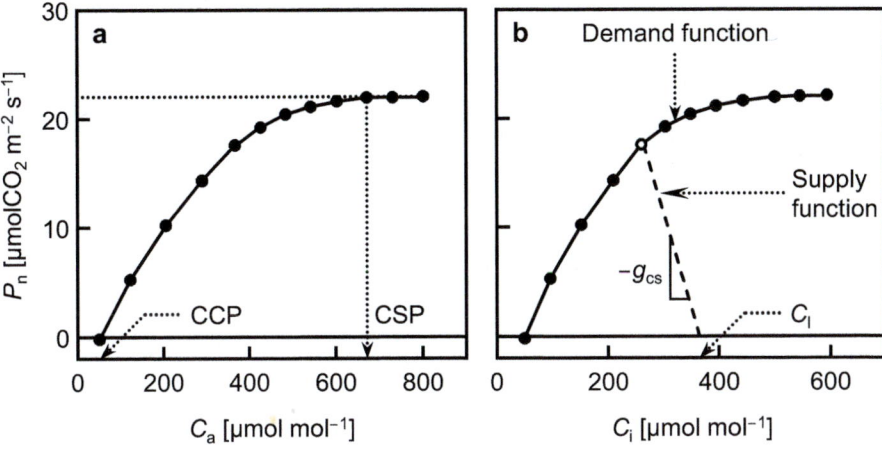

Fig. 12.4 (a) Net photosynthetic rate (P_n) at different atmospheric CO_2 concentrations (C_a) in a tomato leaf (Matsuda et al. unpublished). Measurements were made at a PPFD of 1,500 μmol m^{-2} s^{-1}, a leaf temperature of 25 °C, and a leaf-to-air water vapor pressure deficit of 1.1 kPa. Light was provided by blue and red LEDs. CCP: CO_2 compensation point; CSP: CO_2 saturation point. (b) P_n as a function of intercellular CO_2 concentration (C_i). Dashed line: the supply function [$P_n = g_{cs}(C_1 - C_i) = -g_{cs}(C_i - C_1)$, where $g_{cs} = 1.66$ molCO$_2$ m^{-2} s^{-1} and $C_1 = 367$ μmol mol^{-1}]; solid line: the demand function (C_i-response curve of P_n). The open circle represents the realized P_n and C_i of the leaf

12.4.3 Effects of Light Intensity

Figure 12.5 shows an example of the response of leaf P_n to PPFD. Under the dark conditions of 0 mol m^{-2} s^{-1} PPFD, P_n equals to $-R_d$. With increasing PPFD, P_n initially increases almost linearly. The increment of P_n gradually decreases with further increase in PPFD. Similarly to the CO_2-response curve, the light saturation point (LSP) and light compensation point (LCP) are defined as the PPFDs at which P_n becomes maximum and zero, respectively. The LSP of mature leaves is approximately 1,000–1,500 μmol m^{-2} s^{-1} for species favoring high PPFDs, such as tomato and cucumber, whereas it is in general less than 1,000 μmol m^{-2} s^{-1} for leafy vegetables. P_n measured at LSP, at a normal CO_2 concentration, and at an optimal leaf temperature is often referred to as the photosynthetic capacity of the leaf. It is well known that leaves grown under high PPFD ("sun" leaves) have higher LSP, LCP, photosynthetic capacity, and R_d than those grown under low PPFD ("shade" leaves). The initial slope of the curve at very low PPFDs reflects the maximal light-use efficiency for photosynthesis. When the efficiency is evaluated on an absorbed PPFD basis (i.e., the quantum yield of photosynthesis), the theoretical maximum for C_3 leaves is 0.125 mol mol^{-1} (1 mol of CO_2 is fixed per 8 mol of absorbed photons).

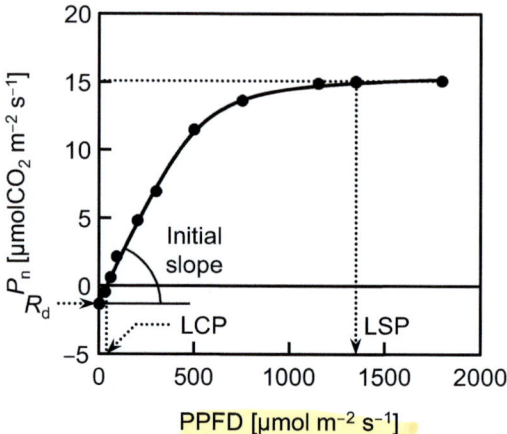

Fig. 12.5 Net photosynthetic rate (P_n) at different incident PPFDs in a spinach leaf (Matsuda et al. unpublished). Measurements were made at an atmospheric CO_2 concentration of 360 μmol mol^{-1}, a leaf temperature of 25 °C, and a leaf-to-air water vapor pressure deficit of 1.1 kPa. Light was provided by a white halogen lamp. R_d: dark respiratory rate; LCP: light compensation point; LSP: light saturation point

12.4.4 Effects of Temperature

The P_g and P_n are low at extremely low and high temperatures and show optimum levels at intermediate temperatures: the temperature dependences can thus be fitted with concave functions (Fig. 12.3). The optimal temperature for P_n is in general slightly lower than that for P_g because of the exponential increase in R_d with increasing temperature (Sect. 12.3.2).

In plants grown at higher temperatures, the optimal temperature for P_n is increased: it increases by 0.4–0.5 °C on average with an increase in growth temperature by 1 °C in C_3 leaves (Yamori et al. 2014). Several physiological and biochemical mechanisms are thought to be involved in this change in temperature dependence of P_n, which are being extensively studied.

Acknowledgment The author would like to thank Keach Murakami for helpful discussions and critical reading of the manuscript.

References

Brooks A, Farquhar GD (1985) Effect of temperature on the CO_2/O_2 specificity of ribulose-1,5-bisphosphate carboxylase/oxygenase and the rate of respiration in the light. Estimates from gas-exchange measurements on spinach. Planta 165:397–406

Campbell GS, Norman JM (1998) An introduction of environmental biophysics, 2nd edn. Springer, New York

Evans JR, Loreto F (2000) Acquisition and diffusion of CO_2 in higher plant leaves. In: Leegood RC, Sharkey TD, von Caemmerer S (eds) Photosynthesis: physiology and metabolism. Kluwer, Dordrecht, pp 321–351
Farquhar GD, Sharkey TD (1982) Stomatal conductance and photosynthesis. Annu Rev Plant Physiol 33:317–345
Jones HG (2014) Plants and microclimate, 3rd edn. Cambridge University Press, Cambridge
Lambers H, Chapin III FS, Pons TL (2008) Plant physiological ecology, 2nd edn. Springer, New York
Shimazaki K, Doi M, Assmann SM, Kinoshita T (2007) Light regulation of stomatal movement. Annu Rev Plant Biol 58:209–247
Yamori W, Hikosaka K, Way DA (2014) Temperature response of photosynthesis in C_3, C_4, and CAM plants: temperature acclimation and temperature adaptation. Photosynth Res 119:101–117

Chapter 13
Air Current Around Single Leaves and Plant Canopies and Its Effect on Transpiration, Photosynthesis, and Plant Organ Temperatures

Yoshiaki Kitaya

Abstract Carbon dioxide (CO_2) and water vapor exchanges between plants and the atmosphere are regulated by resistance to gas diffusion from the atmosphere to the chloroplast for CO_2 diffusion (photosynthesis) and from the stomatal cavity to the atmosphere for water vapor diffusion (transpiration). Photosynthesis and transpiration are commonly controlled by stomatal and boundary layer resistances. Several environmental factors have been reported to affect photosynthesis and transpiration through the stomatal aperture. In the present chapter, the boundary layer resistance as affected by air movement is focused, and the effects of air current speeds lower than 1 m s^{-1} on net photosynthetic (P_n) and transpiration (T_r) rates of single leaves and plant seedling canopies through the boundary layer were mainly assessed. The P_n and T_r of leaves doubled as air current speeds increased from 0.01 to 0.3 m s^{-1} and were almost constant at air current speeds of 0.4–1.0 m s^{-1}. The increase in P_n and T_r was greater in the plant canopy than in the single leaf. Air movement is also important to ensure that the environmental variables remain uniform inside the plant canopy. Controlled air movement is important for enhancing the gas exchange between plants and the ambient air and is consequently important for promoting plant growth especially in semi-closed plant production facilities.

Keywords Boundary layer resistance • CO_2 • Humidity • Leaf temperature • Photosynthesis • Transpiration

---

Y. Kitaya (✉)
Graduate School of Life and Environmental Sciences, Osaka Prefecture University,
1-1 Gakuen-cho, Sakai, Osaka 599-8531, Japan
e-mail: kitaya@envi.osakafu-u.ac.jp

© Springer Science+Business Media Singapore 2016
T. Kozai et al. (eds.), *LED Lighting for Urban Agriculture*,
DOI 10.1007/978-981-10-1848-0_13

13.1 Introduction

The exchanges of carbon dioxide (CO_2) and water vapor between plants and the atmosphere are regulated by the resistance to gas diffusion from the atmosphere to the chloroplast for CO_2 diffusion (photosynthesis) and from the stomatal cavity to the atmosphere for water vapor diffusion (transpiration). In general, photosynthesis and transpiration are commonly controlled by stomatal and boundary layer resistance. Several environmental factors, including air temperature, humidity, atmospheric CO_2, light intensity, soil moisture, soil temperature, and soil salinity, have been shown to affect photosynthesis and transpiration mainly through the stomatal aperture. However, there are fewer reports on the effects of environmental factors on photosynthesis and transpiration through the leaf boundary layer. Leaf boundary layer resistance is usually predominantly controlled by air movement.

In agriculture, the utilization of semi-closed plant production facilities such as greenhouses and plant factories has become increasingly popular. In plant production in such facilities without adequate air circulation systems, air movement is extremely restricted as compared to that under field conditions. Insufficient air movement around plants increases the resistance to gas diffusion in the leaf boundary layer and thus limits photosynthesis and transpiration (Yabuki and Miyagawa 1970; Monteith and Unsworth 1990; Jones 1992; Yabuki 2004), resulting in the suppression of plant growth and development. Therefore, the enhancement of the gas exchange in leaves and the growth of plants depend on appropriate control of air movement. Thus, the aim of the present chapter is to emphasize the importance of air movement for promoting photosynthesis and transpiration and, hence, plant growth.

13.2 Effects of Air Current Speed on Boundary Layer Resistance, Photosynthesis, and Transpiration of Single Leaves

A boundary layer occurs on the surface of a plate in a fluid. The boundary layer on the leaf surface in the air is the leaf boundary layer. Inside the boundary layer, laminar airflow is more dominant than turbulent airflow and air movement is more restricted than the outside of the boundary layer. Gas and heat transfer from the surface to the surrounding air is, therefore, significantly restricted. The leaf boundary layer is thinner at higher air current speeds and over the forward edge than the following region of the leaf as shown in Fig. 13.1. The thinner boundary layer promotes more efficient gas and heat transfer.

Leaf boundary layer resistance against water vapor transfer decreased significantly as air current speeds increased from 0.01 to 0.2 m s^{-1} and decreased gradually at air current speeds ranging from 0.3 to 1.0 m s^{-1} (Fig. 13.2). The leaf boundary layer resistance at the air current speed of 0.2 m s^{-1} was a half of that at

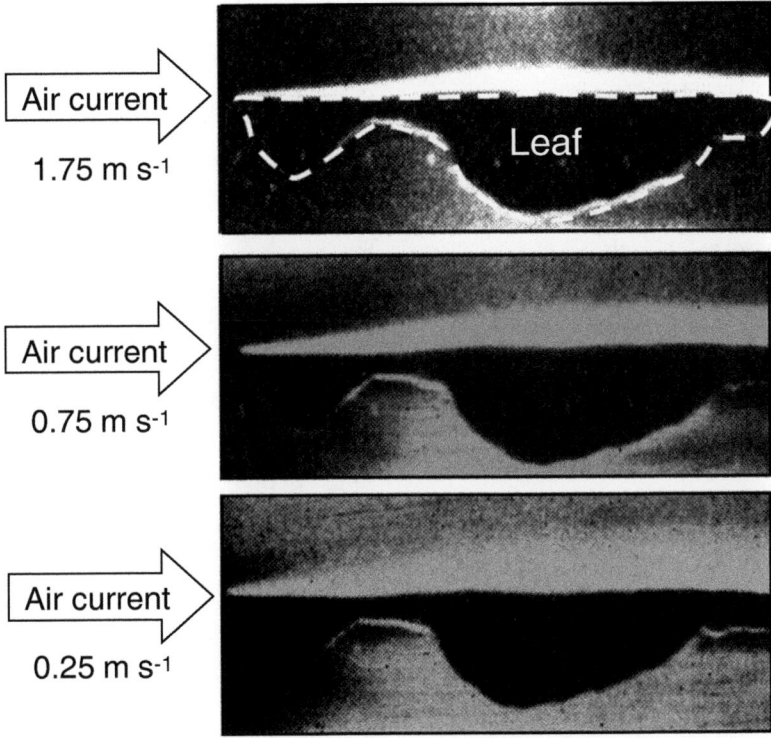

Fig. 13.1 Boundary layers (*white* parts above the leaf) on a cabbage leaf visually identified with a Schlieren optical system at different air current speeds (Yabuki 2004)

0.01 m s^{-1}. The leaf boundary layer resistance was approximately proportional to the minus 0.37 power of the air current speed. Yabuki et al. (1970) reported the same result with cucumber and cabbage leaves at air current speeds lower than 2.0 m s^{-1}. Martin et al. (1999) estimated the boundary layer conductance from energy balance measurements in the field, which is the reciprocal of the boundary layer resistance. The boundary layer conductance increased linearly from 0.01 to 0.15 m s^{-1} as air current speeds increased from 0.1 to 2.0 m s^{-1}.

The net photosynthetic (P_n) and transpiration (T_r) rates increased significantly as the air current speed increased from 0.01 to 0.2 m s^{-1} (Fig. 13.2). The T_r increased gradually at air current speeds ranging from 0.2 to 1.0 m s^{-1}, and the net photosynthetic rate reached an almost constant level at air current speeds ranging from 0.4 to 1.0 m s^{-1}.

P_n and T_r were 1.2 and 1.3 times higher, respectively, at an air current speed of 0.9 m s^{-1} than at an air current speed of 0.1 m s^{-1}, corresponding to a decrease in the leaf boundary layer resistance from 3 to 2 m^2 s mol^{-1}. Increases in air current speeds induced greater increases in T_r than in P_n.

The air current speed inside the plant canopy decreased to 30 % of that above the plant canopy (Fig. 13.3). P_n of the plant canopy increased with increasing air

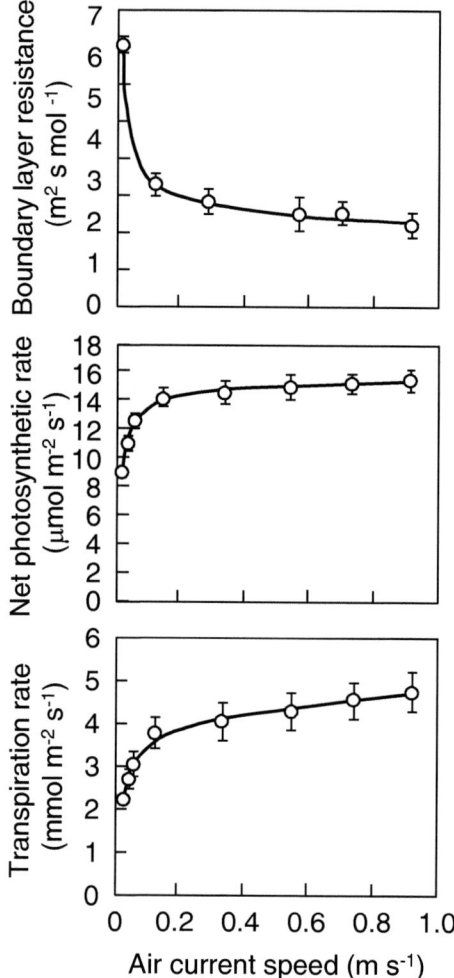

Fig. 13.2 Effect of air current speed on leaf boundary layer resistance, net photosynthetic, and transpiration rates in sweet potato leaves (Kitaya et al. 2003). Measurements were conducted with a leaf chamber method. PPFD, 1000 μmol m^{-2} s^{-1}; air temperature, 28 °C; relative humidity, 65 %; CO_2 concentration, 400 μmol mol^{-1}. *Bars* indicate standard deviations

current speeds to 1.0 m s^{-1} above the plant canopy (Fig. 13.4). P_n in the plant canopy at the air current speed of 1.0 m s^{-1} was two times higher than that at an air current speed of 0.1 m s^{-1} above the plant canopy.

Therefore, forced air movement inside plant production facilities is essential for enhancing gas exchange in leaves and thereby promoting plant growth. Forced air movement is more significant for plant canopies than for single leaves because of a significant reduction of air current speeds inside the canopy (Fig. 13.3). The P_n of plant canopies was also reported to be 1.4 times greater at an air current speed of 0.5 m s^{-1} than at 0.1 m s^{-1} for tomato seedlings (Shibuya and Kozai 1998) and two times greater at an air current speed of 1.0 m s^{-1} than at 0.1 m s^{-1} for rice seedlings (Kitaya et al. 2000). These results confirm that air movement is more important for enhancing gas exchange in plant canopies than in single leaves. The retardation of

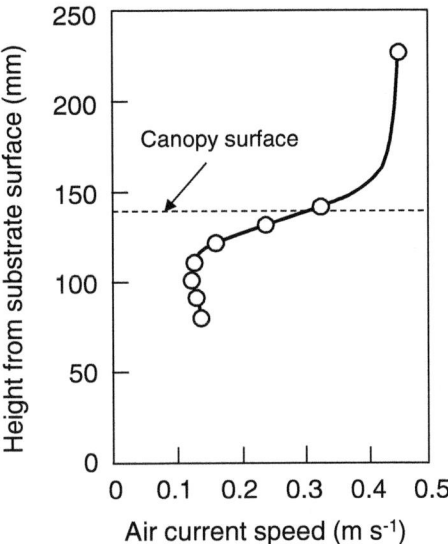

Fig. 13.3 Profile of air current speed inside and outside a tomato seedling canopy (Kitaya et al. 2003)

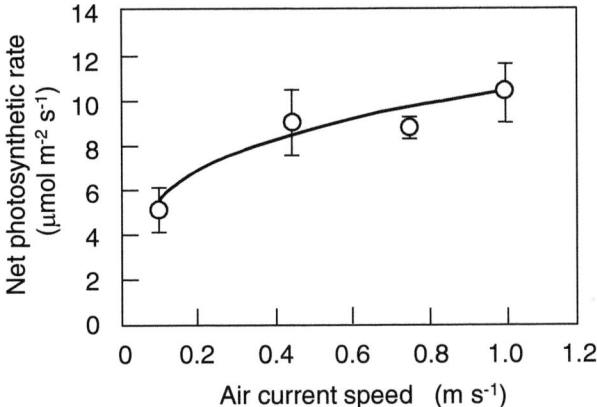

Fig. 13.4 Effect of air current speed on the net photosynthetic rate based on the rooting bed area of a tomato plant canopy (Kitaya et al. 2003). The canopy height was 0.14 m and the leaf area index was 2.1 $m^2\ m^{-2}$. The air current speed was measured 0.1 m above the canopy. PPFD, 250 $\mu mol\ m^{-2}\ s^{-1}$; air temperature, 28 °C; relative humidity, 65 %; CO_2 concentration, 400 $\mu mol\ mol^{-1}$. *Bars* indicate standard deviations

gas exchange in plant canopies is due to an increased leaf boundary layer resistance (Fig. 13.2). Precise control of air movement inside plant canopies promotes gas exchange in leaves and thereby enables the growth of plants to be controllable. The greater effect of the air current speed on transpiration than on photosynthesis in leaves (Fig. 13.2) indicates that air movement affects transpiration more directly than photosynthesis.

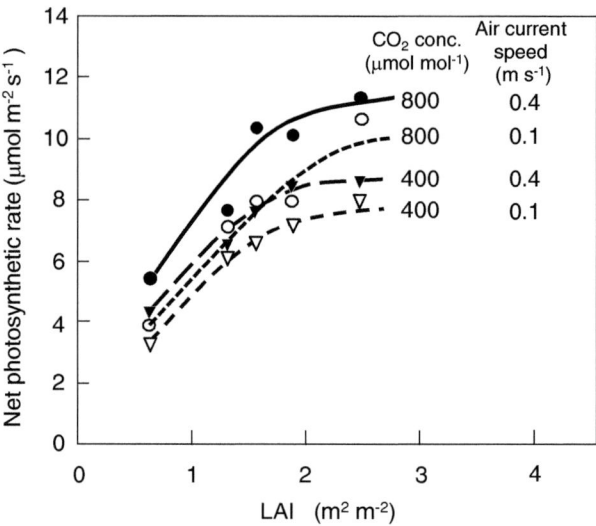

Fig. 13.5 Effects of air current speed, LAI, and CO_2 concentrations on the net photosynthetic rate of a tomato plant canopy (Kitaya et al. 2004). The canopy height was 0.14 m. The air current speed was measured 0.1 m above the canopy. PPFD measured on the canopy, 250 µmol m^{-2} s^{-1}; air temperature, 28 °C; relative humidity, 65 %. *Bars* indicate standard deviations

Figure 13.5 shows P_n of the tomato seedling canopies having different leaf area indexes (LAIs) and being affected by air current speeds and atmospheric CO_2 concentrations. The P_n based on the rooting bed area increased with increasing LAI from 0.6 to 2.0 and then tended to approach a fixed value regardless of the air current speed and atmospheric CO_2 concentrations. Increases in the air current speed and the CO_2 concentration promote photosynthesis. At an air velocity of 0.4 m s^{-1}, P_n at 0.8 mmol mol^{-1} CO_2 were 1.2 times and 1.3 times those at 0.4 mmol mol^{-1} CO_2 when the LAI was 0.6 and 2.5, respectively. At the air current speed of 0.1 m s^{-1}, P_n at 0.8 mmol mol^{-1} CO_2 were 1.2 times and 1.4 times those at 0.4 mmol mol^{-1} CO_2 when the LAIs were 0.6 and 2.5, respectively. The data showed that P_n increased more significantly with elevated CO_2 levels in the plant canopy having higher LAIs and at lower air current speeds.

Figure 13.6 shows the effect of the air current speed inside the plant canopy on P_n based on the leaf area under different LAI and CO_2 conditions. The P_n of the plant canopy increased with increasing air current speeds inside the plant canopy and saturated at 0.2 m s^{-1} (0.5 m s^{-1} above the plant canopy). P_n at the air current speed of 0.2 m s^{-1} was 1.3 times that at 0.05 m s^{-1} at 0.4 and 0.8 mmol mol^{-1} CO_2. The P_n at 0.8 mmol mol^{-1} CO_2 was 1.2 times that at 0.4 mmol mol^{-1} CO_2 at air current speeds ranging from 0.1 to 0.8 m s^{-1}. Forced air movement is more significant for plant canopies than for single leaves because of significant reduction of the air current speed inside the plant canopy compared with that above the plant canopy (Fig. 13.3).

The evapotranspiration rate of a tomato seedling canopy was also reported to increase with increasing air current speeds as well as P_n (Shibuya and Kozai 1998). The appropriate air current speeds for enhancing gas exchanges in leaves must be kept higher than 0.2 m s^{-1} in the vicinity of the leaves. Forced air movement is essential for plant canopies in closed plant culture chambers, and the air current

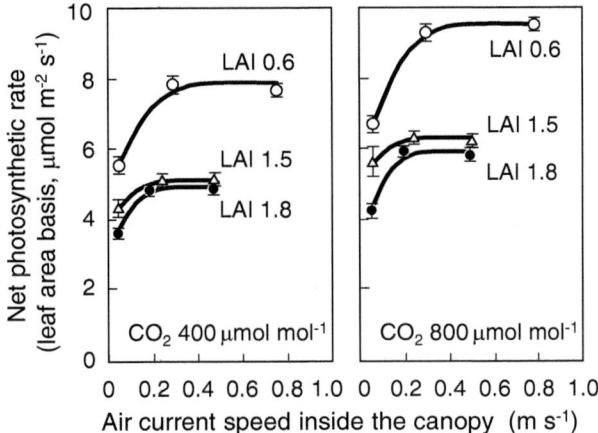

Fig. 13.6 Effect of air current speeds inside tomato seedlings canopies with different LAIs on the net photosynthetic rate based on the leaf area under different CO_2 levels (Kitaya et al. 2004). The measurement condition was the same as Fig. 13.5

speed above the canopy should be higher than 1.0 m s^{-1} to obtain maximal gas exchange rates of the dense plant canopy.

13.3 Effect of Air Current Speed on the Surface Temperatures of Plant Organs

The thermal images of sweet potato leaves showed a decrease in temperatures with increasing air current speeds under a lighting condition (Fig. 13.7). The decrease in temperature is mainly due to the promotion of transpiration and thus an increase in latent heat transfer. The surface temperature was lower at the forward edge than the following regions depending on the thickness of the boundary layer (Fig. 13.1).

Figure 13.8 shows the effect of air movement on the temperature of leaves and reproductive organs of strawberry plants. Leaf temperatures were higher than the temperatures of other organs. The difference in temperatures among the organs may depend on the boundary layer resistance and the heat capacity per surface area of each organ. Temperatures of reproductive organs and leaves decreased with increasing air current speeds. The temperatures of petals, stigmas, anthers, and leaves decreased by 12.8, 11.9, 13.1, and 14.1 °C, respectively, when the air current speed increased from 0.1 to 1.0 m s^{-1}.

The decrease in leaf temperatures was similar when affected by air current speeds at different relative humidity levels in the surrounding air (Fig. 13.9). Leaf temperatures decreased with decreasing relative humidity from 92 % to 66 % due to an increase in differences of water vapor pressures between leaves and air. However, leaf temperature increased with decreasing relative humidity to 52 %. This phenomenon is due to an increase in stomatal resistance caused by stomatal closure to avoid excess water loss.

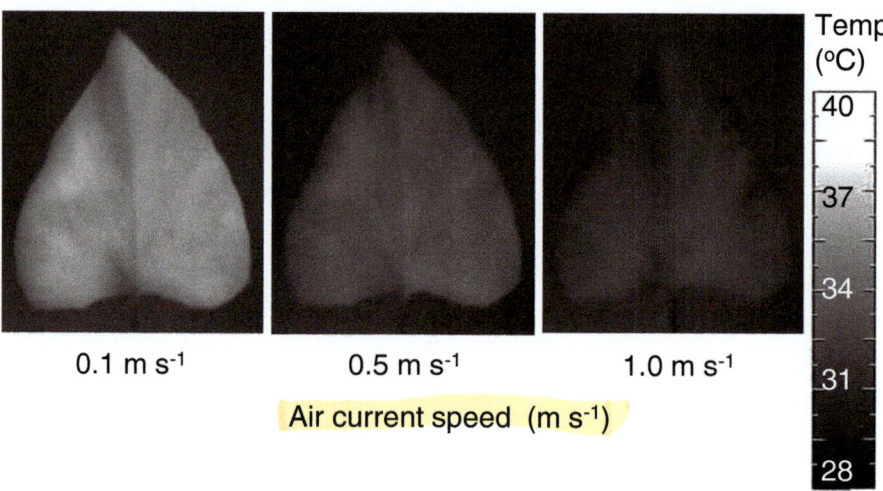

Fig. 13.7 Thermal images of the sweet potato leaf as affected by air movement. Airflow direction was from the *top* to the *bottom* in each figure. Air temperature, 29 °C; relative humidity, 50 %; irradiance, 350 W m^{-2}

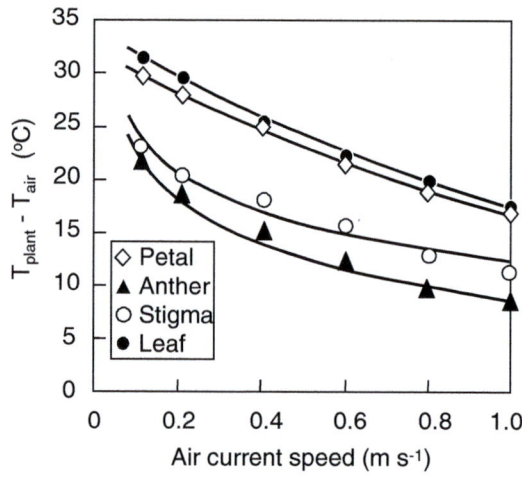

Fig. 13.8 Effect of air current speed on temperatures (T_{plant}) in reproductive organs and leaves of strawberry plants (Kitaya and Hirai 2007). Mean values ($n = 3$) and logarithmic approximation curves are shown. Light source, metal halide lamp; irradiance, 100 W m^{-2}; air temperature (T_{air}), 10 °C; relative humidity, 75 %

13.4 Effects of Light Intensity and Air Current Speed on the Air Temperature, Water Vapor Pressure, and CO$_2$ Concentration Inside Plant Canopies

Restricted air movement induced spatial variations of environmental variables such as air temperature, water vapor pressure, and CO$_2$ concentration inside eggplant seedling canopies (Kim et al. 1996; Kitaya et al. 1998). Under a photosynthetic photon flux density (PPFD) of 500 μmol m^{-2} s^{-1}, 2–3 °C higher air temperatures,

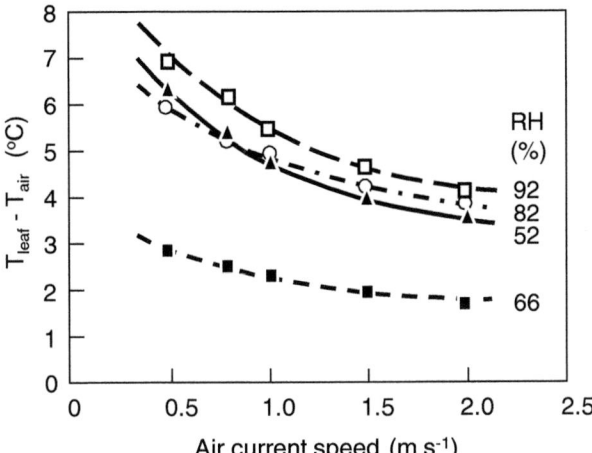

Fig. 13.9 Effects of air current speed and relative humidity (RH) on temperature differences between the leaf (T_{leaf}) of sweet potato and air (T_{air}). Mean values ($n = 3$) and logarithmic approximation curves are shown. Light source, metal halide lamp; irradiance, 360 W m^{-2}, air temperature (T_{air}), 25 °C

0.6 kPa higher water vapor pressures, and 25–35 µmol mol^{-1} lower CO_2 concentrations were observed at around the canopy height than at a height 60 mm above the canopy (Fig. 13.10). Air temperatures and water vapor pressures increased and CO_2 concentrations decreased inside the canopy with increasing PPFD. Lower air temperature and higher CO_2 concentration inside the canopy were observed at an air current speed of 0.3 m s^{-1} than at that of 0.1 m s^{-1} (Fig. 13.11).

Considerable differences in the levels of environmental variables were observed between the inside and outside of plant canopies. Control of environmental variables surrounding plants is important for scheduling crop production and obtaining high yields with a rapid turnover rate in plant production. Precise control of environmental variables inside plant canopies, with sufficient air movement, is necessary for enabling the growth of plants to be controllable.

13.5 Concluding Remarks

Control of air movement is important for enhancing gas exchange between plants and the ambient air and is consequently important for promoting plant growth. Environmental variables surrounding plants must be controlled precisely under adequate air currents for scheduling plant production associated with rapid growth and a high turnover rate. Effective air movement is essential to promote plant growth and development during vegetative and reproductive growth stages. More suitable air movement can also ensure that the environmental variables remain more homogeneous inside the plant canopy.

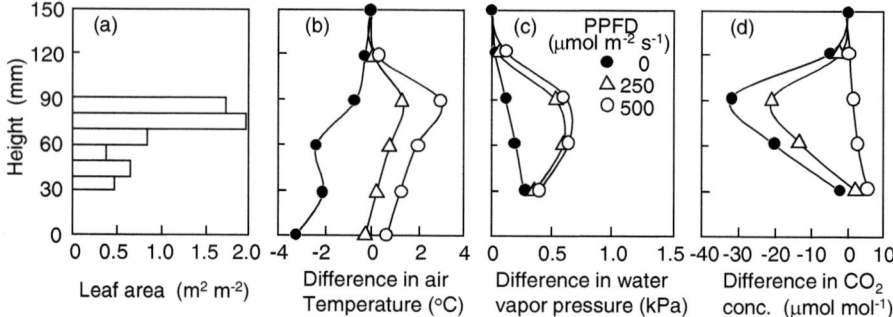

Fig. 13.10 Profiles of leaf area (**a**), air temperature (**b**), water vapor pressure (**c**), and CO_2 concentration (**d**) inside and outside the canopy of eggplant seedlings as affected by PPFD at an air current speed of 0.1 m s^{-1} (Kitaya et al. 1998)

Fig. 13.11 Effects of PPFD and air current speed on differences in the air temperature (**a**) and CO_2 concentration (**b**) between two heights at around the canopy height and 60 mm above the canopy (Kitaya et al. 1998)

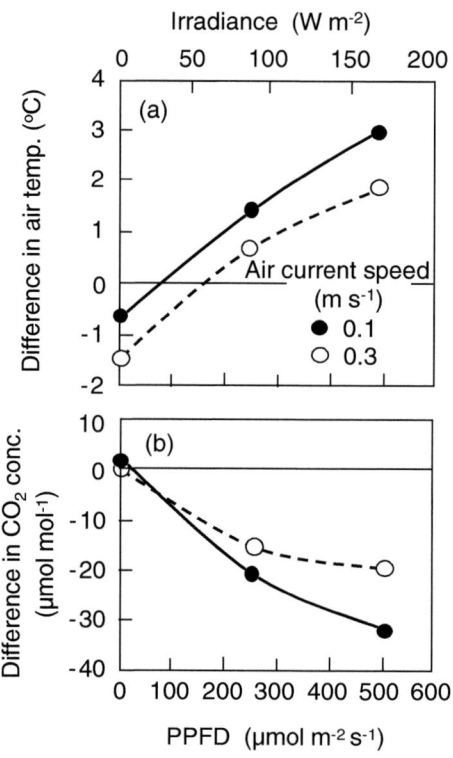

References

Jones HG (1992) Plants and microclimate. Cambridge University Press, Cambridge
Kim YH, Kozai T, Kubota C et al (1996) Effects of air velocities on the microclimate of plug stand under artificial lighting. Acta Hortic 440:354–359

Kitaya Y, Hirai H (2007) Thermal situation of plant reproductive organs affected by gravity and light intensity. J Jpn Soc Micrograv Appl 24(4):325–329

Kitaya Y, Shibuya T, Kozai T et al (1998) Effects of light intensity and air current speed on air temperature, water vapor pressure and CO_2 concentration inside a plants stand under an artificial lighting condition. Life Support Biosph Sci 5:199–203

Kitaya Y, Tsuruyama J, Kawai M et al (2000) Effects of air current on transpiration and net photosynthetic rates of plants in a closed plant production system. In: Kubota C, Chun C (eds) Transplant production in the 21st century. Kluwer Academic Publishers, Dordrecht, pp 83–90

Kitaya Y, Tsuruyama J, Shibuya T et al (2003) Effects of air current speed on gas exchange in plant leaves and plant canopies. Adv Space Res 31:177–182

Kitaya Y, Shibuya T, Yoshida M et al (2004) Effects of air velocity on photosynthesis of plant canopies under elevated CO_2 levels in a plant culture system. Adv Space Res 34:1466–1469

Martin TA, Hinckley TM, Meinzer FC et al (1999) Boundary layer conductance, leaf temperature and transpiration of *Abies amabilis* branches. Tree Physiol 19:435–443

Monteith JL, Unsworth HM (1990) Principles of environmental physics. Edward and Arnold Publishing Co, London

Shibuya T, Kozai T (1998) Effects of air velocity on net photosynthetic and evapotranspiration rates of a tomato plug sheet under artificial light. Environ Control Biol 36:131–136 (in Japanese with English summary)

Yabuki K (2004) Photosynthetic rate and dynamic environment. Kluwer Academic Publishers, Dordrecht

Yabuki K, Miyagawa H (1970) Studies on the effect of wind speed on photosynthesis (2) the relation between wind speed and photosynthesis. J Agric Methods 26:137–141 (in Japanese with English summary)

Yabuki K, Miyagawa H, Ishibashi A (1970) Studies on the effect of wind speed on photosynthesis (1) boundary layer near leaf surfaces. J Agric Methods 26:65–70 (in Japanese with English summary)

Part IV
Greenhouse Crop Production with Supplemental LED Lighting

Chapter 14
Control of Flowering Using Night-Interruption and Day-Extension LED Lighting

Qingwu Meng and Erik S. Runkle

Abstract Flowering of photoperiodic plants is regulated by the duration of the continuous night (dark) period during each 24-h period. When the natural photoperiod is short, longer days (shorter nights) may be desired by commercial growers of ornamentals and other specialty crops to promote flowering of long-day plants or inhibit flowering of short-day plants. To create short nights, electric lighting can extend the daylength (day extension, DE) or interrupt the night (night interruption, NI). Conventional lamps such as incandescent (INC), halide, and compact fluorescent (CFL) can serve this purpose, but they are energy inefficient, have a short life span, and/or emit photons at wavelengths that have little or no effect on regulating flowering. Recent advancements in solid-state lighting enable horticultural applications including regulation of flowering, especially in (semi-) controlled environments. Light-emitting diodes (LEDs) with customized spectra suitable for control of flowering are at least as effective as conventional lamps, last longer, and are more energy efficient. Narrowband radiation from LEDs facilitates research on the role of specific wavelengths in mediating flowering and plant morphology, which are important in commercial production of many specialty crops produced in controlled environments. In addition, applied lighting research helps elucidate how photoreceptors, such as phytochromes and cryptochromes, mediate these physiological processes in plants. LEDs will increasingly replace conventional lamps to regulate flowering of commercial photoperiodic crops as their energy efficiency increases and manufacturing costs decrease.

Keywords Cryptochrome • Far-red radiation • Long-day plants • Photoperiod • Phytochrome • Regulation of flowering • Short-day plants

---

Q. Meng • E.S. Runkle (✉)
Department of Horticulture, Michigan State University, 1066 Bogue Street, East Lansing, MI 48824-1325, USA
e-mail: runkleer@msu.edu

14.1 Introduction

Flowering of a wide range of ornamental crops, including annuals (bedding plants) and herbaceous perennials, is sensitive to the photoperiod (Thomas and Vince-Prue 1997). Long-day plants flower earlier when the dark period is shorter than a critical length, whereas short-day plants flower earlier when the dark period exceeds a critical duration. The critical photoperiod is species and sometimes cultivar specific. Unlike the conversion of light energy to chemical energy for photosynthesis, photoperiodic signaling in plants has a very low threshold light intensity (<2 μmol m^{-2} s^{-1}) (Whitman et al. 1998). When the natural photoperiod is short, the long night can be truncated using electric lights to promote flowering of long-day plants and inhibit flowering of short-day plants. This technique, known as the photoperiodic regulation of flowering, is an important strategy for commercial growers to produce crops efficiently and to schedule them in flower for specific, predetermined market dates. Besides manipulating flowering time, photoperiodic lighting can alter other characteristics such as morphology, vegetative growth, and pigmentation. For example, a delay in flowering of short-day plants is often accompanied by a desired increase in vegetative growth, which is known as "bulking".

Photoperiodic lighting is often delivered during one of two periods during the night: following sunset [day-extension (DE) or end-of-day (EOD) lighting] or during the middle of the night [night-interruption (NI) lighting]. A DE creating a 16-h photoperiod often ensures a long-day response for a wide range of ornamentals (Whitman et al. 1998). Because flowering of photoperiodic plants is determined by the night length, a brief (e.g., several seconds to minutes) pulse of NI light that divides a long night into two short dark periods can regulate flowering of some model crops that need only one or a few inductive cycles (Thomas and Vince-Prue 1997). However, most plants, particularly most ornamental crops, require a longer ($>$30 min) NI lighting duration to be effective. Generally, a 4-h NI is sufficiently long to saturate the promotion of flowering for long-day crops and inhibit flowering of short-day crops (Runkle et al. 1998). When comparing the efficacy of DE and NI lighting, a 4-h NI generated a slightly stronger long-day signal than a 5.5-h DE using the same light source (Meng and Runkle 2016a), although several studies with herbaceous perennials (e.g., *Rudbeckia fulgida*) have reported a similar response to NI and DE lighting (Runkle et al. 1999).

14.2 Conventional Lamps

The application of electric lights in photoperiodic control of flowering has evolved rapidly over the past decade. A wide array of light sources, including incandescent (INC), high-pressure sodium (HPS), and fluorescent lamps, have been extensively researched and used commercially. Although these lamps were designed for general illumination, in many instances they are effective at creating long days for photoperiodic plants. The selection of an appropriate light source for commercial applications depends on factors such as the spectral distribution, intensity, energy efficiency, rated lifetime, annual hours of operation, and costs for installation and operation.

With a spectral distribution similar to a blackbody radiator, INC lamps convert electric energy to photons mostly with long wavelengths. The primary spectral emission of INC lamps in the near-visible range is red (R, 600–700 nm) and far-red (FR, 700–800 nm) radiation, but this accounts for only about 8 % of the total energy emitted (Thimjan and Heins 1983). Despite their energy inefficiency and short life span, INC lamps gained popularity in greenhouses and growth chambers for photoperiodic control because of their low cost. However, they have been phased out of production in compliance with increased energy standards being enforced worldwide and, to some extent, have been replaced by slightly more energy-efficient halide lamps, which emit a very similar spectrum. Compact fluorescent (CFL) lamps are more energy efficient and last longer than INC lamps. However, flowering of some long-day plants, such as petunia (*Petunia* × *hybrida*), was delayed when INC lamps were replaced with CFL lamps (Runkle et al. 2012). CFL lamps emit little FR radiation, which is required to accelerate flowering in some long-day crops. As the light source most commonly used for greenhouse supplemental lighting, HPS lamps can also provide long days to inhibit flowering of short-day plants (Blanchard and Runkle 2009) and promote flowering of long-day plants (Blanchard and Runkle 2010; Whitman et al. 1998).

14.3 Light-Emitting Diodes

14.3.1 Critical Wavebands for Regulation of Flowering of Long-Day Plants

Phytochromes are a class of photoreceptors that primarily absorb R and FR radiation and mediate flowering and photomorphogenesis (Fig. 14.1). In plants, phytochromes exist as an active form absorbing FR radiation (P_{FR}) and an inactive form absorbing R radiation (P_R), the ratio of which depends on the incident spectrum. A phytochrome photoequilibrium (PPE) is established based on the proportion of P_{FR} in the total pool of phytochromes ($P_{FR}+P_R$). An estimated PPE can be calculated using the spectral data of a light source and relative absorption of P_R and P_{FR} (Sager et al. 1988). Using conventional lamps, a mixture of R and FR radiation was more promotive of flowering in long-day plants than either R or FR radiation alone (Thomas and Vince-Prue 1997). Likewise, R and FR LEDs, establishing an intermediate PPE of 0.63 or 0.72, usually elicited the most rapid flowering of the long-day plants tested (Craig and Runkle 2016). The efficacy of INC lamps is not

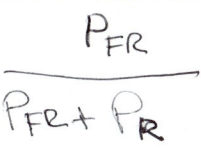

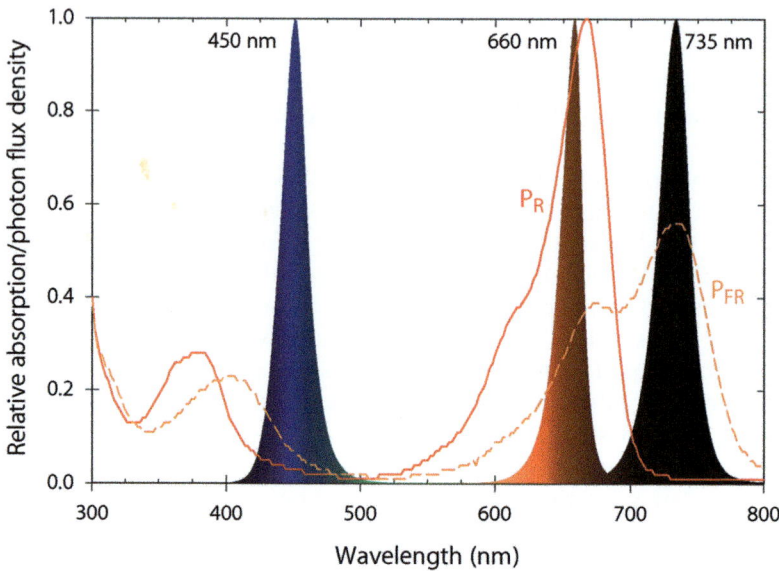

Fig. 14.1 The relative absorption of the two forms of phytochrome and the relative spectral distribution of three LED types used for photoperiodic lighting research at Michigan State University. The names for the two forms of phytochrome are based on their peak absorption of radiation: P_R = the red-absorbing form and P_{FR} = the far-red-absorbing form (Sager et al. 1988). Phytochrome can be manipulated by red (peak = 660 nm) or far-red (peak = 735 nm) LEDs at a low intensity, while a higher intensity of blue radiation (peak = 450 nm) is required, possibly because of its relatively low absorption by phytochrome

surprising because the estimated PPE established by INC lamps is 0.64, which is within this effective range.

The addition of FR to R radiation can accelerate flowering of long-day plants, but typically promotes undesired extension growth of ornamental plants such as calibrachoa (*Calibrachoa* × *hybrida*) and coreopsis (*Coreopsis grandiflora*). The shade-avoidance response triggered by FR radiation, even at a low intensity, modifies physiological and morphological characteristics of plants (Cerdán and Cory 2003). A low R-to-FR ratio (R/FR) can increase the biosynthesis of gibberellins, which are plant hormones mediating stem elongation (Kurepin et al. 2012). This led to a question of whether LEDs that do not emit FR radiation could provide an effective long-day signal while maintaining the compactness of plants. For some long-day plants such as ageratum (*Ageratum houstonianum*) and dianthus (*Dianthus chinensis*), R+white (W) LEDs controlled flowering as effectively as R +W+FR LEDs but produced shorter plants at flowering, showing that R radiation by itself was sufficient for photoperiodic control of flowering of some crops (Kohyama et al. 2014). However, for other long-day plants, the most rapid flowering occurred when both R and FR radiation were delivered. For example, NIs essentially devoid of FR radiation [i.e., R, blue (B, 400–500 nm)+R, and W LEDs] were not perceived as long days for snapdragon (*Antirrhinum majus*), while flowering was accelerated under R+W+FR LEDs (Meng and Runkle 2016b). Therefore, long-day plants can be classified into FR-dependent and FR-neutral varieties based on their flowering responses to FR radiation. Within the FR-dependent category, FR radiation is either required for promotion of flowering (an obligate response) or is not required but promotes flowering if added to R radiation (a facultative response). Examples of obligate FR-dependent long-day plants are snapdragon and pansy (*Viola* × *wittrockiana*). Examples of facultative FR-dependent plants are petunia and coreopsis. In contrast, flowering of FR-neutral plants, such as ageratum, rudbeckia (*Rudbeckia hirta*), and calibrachoa, is primarily regulated by R radiation, and adding FR radiation has no effect on flowering time.

Because LED arrays without FR radiation can control flowering of some long-day plants without promoting extension growth, the application of W LEDs for photoperiodic lighting was explored. LED arrays emitting W radiation are usually B LEDs covered with a phosphor coating, which scatters most photons to longer wavelengths, but can also be created by mixing R, green (G, 500–600 nm), and B LEDs that, when combined, appear W. W LEDs emit little or no FR radiation and thus, cannot necessarily replace INC lamps or R+FR LEDs for some FR-dependent long-day plants. Various types of W LEDs are available including cool-, warm-, and neutral-W LEDs, which depend on the phosphor coating and the resulting spectral distribution and correlated color temperature. Cool- and warm-W LEDs have the same PPE of 0.84 but different B-to-R ratios (0.67 and 0.27, respectively) (Table 14.1). Despite the spectral differences, the effectiveness of cool- and warm-W LEDs at regulating flowering was generally equivalent to that of R and B+R LEDs (Meng and Runkle 2016b).

B radiation is absorbed by the cryptochrome and phototropin families of photoreceptors, but can also be weakly absorbed by phytochromes. Both cryptochromes and phytochromes mediate flowering, whereas phototropins regulate phototropism in plants. The efficacy of B radiation at regulating flowering of photoperiodic crops is dependent on the intensity delivered. A threshold intensity greater than that usually sufficient for R+FR photoperiodic lighting (i.e., 2 µmol m^{-2} s^{-1}) is required to establish a B-mediated flowering response (Meng and Runkle 2016a). At 2–3 µmol m^{-2} s^{-1}, a 4-h NI from B LEDs was not perceived as a long-day signal by any long-day plants tested (Craig 2012; Meng and Runkle 2015). Furthermore, the addition of this low-intensity B radiation to R, FR, or R+FR radiation did not influence flowering (Meng and Runkle 2015). However, at a higher intensity of 30 µmol m^{-2} s^{-1}, B radiation delivered alone as a 4-h NI was perceived as a long day for all long-day plants tested (Meng and Runkle 2016a). Furthermore, this B radiation further promoted flowering of some plants (e.g., petunia) grown under an NI at 2 µmol m^{-2} s^{-1} from R+W+FR LEDs. Collectively, these experiments challenge the notion that the PPE is an accurate predictor of the efficacy of a light source at regulating flowering. First, the PPE only changed from 0.53 to 0.48 when the intensity of B radiation increased from 2 to 30 µmol m^{-2} s^{-1}, but the capacity to control flowering was activated. Second, low-intensity B radiation created an intermediate PPE of 0.53, which should have been at least somewhat

Table 14.1 Spectral characteristics of incandescent (INC), compact fluorescent (CFL), high-pressure sodium (HPS) lamps, and blue (B)+red (R)+far-red (FR), R+white (W), R+W+FR, cool-W (CW), and warm-W (WW) light-emitting diodes (LEDs)

Parameter	INC	CFL	HPS	LEDs				
				B+R+FR[a]	R+W[b]	R+W+FR[c]	CW[d]	WW[e]
Percentage (%) of photon flux (400–800 nm)								
Blue (400–500 nm)	3	14	5	11	6	6	20	12
Green (500–600 nm)	14	37	51	2	14	13	46	39
Red (600–700 nm)	30	42	38	60	78	36	30	43
Far red (700–800 nm)	54	7	6	27	1	44	4	6
Light ratio								
Red:far red	0.56	6.19	5.90	2.24	55.08	0.82	7.47	7.18
Blue:red	0.09	0.32	0.12	0.19	0.08	0.18	0.67	0.27
PPE	0.64	0.83	0.86	0.76	0.88	0.67	0.84	0.84

Phytochrome photoequilibria (PPE) are estimated according to Sager et al. (1988)
[a]TotalGrow Day & Night Management Light
[b]Philips GreenPower LED flowering DR/W
[c]Philips GreenPower LED flowering DR/W/FR
[d]Philips, model 9290002296
[e]Philips, model 9290002204

effective at stimulating flowering of at least some species (Craig and Runkle 2013, 2016). Third, there were no consistent correlations between the estimated PPE of a light source and a flowering index for long-day plants such as dianthus, petunia, and rudbeckia (Meng and Runkle 2015). Therefore, factors such as the radiation intensity, duration, and spectral distribution should be considered with the PPE to predict the photoperiodic efficacy of a light source.

14.3.2 Critical Wavebands for Regulation of Flowering of Short-Day Plants

The spectral requirements to inhibit flowering of short-day plants are slightly different from those to promote flowering of long-day plants. During a long night, low-intensity R radiation delivered as a DE or NI generally inhibits flowering in short-day plants (Thomas and Vince-Prue 1997). For example, R LEDs alone delivered as a 4-h NI inhibited flowering of chrysanthemum (*Chrysanthemum morifolium*) and marigold (*Tagetes erecta*) compared with the 9-h short-day control (Craig and Runkle 2013; Meng and Runkle 2016b). Using R+FR LEDs, a high R/FR (or PPE) was often more effective than a low R/FR (or PPE) at delaying flowering of several short-day plants studied (Craig and Runkle 2013). In addition, FR radiation alone was not perceived as a long day. For some plants that only require one or a few photoinductive cycles, flowering can be at least somewhat influenced by R/FR photoreversibility; an inhibition of flowering by R radiation can be fully or partially reversed by subsequent exposure to FR radiation (Thomas and Vince-Prue 1997). The delivery of R radiation establishes a high PPE that inhibits the signaling pathway for flowering of short-day plants, but subsequent FR radiation attenuates this inhibition by converting some P_{FR} back to P_R.

Similar to long-day plants, the efficacy of B radiation at regulating flowering of short-day plants depends on its intensity. To inhibit flowering of the short-day plant duckweed (*Lemna paucicostata*) by 50%, B, G, R, and FR radiation needed to be 10, 0.5, 0.1, and 3 µmol m^{-2} s^{-1}, respectively (Saji et al. 1982). Similarly, only at a sufficiently high intensity (e.g., 30 µmol m^{-2} s^{-1}) did short-day plants perceive B radiation as a long-day signal (Meng and Runkle 2015, 2016a). Compared with R +W+FR radiation at 2 µmol m^{-2} s^{-1}, B radiation at 30 µmol m^{-2} s^{-1} was similarly effective for marigold but less effective for chrysanthemum.

Relatively few studies have explored the efficacy of G radiation at regulating photoperiodic flowering. As noted previously, G radiation alone was an effective long-day signal at a low intensity for the model plant duckweed (Saji et al. 1982), although the capacity of G radiation to create long days was questionable in other studies (Thomas and Vince-Prue 1997). More research on the efficacy of G radiation has been performed recently with the advancements of LEDs. Under short days, NIs or DEs with low- or high-intensity G LEDs (peak wavelength = 518, 520, or 530 nm) inhibited flowering of the short-day plants cosmos (*Cosmos bipinnatus*), perilla (*Perilla ocymoides*), okra (*Abelmoschus esculentus*), and chrysanthemum (Hamamoto et al. 2003; Hamamoto and Yamazaki 2009; Jeong et al. 2012). G radiation was as effective as R radiation at inhibiting flowering in some of these studies, but was less effective in others. This indicates that the degree of a long-day response activated by G radiation could depend on its intensity, duration, and spectral characteristics and could vary among species. G radiation emitted from W LEDs could also play a role in photoperiodic regulation of flowering. A 4-h NI from W LEDs emitting comparable amounts of G and R radiation inhibited flowering of chrysanthemum more than that from R LEDs alone (Meng and Runkle 2016b). Because G radiation can exert an inhibitory effect similar to R radiation at a low intensity in some species, a combination of these two wavebands could be more effective at inhibiting flowering of short-day plants than either waveband alone.

14.3.3 Comparisons Between Conventional Lamps and Light-Emitting Diodes

Traditional broad-spectrum light sources can effectively create long days for most photoperiodic crops; however, much of the radiation emitted – and therefore energy consumed – is not necessary for photoperiodic lighting. LED arrays developed for

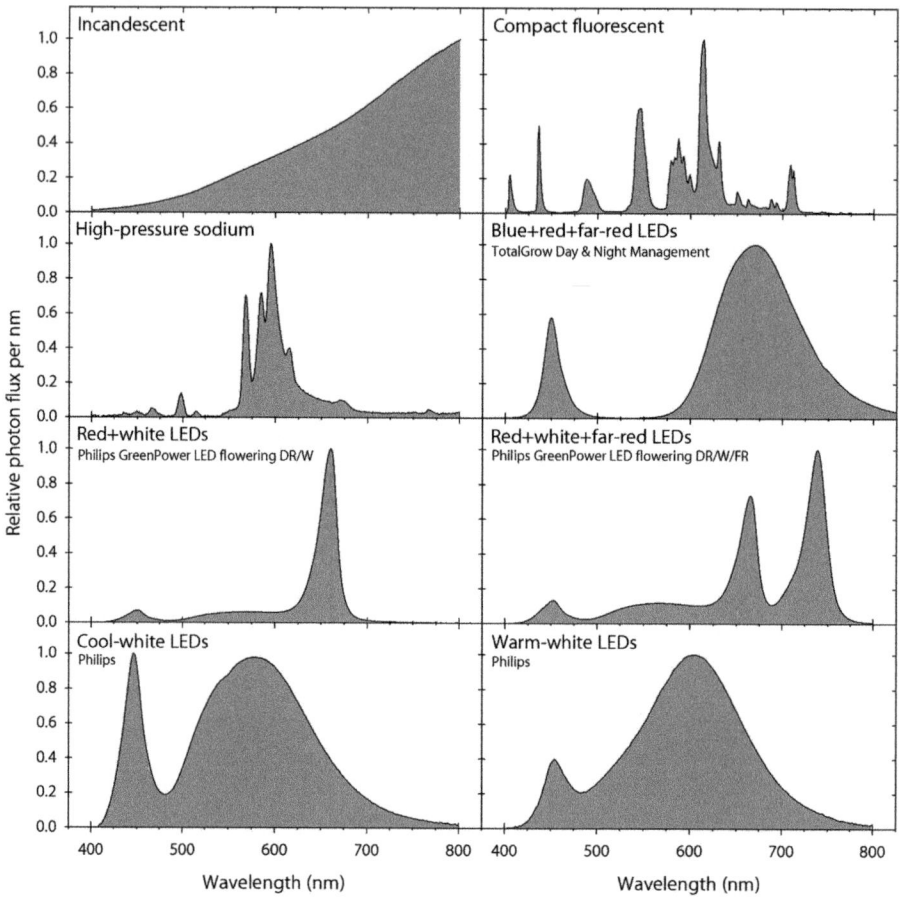

Fig. 14.2 The spectra of several light sources used for regulation of flowering, including lamps traditionally used by commercial growers (incandescent, compact fluorescent, and high-pressure sodium) and newly developed light-emitting diodes (LEDs). A portable spectroradiometer measured photon fluxes every 1 nm from 400 to 800 nm

plant lighting applications can be at least as energy efficient as, and last longer than, conventional lamps (Nelson and Bugbee 2014; Pimputkar et al. 2009; Schubert and Kim 2005). The use of LEDs also enables specification of spectral composition. For example, LEDs can be customized and tailored to emit a spectrum that controls flowering effectively and efficiently. The spectral distributions and characteristics of several conventional lamps and commercial LEDs used for photoperiodic control of plants are in Fig. 14.2 and Table 14.1.

In a coordinated commercial greenhouse grower trial, LED lamps emitting primarily R and FR radiation, plus a little W, were compared with lamps traditionally used by greenhouse growers to create long days, including INC, HPS, and CFL lamps (Meng and Runkle 2014). Flowering of most herbaceous ornamental crops tested was similar under NIs from the LED, INC, and HPS lamps, showing that the LEDs were at least as effective as traditional lamps at regulating flowering. Although the spectral distribution of these commercial LEDs differed from that of INC lamps, the intensity of each 100-nm waveband, from 400 to 800 nm, was similar between the two lamp types. The R/FR emitted from the LEDs and INC lamps was similar (0.8 and 0.6, respectively), so their comparable efficacy was not surprising. Experimental LED arrays delivering a similar R/FR were also as effective as INC lamps at creating long days for long-day and short-day plants (Craig and Runkle 2013, 2016).

14.4 Concluding Summary

In long-day plants, R radiation at a low intensity can regulate flowering of some ornamental crops, while the inclusion of FR radiation can promote flowering of other crops. In contrast, R radiation alone is effective at inhibiting flowering of a wide range of short-day plants. The threshold intensity of B radiation, above which it can regulate flowering of photoperiodic plants, is much greater than that of R radiation. LED products designed specifically to regulate flowering can be at least as effective as conventional INC, HPS, and CFL lamps. Table 14.2 summarizes the efficacy of conventional lamps and LEDs commonly used for photoperiodic control of long-day and short-day ornamental crops. A simplified economic analysis revealed that, in the long term, the total operating cost of LEDs could be less than that of INC or HPS lamps to deliver NIs because of the greater energy efficiency and longer life span of LEDs (Meng and Runkle 2014). The cost of LED products is expected to continue decreasing as the technology matures. As a result, flowering applications using LEDs should become more prevalent as we gain a better understanding of photocontrol of flowering.

Table 14.2 Summary of the efficacy of different lamp types at regulating flowering of photoperiodic crops when delivered during the night at a low intensity (1–3 µmol m^{-2} s^{-1})

Lamp type		Short-day plants	Long-day plants
Incandescent, halogen		✓	✓
Fluorescent (including CFLs)[a]		✓	Some
Mix incandescent + CFL[a]		✓	✓
High-intensity discharge (HPS, MH, mercury)[b]		✓	✓
LEDs	White	✓	Some
	Red	✓	Some
	Red + far red	✓	✓
	Far red	–	–
	Blue	–	–
	Green	Varies	–

✓ generally effective, – generally not effective
[a]*CFL* Compact fluorescent lamps
[b]*HPS* High-pressure sodium, *MH* Metal halide

References

Blanchard MG, Runkle ES (2009) Use of a cyclic high-pressure sodium lamp to inhibit flowering of chrysanthemum and velvet sage. Sci Hortic 122:448–454

Blanchard MG, Runkle ES (2010) Intermittent light from a rotating high-pressure sodium lamp promotes flowering of long-day plants. HortScience 45:236–241

Cerdán PD, Chory J (2003) Regulation of flowering time by light quality. Nature 423:881–885

Craig DS (2012) Determining effective ratios of red and far-red light from light-emitting diodes that control flowering of photoperiodic ornamental crops. MS Thesis, Michigan State University

Craig DS, Runkle ES (2013) A moderate to high red to far-red light ratio from light-emitting diodes controls flowering of short-day plants. J Am Soc Hort Sci 138:167–172

Craig DS, Runkle ES (2016) An intermediate phytochrome photoequilibria from night-interruption lighting optimally promotes flowering of several long-day plants. Environ Exp Bot 121:132–138

Hamamoto H, Yamazaki K (2009) Reproductive response of okra and native rosella to long-day treatment with red, blue, and green light-emitting diode lights. HortScience 44:1494–1497

Hamamoto H, Shimaji H, Higashinde T (2003) Budding and bolting responses of horticultural plants to night-break treatments with LEDs of various colors. J Agric Meteorol 59:103–110

Jeong SW, Park S, Jin JS, Seo ON, Kim GS, Kim YH, Bae H, Lee G, Kim ST, Lee WS, Shin SC (2012) Influences of four different light-emitting diode lights on flowering and polyphenol variations in the leaves of chrysanthemum (*Chrysanthemum morifolium*). J Agric Food Chem 60:9793–9800

Kohyama F, Whitman C, Runkle ES (2014) Comparing flowering responses of long-day plants under incandescent and two commercial light-emitting diode lamps. HortTechnology 24:490–495

Kurepin LV, Joo SH, Kim SK, Pharis RP, Back TG (2012) Interaction of brassinosteroids with light quality and plant hormones in regulating shoot growth of young sunflower and *Arabidopsis* seedlings. J Plant Growth Regul 31:156–164

Meng Q, Runkle ES (2014) Controlling flowering of photoperiodic ornamental crops with light-emitting diode lamps: a coordinated grower trial. HortTechnology 24:702–711

Meng Q, Runkle ES (2015) Low-intensity blue light in night-interruption lighting does not influence flowering of herbaceous ornamentals. Sci Hortic 186:230–238

Meng Q, Runkle ES (2016a) Moderate-intensity blue radiation can regulate flowering, but not extension growth, of several photoperiodic ornamental crops. Environ Exp Bot (in press)

Meng Q, Runkle ES (2016b) Investigating the efficacy of white light-emitting diodes at regulating flowering of photoperiodic ornamental crops. Acta Hortic (under review)

Nelson JA, Bugbee B (2014) Economic analysis of greenhouse lighting: light emitting diodes vs. high intensity discharge fixtures. PLoS One 9(6):e99010

Pimputkar S, Speck JS, DenBaars SP, Nakamura S (2009) Prospects for LED lighting. Nat Photonics 3:180–182

Runkle ES, Heins RD, Cameron AC, Carlson WH (1998) Flowering of herbaceous perennials under various night interruption and cyclic lighting treatments. HortScience 33:672–677

Runkle ES, Heins RD, Cameron AC, Carlson WH (1999) Photoperiod and cold treatment regulate flowering of *Rudbeckia fulgida* 'Goldsturm'. HortScience 34:55–58

Runkle ES, Padhye SR, Oh W, Getter K (2012) Replacing incandescent lamps with compact fluorescent lamps may delay flowering. Sci Hortic 143:56–61

Sager JC, Smith WO, Edwards JL, Cyr KL (1988) Photosynthetic efficiency and phytochrome photoequilibria determination using spectral data. Trans Am Soc Agric Eng 31:1882–1889

Saji H, Masaki F, Takimoto A (1982) Spectral dependence of night-break effect on photoperiodic floral induction in *Lemna paucicostata* 441. Plant Cell Physiol 23:623–629

Schubert EF, Kim JK (2005) Solid-state light sources getting smart. Science 308:1274–1278

Thimijan RW, Heins RD (1983) Photometric, radiometric, and quantum light units of measure: a review of procedures for interconversion. HortScience 18:818–822

Thomas B, Vince-Prue D (1997) Photoperiodism in plants, 2nd edn. Academic Press, San Diego

Whitman CM, Heins RD, Cameron AC, Carlson WH (1998) Lamp type and irradiance level for daylength extensions influence flowering of *Campanula carpatica* 'Blue Clips', *Coreopsis grandiflora* 'Early Sunrise', and *Coreopsis verticillata* 'Moonbeam'. J Am Soc Hort Sci 123:802–807

Chapter 15
Control of Morphology by Manipulating Light Quality and Daily Light Integral Using LEDs

Joshua K. Craver and Roberto G. Lopez

Abstract In northern latitudes, supplemental lighting is utilized to increase the photosynthetic daily light integral in greenhouses during the winter months, which can fall as low as 1–5 mol m^{-2} d^{-1}. Traditionally, supplemental lighting has been provided by high-intensity discharge (HID) lamps, but light-emitting diode (LED) technologies are now available for many greenhouse applications. The use of LEDs for supplemental lighting can be beneficial because wavelengths of light can be selected for applications such as the control of plant growth, development, morphology, and leaf color. However, delivering these precise wavelengths at moderately low intensities with ambient light already present in the greenhouse may prove ineffective at eliciting desired morphological characteristics. Regardless, LEDs have proven to be a viable option to provide supplemental lighting in the many controlled environments.

Keywords Extension growth • Floriculture crops • Light-emitting diodes • Photomorphogenesis • Phytochrome • Seedling growth • Sole-source lighting • Supplemental lighting

15.1 Introduction

Light-emitting diodes (LEDs) provide a novel approach to greenhouse lighting that has yet to be fully researched (Mitchell et al. 2012; Morrow 2008). To date, little research has been published involving the use of LEDs in greenhouses where plants are also subjected to solar radiation. The use of LEDs in a greenhouse production scenario could involve supplemental, photoperiodic, or photomorphogenic lighting

---

J.K. Craver
Department of Horticulture and Landscape Agriculture, Purdue University, 625 Agriculture Mall Drive, West Lafayette, IN 47907-2010, USA

R.G. Lopez (✉)
Department of Horticulture, Michigan State University, 1066 Bogue Street, East Lansing, MI 48824-1325, USA
e-mail: rglopez@msu.edu

15.2 Effects of DLI on Plant Morphology

Daily light integral (DLI) describes the cumulative number of photosynthetically active photons (400–700 nm) received during a 24-h period. Numerous studies have reported that DLI influences plant growth and morphology measured in terms of biomass accumulation, leaf area, branch and flower number, and height. For example, Currey and Lopez (2015) reported that leaf, stem, and root biomass accumulation increased linearly with DLI by 122 %, 118 %, and 211 % for geranium (*Pelargonium* ×*hortorum*), New Guinea impatiens (*Impatiens hawkeri*), and petunia (*Petunia* ×*hybrida*) cuttings, respectively, as DLI during propagation increased from ≈ 2 to 13 mol m^{-2} d^{-1}. Additionally, as DLI during root development increased, the leaf area ratio and specific leaf area of New Guinea impatiens cuttings decreased by 41 and 34 %, respectively. In a separate study, petunia plants were 6 cm shorter when DLI increased from 6.5 to 13.0 mol m^{-2} d^{-1} (Kaczperski et al. 1991). Faust et al. (2005) observed that the number of lateral shoots in ageratum (*Ageratum houstonianum*) and petunia increased by 7.1 and 7.0 as the DLI increased from 5 to 43 mol m^{-2} d^{-1}, respectively.

15.3 Effects of Light Quality on Plant Morphology

Light quality is detected by plants using photoreceptors such as phytochromes, cryptochromes, and ultraviolet light receptors (Runkle and Heins 2001). The detection of these wavelengths by the plant can elicit a wide variety of developmental and morphological responses. One of the benefits of utilizing LEDs in a greenhouse setting is the ability to control the spectrum of wavelengths emitted from the arrays to potentially elicit these desired morphological and physiological responses in the plant (Morrow 2008). Specifically, LEDs can be designed to emit wavelengths of light that match the peak absorption of these critical photoreceptors and plant pigments. This not only provides the benefit to manipulate a desired plant response, but it also potentially saves energy by not providing wavelengths of light less or not necessary for production (Mitchell et al. 2012). Additional potential benefits from targeting specific wavelengths of light using LEDs include reduced pest and disease occurrence and increased concentrations of vitamins, minerals, pigments, or phenolic compounds in the plant tissue (Massa et al. 2008). In the next few sections, we will discuss many of these responses to light quality in detail with an emphasis on regulating plant morphology.

15.3.1 Red Light

Red photons of light have a wavelength between 600 and 700 nm. One of the most common roles for red light is to participate in the physiological process of photosynthesis. Specifically, many LED arrays emit red wavelengths at 660 nm, which is very close to the absorption peak of chlorophyll (Massa et al. 2008). Thus, red LEDs can be used to efficiently drive photosynthetic activity, resulting in increased biomass and overall plant productivity. However, red light alone is not sufficient for the optimum production and quality of most crops. When exposed to solely red light, many dicotyledonous crops develop extensive hypocotyl elongation (Hoenecke et al. 1992). Additionally, *Arabidopsis* plants grown under only red light develop abnormal morphological characteristics (Goins et al. 1998). However, both red and blue light (400–500 nm) control stem elongation (Kigel and Cosgrove 1991). Specifically, blue light, when combined at a low irradiance with red light, can prevent excessive elongation of hypocotyls, stems, and petioles and deter other morphological abnormalities observed under solely red wavelengths (Goins et al. 1998; Hoenecke et al. 1992).

Red light is involved in much more than simply photosynthetic activity. Phytochrome is one of the primary families of photoreceptors that absorb red light as well as far-red (700–800 nm) radiation. When exposed to light, phytochromes exist in two interconvertible forms, the red-absorbing (P_r) and far-red-absorbing (P_{fr}) forms (Smith and Whitelam 1990). The relative proportion of P_{fr} to the total amount of phytochrome (phytochrome photoequilibrium) regulates a variety of photomorphogenic responses including stem extension (Runkle and Heins 2001; Stutte 2009) and flowering (see Chap. 14). These photomorphogenic responses are known to vary with plant species and cultivar, age, light quantity and quality, and temperature. Plants are generally more compact when exposed to light with a high red to far-red ratio (R:FR). They are also more sensitive to red and far-red light at the end of the day (EOD), and 10–60 min of EOD red light may be as effective as a high R:FR during the entire photoperiod to inhibit extension growth (Ilias and Rajapakse 2005).

15.3.2 Blue Light

Blue photons of light have a wavelength between 400 and 500 nm and mediate stem extension, thus inhibiting extension growth for a variety of crops (Cosgrove 1981; Kigel and Cosgrove 1991; Runkle and Heins 2001). However, subjecting plants to solely blue wavelengths can result in elongation responses similar to that under monochromatic red light (van Ieperen et al. 2012). Thus, combinations of both red and blue wavelengths are usually necessary to produce compact plants without excessive elongation. For example, supplemental and sole-source lighting combinations of red and blue light from LEDs produce more compact bedding plant seedlings (Randall and Lopez 2014; Wollaeger and Runkle 2014). Results from studies such as these will be discussed later in this chapter.

Blue light with or without red light can also affect stomatal density and aperture (Kinoshita et al. 2001; van Ieperen et al. 2012; Zeiger et al. 2002). Specifically, when blue light is added in small quantities to red light, stomatal opening increases significantly compared to solely red light (Kinoshita et al. 2001; van Ieperen et al. 2012). This regulation of stomatal opening by blue light is believed to be mediated through the blue light receptors phot1 and phot2. This increase in stomatal opening ultimately leads to increased carbon dioxide (CO_2) uptake, which further stimulates the process of photosynthesis (Kinoshita et al. 2001).

15.3.3 Far-Red Light

Far-red photons have a wavelength from 700 to 800 nm and thus, by definition, are outside the photosynthetically active radiation (PAR, 400–700 nm) wave band. Thus, the utility of far-red light has predominately been connected to flowering and morphological responses for greenhouse production. One such manipulation involves the phytochrome photoequilibrium, which can be manipulated using LEDs to either initiate earlier flowering in some long-day plants or promote continued growth in the vegetative state (Downs and Thomas 1982; Stutte 2009). A deficiency in far red can delay flower initiation or development in select plants with a long-day photoperiodic response (Runkle and Heins 2001). Thus, a lack of ample far-red light in a production environment may lead to a delay in flowering for some species.

15.4 Supplemental Lighting

The DLI measured inside a greenhouse is often 40–50 % lower than outside due to reflection and absorption of photons by the greenhouse infrastructure and glazing (Hanan 1998). Shading materials and energy curtains, which are used to manage temperature during periods of high solar radiation and low outdoor temperature, respectively, can further reduce the DLI by 40–80 % (Faust et al. 2005). Therefore, supplemental lighting is primarily utilized in commercial greenhouses located in temperate latitudes during the darker months of the year, when solar radiation limits production. For example, more than 2150 ha of glasshouses in the Netherlands use supplemental lighting for ornamental and vegetable propagation and production (Heuvelink et al. 2006). Additionally, supplemental lighting enables producers to meet consumer demand for local and year-round production of specialty crops including fruits, vegetables, and ornamentals.

High-intensity discharge lamps, especially high-pressure sodium (HPS) lamps, have been traditionally used for supplemental lighting in commercial greenhouses to increase photosynthesis. High-pressure sodium lamps are the most widely used lamp type because of their relatively high efficacy (conversion of electricity into photosynthetic light) and lifespan of 10,000–12,000 h. However, approximately 70 % of energy consumed by the fixtures is not converted into PAR and, instead, is emitted as radiant heat energy. The surface temperature of HPS lamps can reach as high as 450 °C, which requires the separation of lamps from plants (Fisher and Both 2004; Nelson 2012; Spaargaren 2001). Additionally, HPS lamps primarily emit light in the range of 565–700 nm, which is predominately yellow (565–590 nm) and orange (590–625 nm) light. They only emit 5 % blue light, which is low compared to solar radiation that contains 18 % blue light (Islam et al. 2012).

15.4.1 LED Supplemental Lighting for Ornamental Seedling and Cutting Propagation

The production of ornamental young plants, such as seedlings (plugs) and rooted cuttings (liners) (also called transplants), often occurs during winter and early spring (Styer 2003). However, the DLI during this time is at seasonally low levels of 5 mol m^{-2} d^{-1} or less inside many greenhouses located in northern latitudes (Lopez and Runkle 2008; Pramuk and Runkle 2005). This low-average DLI during seedling and cutting propagation can delay rooting and subsequent performance; previous research has shown that a target DLI of 10–12 mol m^{-2} d^{-1} is recommended to produce high-quality young plants (Pramuk and Runkle 2005; Randall and Lopez 2014). Generally, increases in DLI increase the quality of both plugs and liners. A quality young plant is one that has a compact growth habit without excessively large leaves, a high root and shoot mass, a well-developed root system, and a thick stem (Oh et al. 2010; Pramuk and Runkle 2005; Randall and Lopez 2015). These qualitative parameters ultimately lead to seedlings and rooted cuttings that are more easily processed, shipped, and mechanically transplanted, which are desired by growers (Pramuk and Runkle 2005). These higher-quality young plants are often produced at elevated DLIs, leading to increases in dry mass per unit fresh mass (Faust et al. 2005). This results in thicker tissues that contain more carbohydrates and structural materials for use in growth and development. In contrast, young plants grown under a low DLI possess more water in their tissues, resulting in softer tissues that growers refer to as being less "toned" (Faust et al. 2005).

Providing supplemental lighting is a means by which young plants can be grown under a favorable DLI for uniform and consistent production, quality, and subsequent performance (Hernández and Kubota 2012). Currently, HPS lamps are the industry standard for providing supplemental lighting and commonly deliver a photosynthetic photon flux (PPF) of 50–80 μmol m^{-2} s^{-1} for most crops (Fisher and Both 2004). Although high-intensity LED arrays for greenhouse supplemental lighting are still a relatively new technology, they have the potential to offer greater efficiencies, longer lifetimes, and wavelength specificity for young plant production.

Young plants grown under LED supplemental lighting are typically of equal or higher quality to those grown under HPS lamps. Specifically, Randall and Lopez (2014) placed seedlings of celosia (*Celosia argentea*), geranium, impatiens (*Impatiens walleriana*), marigold (*Tagetes patula*), pansy (*Viola ×wittrockiana*), petunia, salvia (*Salvia splendens*), snapdragon (*Antirrhinum majus*), and vinca (*Catharanthus roseus*) under a 16-h photoperiod of solar radiation plus supplemental lighting providing a PPF of 100 µmol m^{-2} s^{-1} from either HPS lamps or LED arrays with varying proportions (%) of red/blue ratios (100:0, 85:15, or 70:30). Seedlings of some species grown under red + blue LEDs were generally more compact, had a greater stem diameter and relative chlorophyll content, and possessed a higher quality index [a quantitative measurement of quality (Currey et al. 2013)] than those under HPS lamps or only red LEDs. For example, stem extension of celosia, impatiens, marigold, pansy, petunia, salvia, and vinca was suppressed by 29 %, 31 %, 20 %, 35 %, 55 %, 9 %, and 31 %, respectively, for seedlings grown for 28 days under 85:15 red/blue LEDs compared with those grown under HPS lamps (Fig. 15.1). Additionally, stem diameter of geranium, marigold, and snapdragon was 8 %, 13 %, and 18 % greater, respectively, for seedlings grown under the 85:15 red/blue LEDs compared with seedlings grown under HPS lamps. However, root and shoot dry mass were similar between seedlings of the same three species and vinca grown under LED and HPS supplemental lighting. Overall, the quality index was similar for geranium, impatiens, marigold, snapdragon, and vinca grown under LEDs and HPS lamps. In contrast, Randall and Lopez (2014) reported that the quality index was higher for pansy, petunia, and salvia under 100:0, 85:15, and 70:30 red/blue LEDs than under HPS lamps.

In a separate study, Randall and Lopez (2015) compared seedlings grown under a low ambient greenhouse DLI of ≈ 6 mol m^{-2} d^{-1} with those grown under supplemental lighting or with sole-source lighting in a growth chamber with a DLI of ≈ 11 mol m^{-2} d^{-1}. The researchers placed marigold, geranium, impatiens, petunia, and vinca under ambient solar light for 16 h with supplemental light (70 µmol m^{-2} s^{-1}) from HPS lamps or LEDs providing 87:13 red/blue light. Seedlings of the same species were also grown under LEDs providing 185 µmol m^{-2} s^{-1} from either 87:13 or 70:30 red/blue light for 16 h. Root and shoot dry mass, leaf number and area, stem diameter, relative chlorophyll content, and the quality index of most species were generally greater under supplemental and sole-source lighting. Similar to the results of Randall and Lopez (2014), stem extension of all species under the red + blue LED supplemental lighting was less than seedlings grown under HPS lamps. Additionally, stem length of geranium, marigold, and petunia under red + blue LED sole-source lighting was suppressed by 21–26 %, 16–18 %, and 75–79 %, respectively, than seedlings grown under HPS supplemental lighting. Interestingly, leaf area of geranium was 49 %, 20 %, and 24 % greater for seedlings grown under HPS lamps compared with those under ambient solar radiation and sole-source LEDs providing 87:13 and 70:30 red/blue light, respectively (Fig. 15.2).

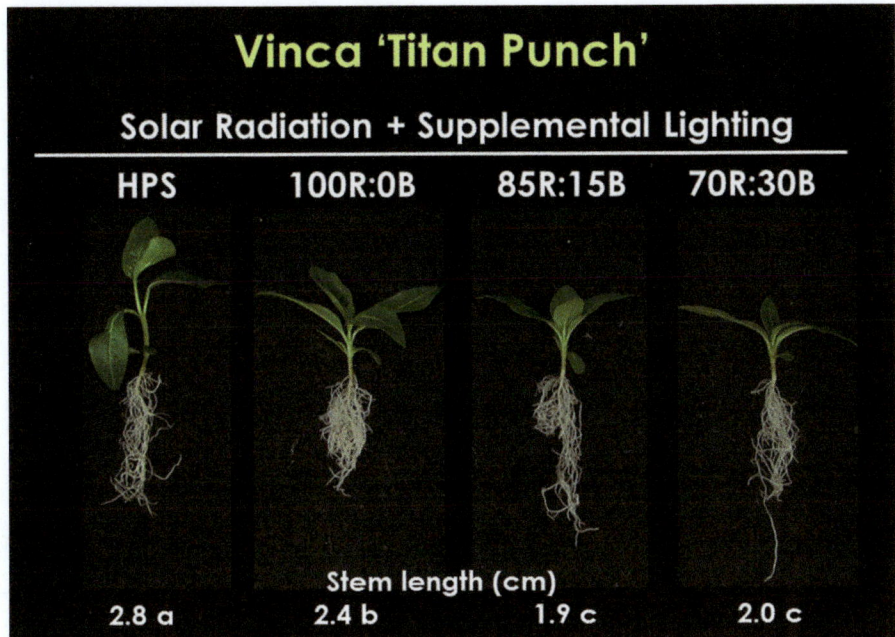

Fig. 15.1 Vinca seedlings grown in a greenhouse at 23 °C for 28 days under ambient solar radiation supplemented with 100 μmol m^{-2} s^{-1} from high-pressure sodium (HPS) lamps or light-emitting diodes (LEDs) with varying red/blue (R/B) light ratios (100:0, 85:15, or 70:30) (Randall and Lopez 2014). Means sharing a letter are not statistically different by Tukey's honestly significant difference at $P \leq 0.05$

Compared to seedlings, Currey and Lopez (2013) observed little effect of supplemental light source during cutting propagation on growth, morphology, and subsequent flowering after transplant. In this study, cuttings of New Guinea impatiens, zonal geranium, and petunia were grown for 7 days under a low DLI of ≈5 mol m^{-2} d^{-1} for callusing. Cuttings were placed under natural days and supplemental lighting as needed to deliver a 16-h photoperiod at 70 μmol m^{-2} s^{-1} from either HPS lamps or LED arrays with red/blue ratios of 100:0, 85:15, or 70:30. The only significant effects observed were that stem length of petunia was shortest for cuttings propagated under 100:0 red/blue LEDs, and leaf and root dry mass of petunia propagated under 70:30 red/blue LEDs were greater than cuttings propagated under HPS lights (Fig. 15.3). However, these differences were not commercially significant for growers. Additionally, supplemental light source during propagation had little or no effect on subsequent flowering of finished plants, including time to flower, stem length, node number below the first flower or inflorescence, flower bud number, and shoot dry mass.

For supplemental lighting in the greenhouse, it is plausible that the spectral quality has little potential to elicit morphological effects given the abundant background radiation provided by natural sunlight. This observation has been

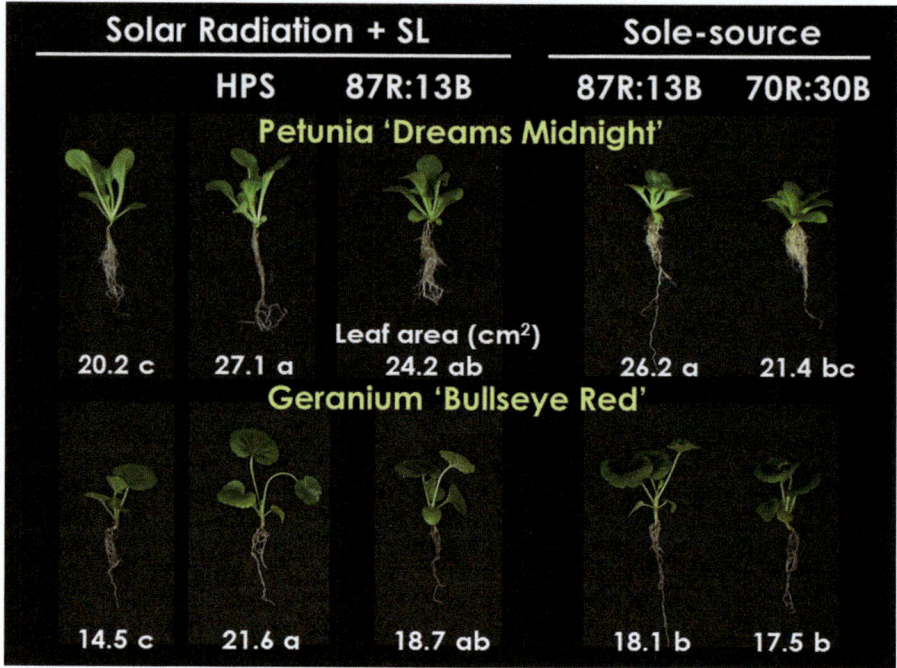

Fig. 15.2 Petunia and geranium seedlings grown at 23 °C for 21 days under ambient greenhouse solar radiation without or with 70 μmol m^{-2} s^{-1} from high-pressure sodium (HPS) lamps or light-emitting diodes (LEDs) for 16 h with an 87:13 red/blue (R:B) light ratio or under 185 μmol m^{-2} s^{-1} of sole-source lighting from LEDs for 16 h with a R/B light ratio of 87:13 or 70:30 (Randall and Lopez 2015). Means sharing a letter are not statistically different by Tukey's honestly significant difference at $P \leq 0.05$

suggested by Hernández and Kubota (2012) when they found that supplemental blue light administered to tomato (*Solanum lycopersicum* 'Komeett') seedlings at red/blue ratios of 96:4 and 84:16 at an intensity of 55 μmol m^{-2} s^{-1} elicited no change to seedling growth, morphology, or photosynthesis. They suggested that supplemental red light is sufficient for the production of tomato seedlings due to the solar background radiation providing sufficient quantities of blue light. Thus, light quality provided to vegetable seedlings may not be as important as reaching the minimum DLI for optimum production.

In addition to supplemental lighting, sole-source lighting with LEDs has become an alternative for seedling production. The commercial production viability of using LEDs in sole-source lighting is related to their low output of radiant heat, allowing the arrays to be placed close to the crop canopy to maximize production space efficiency. As discussed previously, Randall and Lopez (2015) evaluated various popular bedding plant seedlings under sole-source lighting using LEDs providing a red/blue light ratio of either 87:13 or 70:30 at 185 μmol m^{-2} s^{-1}. Generally, seedlings produced under sole-source lighting were more compact (reduced height and leaf area), darker in foliage color (higher relative chlorophyll

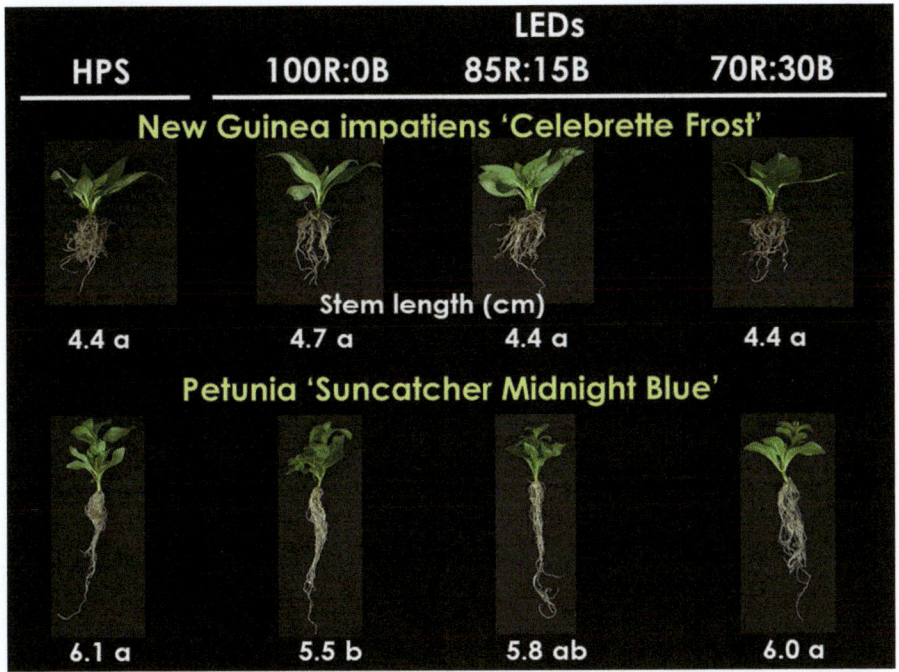

Fig. 15.3 New Guinea impatiens and petunia cuttings propagated for 14 days at 23 °C under ambient greenhouse solar radiation supplemented with 70 µmol m^{-2} s^{-1} from high-pressure sodium (HPS) lamps or light-emitting diodes (LEDs) for 16 h with red/blue (R/B) light ratios of 100:0, 85:15, or 70:30 (Currey and Lopez 2013). Means sharing a letter are not statistically different by Tukey's honestly significant difference at $P \leq 0.05$

content), and had a higher root mass than those produced under supplemental lighting from HPS lamps or ambient lighting conditions in the greenhouse. Additionally, Wollaeger and Runkle (2014) reported that compact bedding plant seedlings could be produced under LED sole-source lighting providing red and blue and/or green wavelengths of light (Fig. 15.4). Ultimately, they believe that this production method may eliminate the need for plant growth regulator applications to inhibit extension growth because height can be manipulated through lighting. Thus, while the effects of light quality may be diminished from background ambient sunlight during supplemental lighting applications in the greenhouse, sole-source lighting enables growers to manipulate light quality to elicit desired morphological and physiological attributes.

Fig. 15.4 Seedlings of impatiens or salvia grown for 4 weeks at 20 °C under sole-source LED lighting treatments that each delivered a PPF of 160 μmol m^{-2} s^{-1} for 18 h·d^{-1} (Wollaeger and Runkle 2014). Peak wavelengths: blue (B), 446 nm; green (G), 516 nm; red (R), 634 nm +664 nm. Means sharing a letter are not statistically different by Tukey's honestly significant difference at $P \leq 0.05$

15.4.2 LED Supplemental Light for Ornamental Crop Finishing

Manipulation of morphology, flowering, and foliage color is important in the greenhouse production of high-quality potted and bedding plants. Light-emitting diodes can be used to manipulate plant morphology and foliage color by delivering specific wave bands of light. For example, Islam et al. (2012) conducted a study to determine whether supplemental lighting from LEDs providing an increased amount of blue light influenced the growth and morphology of poinsettia (*Euphorbia pulcherrima*).

They placed poinsettia 'Christmas Spirit', 'Christmas Eve', and 'Advent Red', under a 10-h photoperiod of solar radiation plus supplemental lighting providing a PPF of 100 µmol m^{-2} s^{-1} from either HPS lamps emitting 5 % blue light or LED arrays emitting a red/blue light ratio of 80:20. The estimated phytochrome photoequilibrium state was 0.85 and 0.89 for the HPS and LEDs, respectively. After 12 weeks, stem, internode, and petiole length; bract area; leaf area; and total dry mass of all cultivars were reduced (by 17–61 %) when grown under the R + B LEDs compared to HPS lamps.

In a separate study, 8–10-day-old petunia 'Tidal Wave' seedlings were placed under frames covered with either a far-red absorbing film (−FR) or two layers of clear polyethylene plastic film (+FR) in a greenhouse (Gautam et al. 2015). Light-emitting diode arrays with a red/blue light ratio of 50:50 were placed underneath the films and provided a PPF of 50 µmol m^{-2} s^{-1}. Additionally, one treatment only contained HPS lamps above the films to provide a PPF of 180 µmol m^{-2} s^{-1} for 16 h. High-pressure sodium lamps were placed above the LED treatments and provided an additional PPF of 100 µmol m^{-2} s^{-1} for 16 h. Under a lower natural irradiance in the spring, petunia shoots were 25 % shorter if they were grown under supplemental lighting from red LEDs compared to plants under the + FR and −FR treatments receiving supplemental light from HPS lamps or blue LEDs. Additionally, under the low natural irradiance, and especially in the presence of FR light, blue light promoted flowering of petunia, but increased shoot elongation and plant height.

Growers sometimes observe that ornamental foliage is not as colorful (e.g., a pale red/purple) during the winter and early spring due to low-light greenhouse conditions. Owen and Lopez (2015a) determined whether a short exposure to end-of-production supplemental lighting from different sources and intensities would influence foliage color. They placed geranium 'Black Velvet' and purple fountain grass (*Pennisetum setaceum* 'Rubrum') plants under a 16-h photoperiod consisting of ambient solar radiation plus day-extension light at 4.5 µmol m^{-2} s^{-1} from LEDs, 70 µmol m^{-2} s^{-1} from HPS lamps, or one of six LED arrays. The six LED arrays were 100 µmol m^{-2} s^{-1} of monochromatic red; 25, 50, or 100 µmol m^{-2} s^{-1} of monochromatic blue; or 100 µmol m^{-2} s^{-1} of 87:13 or 50:50 red/blue light. The supplemental light providing 100 µmol m^{-2} s^{-1} of 0:100 red/blue light enhanced pigmentation of geranium and purple fountain grass leaves the most when plants were grown under a greenhouse DLI of <9 mol m^{-2} d^{-1}. For example, purple fountain grass placed under the control or 100 µmol m^{-2} s^{-1} of 100:0, 0:100, 50:50, or 87:13 red/blue supplemental light had hue angle ($h°$) values of 91°, 57°, 4°, 44°, and 70°, respectively. Therefore, foliage color of plants under blue LEDs at 100 µmol m^{-2} s^{-1} were the darkest red because the $h°$ was the lowest.

15.4.3 *LED Supplemental Light for Vegetable Production*

During the winter months, supplemental lighting for greenhouse vegetable crop production is becoming more commonplace for many species (Chap. 16). Much of the research conducted regarding LED supplemental lighting for vegetable production has naturally focused on how to impact factors such as yield and marketability for a variety of crops (Deram et al. 2014; Gómez et al. 2013; Trouwborst et al. 2010). Light-emitting diodes can also be used for a few days or weeks prior to harvest to elicit specific morphological or physiological effects, such as to increase coloration on lettuce (*Lactuca sativa*) varieties to increase the marketability and quality of the crop. Providing lettuce varieties 'Cherokee', 'Magenta', 'Ruby Sky', and 'Vulcan' with LED red/blue light ratios of 100:0, 50:50, or 0:100 at 100 μmol m^{-2} s^{-1} with a 16-h photoperiod produced darker red foliage compared to those grown under HPS lamps, lower intensities of the same ratios, or no supplemental lighting (Owen and Lopez 2015b; Fig. 15.5). With as little as 5–7 days of supplemental lighting prior to harvest, anthocyanin synthesis and thus pigmentation increased in these lettuce crops. Anthocyanins are a plant pigment involved in the red coloration of plant foliage and have a variety of human health-promoting benefits. Earlier research by Li and Kubota (2009) reported that supplemental blue or UV-A light could be utilized to increase the accumulation of this plant pigment in baby leaf lettuce 'Red Cross'. Thus, the utilization of LEDs for supplemental lighting of lettuce crops has the potential to improve aesthetic quality and nutritional value.

Supplemental lighting provided by red-orange and blue LEDs can significantly inhibit hypocotyl elongation of lettuce seedlings during the early stages of plant growth compared to under HPS lamps (Pinho et al. 2007). This inhibition of elongation was attributed to the 20 % of blue light included in the LEDs. Similarly, the inclusion of blue light in intracanopy supplemental lighting reduced the length of internodes for both tomato and cucumber (*Cucumis sativus*) (Ménard et al. 2006). An increase in the red/far-red ratio also reduced internode length and increased fruit coloration for these species (Ménard et al. 2006). Thus, by manipulating the wavelengths of light provided, the production of many crops can be enhanced by controlling excessive growth and elongation. However, results were obtained using intracanopy supplemental lighting, where ambient light is significantly reduced due to the developing canopy. Therefore, while supplemental light quality effects are negligible at times because of the background radiation, as discussed previously, the manipulation of plant morphology through light quality is more plausible using intracanopy supplemental lighting.

15.5 Concluding Summary

Light-emitting diodes provide an alternative to traditional greenhouse lighting sources for many supplemental lighting applications. While the benefit of targeting specific wavelengths of light may be negligible against ambient light, the relatively high energy efficiency and long lifespan of most LEDs remain desirable attributes. Additionally, with applications such as intracanopy, end of production, and sole-source lighting becoming increasingly popular, the utilization of wavelength

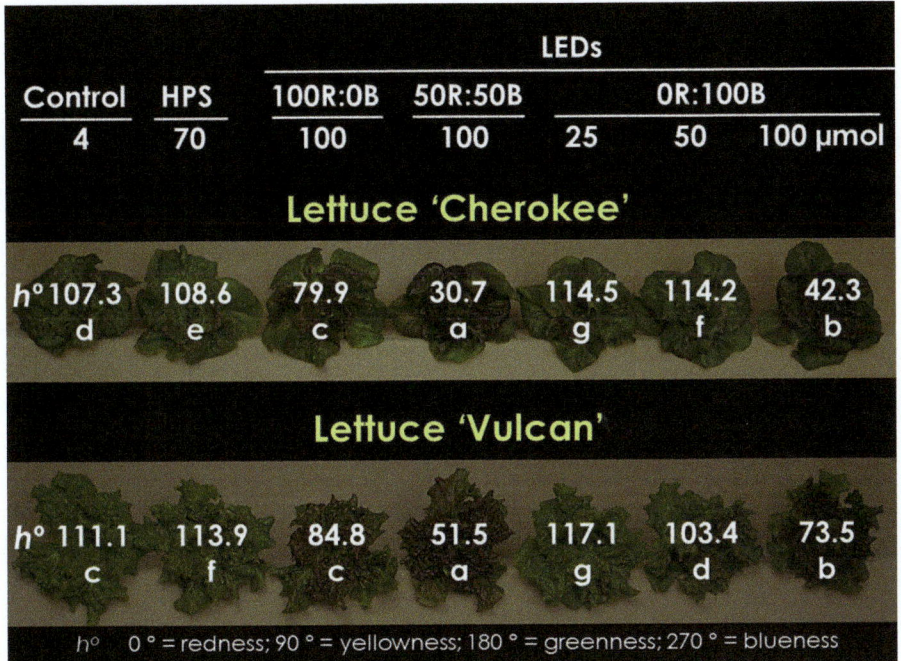

Fig. 15.5 Average hue angle ($h°$) of lettuce grown in a greenhouse at 18 °C and then provided with 7 days of day-extension lighting to deliver a 16-h photoperiod from low-intensity light-emitting diode (LED) lamps or 16 h of supplemental light from high-pressure sodium (HPS) lamps or LED arrays with different red/blue ratios (R:B). Values under each lighting treatment indicate the PPF in μmol m^{-2} s^{-1}. Within-row $h°$ means followed by different letters are significantly different (Owen and Lopez 2015 b)

specificity provided by LEDs will only increase as their manufacturing costs continue to decrease while their electrical efficacy continues to increase.

References

Cosgrove DJ (1981) Rapid suppression of growth by blue light. Plant Physiol 67:584–590

Currey CJ, Lopez RG (2013) Cuttings of impatiens, pelargonium, and petunia propagated under light-emitting diodes and high-pressure sodium lamps have comparable growth, morphology, gas exchange, and post-transplant performance. HortScience 48:428–434

Currey CJ, Lopez RG (2015) Biomass accumulation and allocation, photosynthesis, and carbohydrate status of impatiens, pelargonium, and petunia cuttings are affected by photosynthetic daily light integral during root development. J Am Soc HortScience 140:542–549

Currey CJ, Torres AP, Jacobs DF, Lopez RG (2013) The quality index – a new tool for integrating quantitative measurements to assess quality of young floriculture plants. Acta Hortic 1000:385–392

Deram P, Lefsrud MG, Orsat V (2014) Supplemental lighting orientation and red-to-blue ratio of light-emitting diodes for greenhouse tomato production. HortScience 49:448–452

Downs RJ, Thomas JF (1982) Phytochrome regulation of flowering in the long-day plant, *Hyoscyamus niger*. Plant Physiol 70:898–900

Faust JE, Holcombe V, Rajapakse NC, Layne DR (2005) The effect of daily light integral on bedding plant growth and flowering. HortScience 40:645–649

Fisher P, Both AJ (2004) Supplemental lighting technology and costs. In: Fisher P, Runkle E (eds) Lighting up profits: understanding greenhouse lighting. Meister Media Worldwide, Willoughby, pp 43–46

Gautam P, Terfa MT, Olsen JE, Torre S (2015) Red and blue light effects on morphology and flowering of *Petunia ×hybrida*. Sci Hortic 184:171–178

Goins GD, Yorio NC, Sanwo-Lewandowski MM, Brown CS (1998) Life cycle experiments with *Arabidopsis* grown under red light-emitting diodes (LEDs). Life Support Biosph Sci 5:143–149

Gómez C, Morrow RC, Bourget CM, Massa GD, Mitchell CA (2013) Comparison of intracanopy light-emitting diode towers and overhead high-pressure sodium lamps for supplemental lighting of greenhouse-grown tomatoes. HortTechnology 23:93–98

Hanan JJ (1998) Radiation. In: Greenhouses: advanced technology for protected horticulture. CRC Press, Boca Raton, pp 91–166

Hernández R, Kubota C (2012) Tomato seedling growth and morphology responses to supplemental LED lighting red:blue ratios under varied daily solar light integrals. Acta Hortic 956:187–194

Heuvelink E, Bakker MJ, Hogendonk L, Janse J, Kaarsemaker R, Maaswinkel R (2006) Horticultural lighting in the Netherlands: new developments. Acta Hortic 711:25–34

Hoenecke ME, Bula RJ, Tibbits TW (1992) Importance of 'blue' photon levels for lettuce seedlings grown under red-light-emitting diodes. HortScience 27:427–430

Ilias IF, Rajapakse N (2005) The effects of end-of-the-day red and far-red light on growth and flowering of *Petunia ×hybrida* 'Countdown burgundy' grown under photoselective films. HortScience 40:131–133

Islam MA, Kuwara G, Clarkeb JL, Blystadb D-R, Gisleröda HR, Olsena JE, Torrea S (2012) Artificial light from light emitting diodes (LEDs) with a high portion of blue light results in shorter poinsettias compared to high pressure sodium (HPS) lamps. Scientia Hortic 147:136–143

Kaczperski MP, Carlson WH, Karlsson MG (1991) Growth and development of *Petunia ×hybrida* as a function of temperature and irradiance. J Am Soc Hort Sci 116:232–237

Kigel J, Cosgrove DJ (1991) Photoinhibition of stem elongation by blue and red light. Plant Physiol 95:1049–1056

Kinoshita T, Doi M, Suetsugu N, Kagawa T, Wada M, Shimazaki K (2001) Phot1 and phot2 mediate blue light regulation of stomatal opening. Nature 414:656–660

Li Q, Kubota C (2009) Effects of supplemental light quality on growth and phytochemicals of baby leaf lettuce. Environ Exp Bot 67:59–64

Lopez RG, Runkle ES (2008) Photosynthetic daily light integral during propagation influences rooting and growth of cuttings and subsequent development of new guinea impatiens and petunia. HortScience 43:2052–2059

Massa GD, Kim H, Wheeler RM, Mitchell CA (2008) Plant productivity in response to LED lighting. HortScience 43:1951–1956

Ménard C, Dorais M, Hovi T, Gosselin A (2006) Developmental and physiological responses of tomato and cucumber to additional blue light. Acta Hortic 711:291–296

Mitchell CA, Both A, Bourget CM, Burr JF, Kubota C, Lopez RG, Morrow RC, Runkle ES (2012) LEDs: the future of greenhouse lighting! Chron Hortic 52(1):6–12

Morrow RC (2008) LED lighting in horticulture. HortScience 43:1947–1950

Nelson PV (2012) Greenhouse operation and management, 7th edn. Pearson Prentice Hall, Upper Saddle River

Oh W, Runkle ES, Warner RM (2010) Timing and duration of supplemental lighting during the seedling stage influence quality and flowering in petunia and pansy. HortScience 45:1332–1337

Owen WG, Lopez RG (2015a) Customizing crop foliage color with LEDs: ornamental Crops. Greenh Grow 33(9):76–80

Owen WG, Lopez RG (2015b) End-of-production supplemental lighting with red and blue light-emitting diodes (LEDs) influences red pigmentation of four lettuce varieties. HortScience 50:676–684

Pinho P, Nyrhilä R, Särkkä L, Tahvonen R, Tetri E, Halonen L (2007) Evaluation of lettuce growth under multi-spectral-component supplemental solid state lighting in greenhouse environment. Int Rev Electr Eng 2:854–860

Pramuk LA, Runkle ES (2005) Photosynthetic daily light integral during the seedling stage influences subsequent growth and flowering of *Celosia*, *Impatiens*, *Salvia*, *Tagetes*, and *Viola*. HortScience 40:1336–1339

Randall WC, Lopez RG (2014) Comparison of supplemental lighting from high pressure sodium lamps and light-emitting diodes during bedding plant seedling production. HortScience 49:589–595

Randall WC, Lopez RG (2015) Comparison of bedding plant seedlings grown under sole-source light-emitting diodes (LEDs) and greenhouse supplemental lighting from LEDs and high-pressure sodium lamps. HortScience 50:705–713

Runkle ES, Heins RD (2001) Specific functions of red, far red, and blue light in flowering and stem extension of long-day plants. J Am Soc Hort Sci 126:275–282

Smith H, Whitelam GC (1990) Phytochrome, a family of photoreceptors with multiple physiological roles. Plant Cell Environ 13:695–707

Spaargaren IJ (2001) Supplemental lighting for greenhouse crops, 2nd edn. Hortilux Schreder, Amsterdam

Stutte GW (2009) Light-emitting diodes for manipulating the phytochrome apparatus. HortScience 44:231–234

Styer C (2003) Propagating seed crops. In: Hamrick D (ed) Ball redbook crop production, vol 2, 17th edn. Ball Publishing, Batavia, pp 151–163

Trouwborst G, Oosterkamp J, Hogewoning SW, Harbinson J, van Ieperen W (2010) The responses of light interception, photosynthesis and fruit yield of cucumber to LED lighting within the canopy. Physiol Plant 138:289–300

van Ieperen W, Savvides A, Fanourakis D (2012) Red and blue light effects during growth on hydraulic and stomatal conductance in leaves of young cucumber plants. Acta Hortic 956:223–230

Wollaeger HM, Runkle ES (2014) Growth of impatiens, petunia, salvia, and tomato seedlings under blue, green, and red light-emitting diodes. HortScience 49:734–740

Zeiger E, Talbott LD, Frechilla S, Srivastava A, Zhu J (2002) The guard cell chloroplast: a perspective for the twenty-first century. New Phytol 153:415–424

Chapter 16
Supplemental Lighting for Greenhouse-Grown Fruiting Vegetables

Na Lu and Cary A. Mitchell

Abstract Supplemental lighting (SL) technology has played an important role increasing the productivity of greenhouse crops over the past 30 years and has been more extensively and flexibly employed since LED lights became commercially available for horticultural use. This chapter reviews the applicable regions of the world and conditions for using SL, types of SL, suitable light sources, economic considerations, and current research on each type of SL for fruiting vegetables grown in greenhouses. Important aspects of using SL are summarized.

Keywords Energy • Fruit quality • Fruit yield • Intracanopy lighting • LEDs • Overhead/top lighting • Vertical crops

16.1 Introduction

Supplemental (moderate- to high-intensity) lighting is primarily utilized in greenhouses to increase crop production during periods of low solar radiation. Especially at high latitudes, such as the northern half of the USA, Canada, and northern Europe, during darker months of the year and on overcast days, the amount of solar radiation reaching plants in greenhouses can be insufficient to sustain adequate fruit yields. By providing supplemental lighting (SL) from above and/or to the inner canopy of a crop stand, the light environment can be improved, thereby increasing photosynthesis and yield. In addition to increased fruit production, SL can improve product quality, enable earlier or year-round production, and sustain a more stable workforce (Heuvelink et al. 2006).

There are several types of electric light sources, including high-intensity discharge (HID) lamps and light-emitting diodes (LEDs), which are generally used for

---

N. Lu (✉)
Center for Environment, Health and Field Sciences, Chiba University, 6-2-1 Kashiwa-no-ha, Kashiwa, Chiba 277-0882, Japan
e-mail: na.lu@chiba-u.jp

C.A. Mitchell
Department of Horticulture and Landscape Architecture, Purdue University, 625 Agriculture Mall Drive, West Lafayette, IN 47907-2010, USA

SL of crops grown in greenhouses. The successful use of SL for greenhouse crop production requires careful consideration incorporating such elements as light intensity, photoperiod, light orientation, light distribution and uniformity, cost of installation and operation, and system maintenance. In addition, other environmental factors including carbon dioxide and temperature should be considered as well as crop-specific light responses that can vary during a crop's production period.

16.2 Types of SL and Light Sources for Fruiting Vegetables

16.2.1 Overhead/Top SL

Electric lamps typically are installed above the crop stand to provide overhead lighting (or top lighting).

16.2.1.1 HID Lamps for Overhead SL

Two types of HID lamps, metal halide (MH) and high-pressure sodium (HPS), are capable of delivering adequate photosynthetically active radiation (PAR) to crop surfaces. MH lamps produce a cool-white light with almost as much blue as red light, whereas the light emitted by HPS lamps is yellowish orange with much less blue. However, HPS lamps are somewhat more efficient than MH lamps at converting electrical energy to PAR and have an average rated lifespan up to three times longer (Dorais 2003). HPS lamps have been widely used for greenhouse overhead lighting because they have been the most economically viable mass-produced light source available that can provide adequate intensity and spectrum for plant growth.

Common power consumptions of HPS lamps are 400, 600, or 1000 W per lamp. The radiant thermal energy emitted from HPS lamps can be used to maintain adequate ambient greenhouse and plant temperature during cold weather. Brault et al. (1989) estimated that, in northern climates, the heat emitted from HPS lamps provided as much as 41 % of the heat needed for greenhouse operation, although this value depends substantially on the light intensity delivered and greenhouse location. Thus, heat generation is considered a useful by-product of HPS lamp used in the winter. On the other hand, HPS lamps have a high life-cycle cost and an intense environmental impact (Nelson 2012). These lamps typically require reflectors or luminaires to direct the light beam onto crops, thereby providing satisfactory light distribution and efficiency. However, the fixture blocks some sunlight from reaching crop surfaces. This downside of HPS lamps is minimized by mounting them as high as possible in the greenhouse above the crop stand and minimizing their density, which in turn limits the photosynthetic photon flux (PPF) that they can deliver.

16.2.1.2 LEDs for Overhead SL

LED arrays are an alternative for SL of greenhouse fruiting vegetables. Morrow (2008) pointed out that LEDs are a promising SL source for the greenhouse industry as they surpass many capabilities of commercially available lamps commonly used in horticulture. LEDs can be designed to emit narrow-spectrum light to maximize photosynthetic quantum efficiency for specific crop species (Bourget 2008). The most electrically efficient colors of LEDs, based on micromoles of photosynthetic photons emitted per joule of electricity consumed, are blue, red, and cool white (Nelson and Bugbee 2014). Thus, LED fixtures typically come in combinations of these colors. As of 2015, blue and red LEDs became up to 50 % efficient, meaning that half of their input energy is converted into photon output. The present LED efficacy already is higher than that of HPS lamps, which are up to 34 % efficient (Philips Lighting 2015). LEDs also have a long lifespan. Conventional HPS bulbs typically need to be replaced after 10,000 h, whereas well-designed LEDs still emit at least 90 % of their output after 25,000 h and can last five times longer than conventional HPS light sources (Philips Lighting 2015). With ongoing expected improvements in energy efficiency, availability of photosynthesis-driving wave bands, and higher-output/high-density arrays, LEDs represent potential solutions to profitability and sustainability issues (Gomez et al. 2013; Mitchell 2015). For tall crops, such as high-wire tomato, overhead LED lighting does not take full advantage of their unique properties for plant lighting, especially considering the relatively cool photon-emitting surfaces that allow LEDs to be located in close proximity to plant surfaces (Massa et al. 2008). To avoid dense overhead arrays of LEDs from blocking large amounts of incoming solar radiation within greenhouses, such arrays also would have to be positioned high above the crop canopy, necessitating dense arrays of high-output LEDs. Such arrays not only would be costly, but would fail to capture some of the advantages of using LEDs over other supplemental light sources (Mitchell 2015).

16.2.1.3 Application of Overhead SL

Supplemental greenhouse lighting has been used for several decades in horticultural crop production. Many researchers have shown that the effect of overhead SL to increase tomato yield is pronounced at low natural light levels (Rodriguez and Lambeth 1975; Blom and Ingratta 1984; Grimstad 1987; Dorais et al. 1991).

In the Netherlands, more than 2000 ha of greenhouses are equipped with overhead SL, and the use of SL for vegetable production continues to increase. Tomato production under SL accounted for more than 120 ha, production of sweet peppers for more than 60 ha, and cucumbers about 10 ha (Heuvelink et al. 2006). In Canada, approximately 15 % of tomato growers and 10 % of cucumber growers were producing winter greenhouse crops with SL (Dorais 2003). By applying SL technology in the Netherlands, reported maximum production rates of 110, 64, and

168 kg·m^{-2}·year^{-1} were achieved for tomatoes, sweet peppers, and cucumbers, respectively. For Canada, potential yields were 15 % higher than for the Netherlands at 127, 76, and 199 kg·m^{-2}·year^{-1} for tomatoes, sweet peppers, and cucumbers, respectively. Because Canada receives on average 15 % more natural light annually than the Netherlands (Heuvelink et al. 2006), this corresponds well with the "1 % rule," which states that 1 % more light will result in 1 % more yield (Marcelis et al. 2006). Thus, potential yields of these crops under an appropriate SL strategy could be roughly twofold higher than without SL.

16.2.2 Combination of Overhead and Intracanopy SL

16.2.2.1 Light Distribution in Tall Greenhouse Crops

Crop growth uniformity and productivity are influenced by the homogeneity of light distribution and amount of light delivered to the entire foliar canopy. Overhead lighting alone provides uneven distribution and spectrum of radiation along the vertical profile of tall crops, with the bottom of the canopy receiving much less and different spectrum light than is incident upon upper, outer leaves (Frantz et al. 2000). Such mutually shaded plants within a dense crop canopy are known to benefit from supplemental radiation within the foliar canopy when the intensity of light received is between the light compensation and saturation points of photosynthesis. Otherwise, interior and lower leaves deteriorate rapidly in photosynthetic capability, contributing instead to net respiratory carbon losses, premature senescence, and abscission (Massa et al. 2008).

16.2.2.2 Intracanopy Lighting

One method to improve biomass production and yield of tall fruiting greenhouse crops is to introduce light of suitable quantity and quality to the inner foliar canopy. Such "intracanopy lighting" helps increase the efficiency of radiation use by introducing light directly to the interior of crop stands, thereby utilizing full photosynthetic capacity of otherwise mutually shaded leaves. The intense thermal radiation from HPS lamps makes intracanopy or close-canopy lighting unfeasible for high-wattage fixtures. Unlike traditional HID light sources used in commercial greenhouses, the relative coolness of LED photon-emitting surfaces allows them to operate in close proximity to plants without overheating or scorching them, thereby increasing available PAR at leaf level while using less energy (Massa et al. 2008). Therefore, LEDs can be effectively used to direct highly focused lighting to the inner canopy of closed crop stands. SL system efficiency is the combined effect of efficient fixtures and efficient canopy photon capture (Nelson and Bugbee 2014). A combination of overhead lighting and intracanopy lighting could maximize light distribution over an entire crop canopy. An example of an LED overhead lighting

Fig. 16.1 Overhead and intracanopy SL from LEDs for tomato plants grown in a greenhouse

and LED intracanopy lighting system is shown in Fig. 16.1. The inclusion of LED intracanopy lighting along with adequate overhead lighting can increase photon capture rates to nearly 100 % (Nelson and Bugbee 2014) and may have other beneficial effects such as an increased light sharing with interior leaves and as a replacement for heating pipes near growing plants.

16.2.2.3 Application of Overhead and Intracanopy SL

Different combinations of overhead and/or intracanopy SL for tomato (Gunnlaugsson and Adalsteinsson 2006; Dueck et al. 2012a; Gomez and Mitchell 2016; Moerkens et al. 2016), sweet pepper (Hovi-Pekkanen et al. 2006; Jokinen et al. 2012), and cucumber production (Hovi et al. 2004; Hovi-Pekkanen and Tahvonen 2008; Pettersen et al. 2010; Trouwborst et al. 2010; Hao et al. 2014) have been studied to improve crop productivity. In one study, although intracanopy LED SL alone or hybrid SL involving LED interlighting diminished the top-to-bottom decline in photosynthetic activity of inner-canopy leaves (e.g., net photosynthesis and quantum use efficiency) compared with overhead lighting only (HPS or solar), no yield differences in high-wire tomato occurred among SL treatments (Gomez and Mitchell 2016). Preferential partitioning of photoassimilates to vegetative plant parts likely occurred. Moreover, all SL treatments provided equivalent daily light integral (DLI) and similar increases in fruit yield relative to controls, but LED intracanopy SL provided substantial energy savings compared to SL systems including overhead HPS lamps.

Hao et al. (2014) reported that LED intracanopy lighting of cucumber increased light use efficiency, mainly by increasing light reaching the inner canopy, compared with overhead HPS lamps. Moreover, the response of mini-cucumbers to LED intracanopy lighting could be optimized by using proper crop management (e.g., certain plant density) and ratio of overhead light/intracanopy light. Dueck et al. (2012a) compared the effect of overhead lighting and intracanopy lighting with HPS fixtures and/or LEDs on the growth and production of tomatoes in greenhouses. The amount of energy required per kilogram of tomatoes harvested was highest for the LED treatment and hybrid system with overhead LED lighting. They suggested that a combination of overhead HPS lighting and LEDs was the most promising alternative for their climate, when taking into consideration various production parameters and energy costs of using the different systems. Another experiment compared crop production and energy consumption between HPS and LED SL, and the results showed that while achieving nearly identical production, LED SL saved 30 % of dehumidification and heat energy and 27 % of electricity relative to the crop grown with HPS SL (Dueck et al. 2012b).

16.2.3 Intracanopy SL Alone

16.2.3.1 Conditions for the Use of Intracanopy Lighting Alone

During periods with adequate sunlight, overhead SL may not be needed if the upper plant canopy receives sufficient solar radiation. However, leaves in the lower canopy, especially of high-wire crops, may still receive limited solar light because of mutual shading. Therefore, providing intracanopy SL alone still can be beneficial during specific periods. Even in areas that receive "sufficient" sunlight year-round, intracanopy lighting could be an effective method to increase production and ensure more uniform crop yield to the market year-round. Research-grade LED towers have been developed with adjustable, actively heat-sinked, high-output LEDs that can provide substantial intracanopy lighting along the entire vertical profile of a high-wire crop (Fig. 16.2). Such intracanopy lighting sources do not require additional overhead lighting and give the same yield stimulation as overhead HPS alone or hybrid overhead HPS/intracanopy LED interlights, but at a much lower cost of electrical energy (Gomez and Mitchell 2016).

A low-truss, high-density tomato production system has been developed (Giacomelli et al. 1994; Okano et al. 2001; Lu et al. 2012a, b). This system integrates movable benches, computer-assisted management, and other environmental control systems to achieve year-round, continuous, predictable production of uniform quality tomatoes. In this system, the plant density is normally three to five times higher than for a high-wire tomato system, which also creates limited light distribution to lower leaves. Thus, intracanopy lighting also can be applied in such cultivation systems to improve production.

Fig. 16.2 Intracanopy LED lighting of high-wire tomato plants with the foliar canopy entirely closed around LED towers irradiating red + blue light in both directions along the row and with all three vertical LED zones energized

16.2.3.2 Application of Intracanopy SL Alone

Gomez and Mitchell (2014) and Gomez et al. (2013) compared year-round high-wire tomato production with and without SL and evaluated two different SL positions (HPS lamps as overhead lighting vs. LEDs as intracanopy lighting) on production and energy consumption parameters. SL induced significantly earlier and higher fruit production compared with controls lacking SL. LED intracanopy lighting treatment of the same PPF and DLI as provided by different wattages of overhead HPS lighting achieved yields comparable to those with HPS, but with 55–75 % energy savings. Lu et al. (2012a) applied intracanopy lighting at different developmental stages of single-truss tomato plants and determined that SL during the stage from fruit-set to mature green was most efficient for increasing fruit yield and sugar content up to 20 % and 12 %, respectively, compared to plants without SL. Intracanopy lighting using different colors of LED light was also examined for single-truss tomato crops. The results demonstrated combined-spectrum supplemental light (red + green + blue LEDs) or red light alone was superior to blue light for increasing tomato production (Lu et al. 2012b).

An experiment was conducted from winter to spring using commercialized intracanopy LEDs. Single-truss tomato plants were subjected to supplemental light from immediately after first anthesis until harvest. The results showed that intracanopy lighting increased tomato yield by 15–20 % (Fig. 16.3a) and fruit sugar content by 11–12 % (Fig. 16.3b) compared to plants without SL (data not published). Tewolde et al. (2016) used the same LED system and determined the effects of daytime and nighttime intracanopy lighting on single-truss tomato growth and yield in both winter and summer. They found that nighttime LED intracanopy lighting increased photosynthetic capacity in both winter and summer, whereas yield increased by 24 % in winter and 12 % in summer. Daytime LED intracanopy lighting also increased photosynthetic capacity of middle- and lower-canopy leaves and yield (by 27 %) in winter but not in summer. Therefore, using nighttime LED intracanopy lighting may be more advantageous because of lower off-peak electricity rates.

16.3 Light Intensities, Photoperiods, and DLIs of SL

McAvoy and Janes (1984) reported that yields of tomato were 62, 79, and 90 % greater than unlit plants for SL of 100, 125, or 150 $\mu mol\ m^{-2}\ s^{-1}$, respectively, for an 18-h photoperiod. Dorais et al. (1991) reported that tomato yield increased 27 % at a PPF of 150 $\mu mol\ m^{-2}\ s^{-1}$ compared to that at 100 $\mu mol\ m^{-2}\ s^{-1}$ for a 16-h photoperiod. Their results showed that SL allows for an increase in planting density while obtaining greater fruit production. A higher planting density increased the proportion of light intercepted by the canopy, thereby improving light utilization. Higher light intensity also increased the total number of fruits produced. For sweet pepper, increasing SL intensity to 188 $\mu mol\ m^{-2}\ s^{-1}$ also improved fruit-set compared to that at 125 $\mu mol\ m^{-2}\ s^{-1}$ (Heuvelink et al. 2006). Moerkens et al. (2016) reported a 20 % increase in yield under 169 $\mu mol\ m^{-2}\ s^{-1}$ overhead HPS light with 55 $\mu mol\ m^{-2}\ s^{-1}$ extra intracanopy LED light compared to that without extra LEDs. For cucumber, Hao and Papadopoulos (1999) found SL to increase cucumber marketable yield by 28 % when light equivalent to 10–30 % of ambient radiation was provided by HPS lamps. Trouwborst et al. (2011) compared effects of 220 $\mu mol\ m^{-2}\ s^{-1}$ overhead HPS light with a combination of 140 $\mu mol\ m^{-2}\ s^{-1}$ overhead HPS + 80 $\mu mol\ m^{-2}\ s^{-1}$ intracanopy LEDs on cucumber production for a 20-h photoperiod. They concluded that intracanopy lighting caused an increase in yield that can be mainly explained by an increase in light absorption and more homogeneous light distribution within the canopy. Most of these researchers used an SL intensity between 100 and 220 $\mu mol\ m^{-2}\ s^{-1}$ for 12–20 h per day. Some studies have used higher intensities such as 350 $\mu mol\ m^{-2}\ s^{-1}$ as well as longer photoperiods such as 24 h, but their results showed negative effects on tomato yield from continuous lighting (Demers et al. 1998). Dorais (2003) summarized that SL of 100–150 $\mu mol\ m^{-2}\ s^{-1}$ with a 14- to 18-h photoperiod for tomato, 150–175 $\mu mol\ m^{-2}\ s^{-1}$ with a 12–18-h photoperiod for sweet pepper, and 120–150 $\mu mol\ m^{-2}\ s^{-1}$ with an 18–20-h

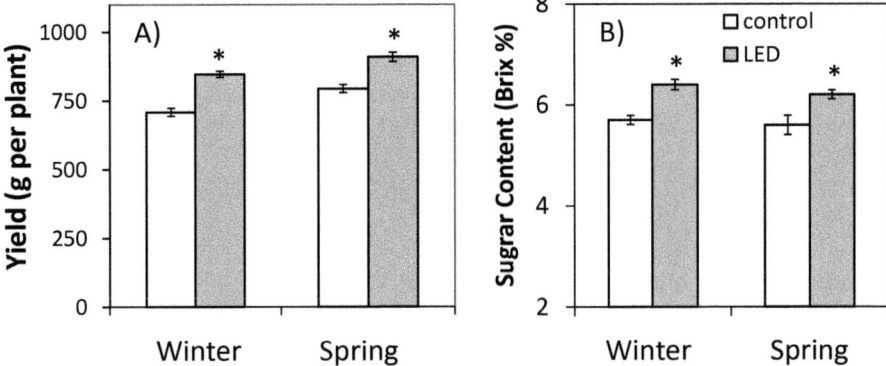

Fig. 16.3 (a) Fruit yield of single-truss tomato plants with and without LED intracanopy lighting in winter and spring. (b) Fruit sugar content of single-truss tomato plants with and without LED intracanopy lighting in winter and spring. A light intensity of 88 μmol m^{-2} s^{-1} was provided by LEDs. Data represent mean ± SE ($n = 20$). Different letters indicate statistically significant differences (Tukey's honest significant difference test at $P < 0.05$) (Data from Na Lu)

photoperiod for cucumber may be suitable for use in Canada during low natural light periods.

DLI is another option for management of crop production. DLI is the amount of PAR received each day as a function of light intensity and duration. It is expressed as moles of light per square meter per day (mol m^{-2} d^{-1}). A DLI of at least 25 mol m^{-2} d^{-1} is suggested for greenhouse tomato production, which can necessitate SL at 150–200 μmol m^{-2} s^{-1} (Dorais 2004; Moe et al. 2006). For cucumber production in temperate climates, Dorais (2004) recommends SL of at least 120–150 μmol m^{-2} s^{-1} for 18–20 h per day, which increases the ambient DLI by 7.8–10.8 mol m^{-2} d^{-1}. The minimum DLI for sweet pepper is 12 mol m^{-2} d^{-1} and SL at 150–175 μmol m^{-2} s^{-1} for 16–20 h per day is recommended (Dorais 2004). Growers should make SL plans based on their own conditions and experience.

16.4 LED SL for Improving Fruit Quality

As growers start to leverage LED technology for SL of fruiting vegetables, the quality of the final product is as important for income as productivity and yield. The demand and acceptance of fresh fruit vegetables are based largely on flavor, fruit sugar content, and color. Ascorbic acid content, lycopene, and oxygen radical absorbance capacity (ORAC) have well-known nutritional benefits for consumers. Increasing such health-related phytochemicals (HRPs) in fruit by using SL may add additional value for consumers.

Verkerke et al. (2014) developed greenhouse cultivation protocols for growers to fine-tune plant growth and create products with higher amounts of HRPs. They used an SL system with red and blue LEDs to irradiate only tomato fruit rather than leaves. It was found that, compared to a control group without LED light, the vitamin C content increased under LED light treatment, increasing logarithmically with increasing radiation exposure. Similarly, total ORAC values also increased under LED light treatment. In a single-truss tomato system, intracanopy LED lighting increased both total soluble solids and ascorbic acid content of tomato fruit grown in winter, compared to that of no SL (Tewolde et al. 2016). In mini-cucumber, fruit color also was improved by LED intracanopy lighting compared to that without SL (Hao et al. 2014). Moreover, Alcock and Bertling (2012) reported that supplemental LED lighting accelerated color changes of green pepper fruit compared to those lacking exposure to LED light. This could improve profitability by enabling producers to exploit early-season prices, particularly for cultivars with pale-green, unripe pepper fruit.

16.5 Other Aspects of Using SL

When considering SL technologies for fruiting vegetables, the following should be taken into account:

1. The average amount and seasonal variation of solar radiation for a given location.
2. The type of greenhouse structure and cover, because they collectively affect the transmission and internal distribution of sunlight. Growers need to determine the extent to which transmission is affected so that they can calculate how much additional light is needed to achieve target DLI and desired production.
3. Light uniformity is an important factor to consider when designing lighting systems for greenhouses. Calculation of light uniformity will determine how much overhead SL is needed and how much light can come from intracanopy lighting or other directions.
4. Consider DLI instead of just instantaneous PPF or lighting duration (photoperiod) alone. DLI determination will help growers manage more stable production over time based on natural variations in solar DLI. Also consider combinations of lighting strategy and plant density since variable planting density could improve the utilization of supplemented light and increase yield per unit area (Dorais et al. 1991; Papadopoulos and Pararajasingham 1997).
5. Different crop types and varieties may require different light levels and photoperiods. The type of crop grown in a greenhouse determines lighting requirements (such as light intensity, duration, and DLI), along with space needed in the greenhouse to install lamps.
6. Use lamps efficiently to minimize cost and maximize benefits. It is sometimes not economical to install lighting systems to provide uniform high-light

intensities in greenhouses because of a high initial investment cost. Thus, SL systems can be designed to provide a specific DLI by increasing lighting hours with relatively lower light intensity, e.g., 18 h at low intensity rather than 12 h at high intensity (Marcelis et al. 2006). However, some fruiting vegetable crops (e.g., tomato) grown under continuous light develop physiological disorders, and thus, 4–6 h of darkness is required each day.
7. Operate lights only during hours and on days when ambient sunlight is low. Switch off lights when sunlight intensity is high, e.g., greater than 300 W·m^{-2} (or 600 µmol m^{-2} s^{-1}) for tomato (Heuvelink et al. 2006). Supplemental lighting on cloudy days or during the night adds the most photosynthesis per unit photons of SL provided, and taking advantage of off-peak electricity rates could improve cost efficiency.
8. Interactions with other environmental factors such as temperature, carbon dioxide, relative humidity, nutrient solution concentration, water status, etc., also should be considered to maximize light use efficiency.

16.6 Economic Considerations for Greenhouse SL

Northern growers of greenhouse fruiting vegetables are increasingly trying to compete in local or regional markets with produce shipped long distances from milder climates, especially during low-light seasons of the year. Since fruiting vegetables typically have high-light requirements for production, growers using or considering supplemental greenhouse lighting must weigh operating costs of a given SL source against capital investment costs.

The standard SL source for greenhouse fruiting vegetable production is the HPS lamp. Because HPS is a long-term, established lighting technology for many applications beyond specialty crop production, it is mass produced and competitively priced. The main economic disadvantage of HPS is its high electrical energy consumption, especially due to its long-wave radiant energy output, which has winter heating cost offset value in greenhouses, but otherwise amounts to copious waste heat generation in addition to useful PAR generation.

LEDs provide the same SL boost to greenhouse high-wire vegetable production that HPS fixtures do, but at substantially lower electrical cost, due mainly to lack of thermal radiation output in the photon emission beam. In 2014, the photosynthetic photon efficiency of the most efficient LED arrays and high-wattage double-ended HPS fixtures was similar at 1.7 µmol PAR/J of electricity consumed (µmol/J) (Nelson and Bugbee 2014), but now, the efficiency of some commercial LEDs is reportedly ≥ 2.5 µmol/J. The lack of thermal emissions in the LED photon beam allows for close-canopy and/or intracanopy placement of LEDs. Thus, a much lower power is needed for LEDs in close proximity to leaves to provide a PPF equivalent to that from HPS sources that require considerable spatial separation from plants and higher electrical power to the fixture. The lower power input to LEDs also means minimal heat-sinking requirements at the back of the diode, although the electrical-to-photon conversion efficiency may be similar for HPS lamps and LEDs.

The present disadvantage of LEDs is that the initial capital fixture cost per photon delivered is as much as five to ten times more than for HPS fixtures (Nelson and Bugbee 2014). Over a 5-year period of use, factoring together the energy use advantage of LEDs as well as their capital investment disadvantage, the electric-plus-fixture cost per mole of PAR photons delivered still is 2.3 times higher for LEDs. However, as the electrical efficacy of LEDs improves with ongoing advances in LED technology, and as mass production of LED arrays for plant growth applications increases, both costs of operation and capital investment are anticipated to decline. Over the long term, the lifespan of LEDs (70–80 % of initial photon output at 50,000 h) is predicted to easily outperform that of HPS (typical bulb replacement at 10,000 h of operation). Cost of replacement of lamps, labor, and disposal are anticipated to weigh in favor of LEDs for greenhouse SL (Zhang et al. 2016). The use of LEDs also has a 38–47 % smaller environmental impact than HPS, all of which have significant economic implications.

Although HIDs and LEDs have their respective economic advantages and disadvantages, regions of the country where electrical energy costs and demand for SL are high also favor LEDs as the light source of choice for greenhouse SL (Burr 2015). Other secondary factors that will influence economy-based choice of SL source include layout of greenhouses as well as crop type, growth habit, and cropping density. Each of these factors contributes to the proportion of SL photons that will be intercepted by plants and converted to useful biomass. Obviously, photons that fall on surfaces not populated by photosynthesizing plant tissues will not contribute to crop productivity. For large greenhouse areas covered by uniform vegetation with few, narrow aisles and relatively short, non-mutually shaded, closed foliar canopies, 1000-W HPS lamps with beam-spreading luminaires mounted high above growth surfaces would be the SL source of choice to provide the most uniform, modest DLI levels with minimal photon wastage and the least blockage of solar PAR. For greenhouses with frequent, wide aisles or moveable benches and high-market-value crops, LED arrays with highly focused photon beams that maintain photon flux right up to the edge of the bench but do not allow light to spill over into the aisles will be most economic. For self-shading high-wire crops, intracanopy LED arrays would be one SL source with potential for nutritional value added because fruits could be irradiated directly. HPS could be used for overhead SL, but would not have the nutritional value-added potential that intracanopy SL might have. In a survey of greenhouse crops for economic effectiveness of SL, relatively low-light, rapidly turning crops like lettuce scored high for profitability potential; high-light-requiring tomato scored intermediate because of its productivity potential and heating cost offset; and strawberry scored lowest in spite of high market value, suggesting that SL might not be needed (Kubota et al. 2016). These findings were independent of SL source per se.

Many considerations factor into the choice of radiation sources for greenhouse SL. Directly or indirectly, economics tends to be one of the most important selection factors, especially for growers.

References

Alcock CM, Bertling I (2012) Light-induced colour change in two winter-grown pepper cultivars (*Capsicum annuum* L.). Acta Hortic 956:275–281

Blom TJ, Ingratta FJ (1984) The effect of high pressure sodium lighting on the production of tomatoes, cucumbers and roses. Acta Hortic 148:905–914

Bourget CM (2008) An introduction to light emitting diodes. HortScience 43:1944–1946

Brault D, Gueymard C, Boily R, Gosselin A (1989) Contribution of HPS lighting to the heating requirements of a greenhouse. Am Soc Agric Eng 89:4039

Burr J (2015) Economics of adoption of LED technology by horticultural industries. In: Mitchell et al. Light-emitting diodes in horticulture, Hortic Rev 43:1–87, Janick J (ed), Wiley Blackwell, Hoboken, New Jersey, USA

Demers DA, Dorais M, Wien CH, Gosselin A (1998) Effects of supplemental light duration on greenhouse tomato (*Lycopersicon esculentum* Mill.) plants and fruit yields. Sci Hortic 74:295–306

Dorais M (2003) The use of SL for vegetable crop production: light intensity, crop response, nutrition, crop management, cultural practices. Can Greenh Conf 9:2003

Dorais M (2004) Lighting greenhouse vegetables. In: Fisher P, Runkle E (eds) Lighting up profits. Meister Media Worldwide, Willoughby, Ohio, USA, pp 93–96

Dorais M, Gosselin A, Trudel MJ (1991) Annual greenhouse tomato production under a sequential intercropping system using supplemental light. Sci Hortic 45:225–234

Dueck TA, Kempkes FLK, Marcelis LFM, Janse J, Eveleens-Clark BA (2012a) Growth of tomatoes under hybrid LED and HPS lighting. Acta Hortic 952:335–342

Dueck T, Nieboer S, Janse J, Valstar W, Eveleens B, Grootscholten M (2012b) Report: LED lights and next generation cultivation tomatoes. http://www.wageningenur.nl/upload_mm/a/7/6/86717a68-0e12-4328-ab91-cd0033ab2115_edepotin_t5028c540_001.pdf

Frantz JM, Joly RJ, Mitchell CA (2000) Intracanopy lighting influences radiation capture, productivity, and leaf senescence in cowpea canopies. J Am Soc HortScience 125:694–701

Giacomelli GA, Ting KC, Mears DR (1994) Design of a single truss tomato production system (STTPS). Acta Hortic 361:77–84

Gomez C, Mitchell CA (2014) SL for greenhouse-grown tomatoes: intracanopy LED towers vs. overhead HPS lamps. Acta Hortic 1037:855–862

Gomez C, Mitchell CA (2016) Physiological and productivity responses of high-wire tomato as affected by supplemental light source and distribution within the canopy. J Am Soc HortScience 141:196–208

Gomez C, Morrow RC, Bourget CM, Massa GD, Mitchell CA (2013) Comparison of intracanopy light-emitting diode towers and overhead high-pressure sodium lamps for SL of greenhouse-grown tomatoes. HortTechnology 23:93–98

Grimstad SO (1987) Supplementary lighting of early tomatoes after planting out in glass and acrylic greenhouses. Sci Hortic 33:189–196

Gunnlaugsson B, Adalsteinsson S (2006) Interlight and plant density in year-round production of tomato at northern latitudes. Acta Hortic 711:71–76

Hao X, Papadopoulos AP (1999) Effects of supplemental lighting and cover materials on growth, photosynthesis, biomass partitioning, early yield and quality of greenhouse cucumber. Sci Hortic 80:1–18

Hao X, Guo X, Chen X, Khosla S (2014) Inter-lighting in mini-cucumbers: interactions with overhead lighting and plant density. Acta Hortic 1107:291–296

Heuvelink E, Bakker MJ, Hogendonk L, Janse J, Kaarsemaker RC, Maaswinkel RHM (2006) Horticultural lighting in the Netherlands: new developments. Acta Hortic 711:25–33

Hovi T, Näkkilä J, Tahvonen R (2004) Interlighting improves production of year-round cucumber. Sci Hortic 102:283–294

Hovi-Pekkanen T, Tahvonen R (2008) Effects of interlighting on yield and external fruit quality in year-round cultivated cucumber. Sci Hortic 116:152–161

Hovi-Pekkanen T, Näkkilä J, Tahvonen R (2006) Increasing productivity of sweet pepper with interlighting. Acta Hortic 711:165–170

Jokinen K, Särkkä LE, Näkkilä J (2012) Improving sweet pepper productivity by LED interlighting. Acta Hortic 956:59–66

Kubota C, Kroggel M, Both A, Burr J, Whalen M (2016) Does SL make sense for my crop? - empirical evaluations. Acta Hortic 1134:403–411

Lu N, Maruo T, Johkan M, Hohjo M, Tsukagoshi S, Ito Y et al (2012a) Effects of SL within the canopy at different developing stages on tomato yield and quality of single-truss tomato plants grown at high density. Environ Control Biol 50:1–11

Lu N, Maruo T, Johkan M, Hohjo M, Tsukagoshi S, Ito Y et al (2012b) Effects of SL with light-emitting diodes (LEDs) on tomato yield and quality of single-truss tomato plants grown at high planting density. Environ Control Biol 50:63–74

Marcelis LFM, Broekhuijsen AGM, Nijs EMFM, Raaphorst MGM (2006) Quantification of the growth response of light quantity of greenhouse grown crops. Acta Hortic 711:97–103

Massa GD, Kim H-H, Wheeler RM, Mitchell CA (2008) Plant productivity in response to LED lighting. HortScience 43:1951–1956

McAvoy R, Janes HW (1984) The use of high pressure sodium lights in greenhouse tomato crop production. Acta Hortic 148:877–888

Mitchell CA (2015) Academic research perspective of LEDs for the horticulture industry. HortScience 50:1293–1296

Moe R, Grimstad SO, Gislerod HR (2006) The use of artificial light in year round production of greenhouse crops in Norway. Acta Hortic 711:35–42

Moerkens R, Vanlommel W, Vanderbruggen R, Van Delm T (2016) The added value of LED assimilation light in combination with high pressure sodium lamps in protected tomato crops in Belgium. Acta Hortic 1134:119–124

Morrow RC (2008) LED lighting in horticulture. HortScience 43:1947–1950

Nelson PV (2012) Greenhouse operation and management, 7th edn. Prentice-Hall, Upper Saddle River, New Jersey, USA

Nelson JA, Bugbee B (2014) Economic analysis of greenhouse lighting: light emitting diodes vs. high intensity discharge fixtures. PLoS One 9(6):e99010

Okano K, Sakamoto Y, Watanabe S-I (2001) Source-sink relationship of ^{13}C-photosynthates in single-truss tomato. Bull Natl Inst Veg Tea Sci 16:351–361 (in Japanese with English synopsis)

Papadopoulos AP, Pararajasingham S (1997) The influence of plant spacing on light interception and use in greenhouse tomato (*Lycopersicon esculentum* Mill.): a review. Sci Hortic 69:1–29

Pettersen RI, Torre S, Gislerod HR (2010) Effects of intracanopy lighting on photosynthetic characteristics in cucumber. Sci Hortic 125:77–81

Philips Lighting BV (2015) Philips horticulture LED toplighting. http://images.philips.com/is/content/PhilipsConsumer/PDFDownloads/Global/ODLI20150701_001-UPD-en_AA-CL_LED_Toplighting_Philips_Horticulture_EN.pdf

Rodriguez BP, Lambeth VN (1975) Artificial lighting and spacing as photosynthetic and yield factors in winter greenhouse tomato culture. J Am Soc HortScience 100:694–697

Tewolde FT, Lu N, Shiina K, Maruo T, Takagaki M, Kozai T, Yamori W (2016) Nighttime supplemental LED inter-lighting improves growth and yield of single-truss tomatoes by enhancing photosynthesis in both winter and summer. Front Plant Sci 7:448

Trouwborst G, Oosterkamp J, Hogewoning SW, Harbinson J, Ieperen WV (2010) The responses of light interception, photosynthesis and fruit yield of cucumber to LED-lighting within the canopy. Physiol Plant 138:289–300

Trouwborst G, Schapendonk AH, Rappoldt K, Pot S, Hogewoning SW, van Ieperen W (2011) The effect of intracanopy lighting on cucumber fruit yield–model analysis. Sci Hortic 129:273–278

Verkerke W, Labrie C, Dueck T (2014) The effect of light intensity and duration on vitamin C concentration in tomato fruits. Acta Hortic 1106:49–54

Zhang H, Burr J, Zhao F (2016) A comparative life cycle assessment (LCA) of lighting technologies for greenhouse crop production. J Clean Prod (In press) http://dx.doi.org/10.1016/j.jclepro.2016.01.014

Chapter 17
Recent Developments in Plant Lighting

Erik S. Runkle

Abstract There is substantial interest among commercial growers and plant hobbyists, academics, and lighting manufacturers in the development of new lighting technologies and lighting applications. LEDs are of particular interest because of the ability to control the light spectrum to regulate crop growth characteristics as well as their increasing energy efficiency. This chapter briefly discusses recent developments in plant lighting including an international light symposium held in 2016, a new book on horticultural lighting, and the development of lighting standards for plant applications, which are especially needed for LEDs.

Keywords ISHS symposium • Lamp efficacy • LEDs • Lighting book • Lighting standards

17.1 Introduction

LED lighting is increasingly being used in the production of specialty crops grown in controlled environments, including in vertical farms and greenhouses. As the technologies for LED lighting advance, they become more economical and also increase opportunities to regulate plant growth, development, and concentration of phytonutrients through control of the light spectrum. Lighting companies, plant producers, and academics are working together to advance the science and application of plant lighting, especially for high-value specialty crops. A key role for academics is to produce unbiased, research-based information and then communicate that to peers and the specialty crops industries. There are several ways in which this can be accomplished, including convening conferences and publishing written works for both scientific and industry audiences. Furthermore, there is a compelling need to define lighting metrics for plant applications because in many situations, metrics developed for general illumination (for human application) are simply not applicable.

E.S. Runkle (✉)
Department of Horticulture, Michigan State University, 1066 Bogue Street,
East Lansing, MI 48824-1325, USA
e-mail: runkleer@msu.edu

17.2 The 8th International Symposium on Light in Horticulture

The 8th International Symposium on Light in Horticulture, held under the auspices of the International Society for Horticultural Science (ISHS), was held in East Lansing, Michigan (USA), in May 2016. Convened by Drs. Erik Runkle and Roberto Lopez from Michigan State University, it attracted over 250 participants from 25 countries around the world, including several chapter coauthors of this book. During the 5-day symposium, a diverse group of faculty, graduate students, and industry professionals delivered 52 oral and 78 poster presentations. The scientific program encompassed themes most relevant to current horticultural lighting research and application including light quality and optimization, lighting technologies, phytonutrients, and growth control, as well as supplemental, sole-source, and photoperiodic lighting. The event was sponsored by 26 leading lighting companies, horticultural companies, growth chamber manufacturers, and horticulture trade magazines. The proceedings of the symposium containing 56 peer-reviewed scientific articles have been published and are available at http://www.actahort.org/books/1134 (Currey et al. 2016).

17.3 New Horticultural Lighting Book

A book entitled *Light Management in Controlled Environments*, edited by Roberto Lopez and Erik Runkle, has been developed primarily for growers, technical industry representatives, and university horticulture students. The book is updated and substantially expanded from the original *Lighting Up Profits* book (Fisher and Runkle 2004). This book presents the underlying biology of how light influences plant growth and development of specialty crops, especially those grown in controlled environments such as greenhouses and indoor production facilities. It contains 18 chapters written by over 20 leading experts in North America on plant lighting. The lighting book is most applicable to floriculture crops (including bedding plants, herbaceous perennials, potted flowering plants, and cut flowers) and vegetables and leafy greens grown in protected environments.

17.4 Standards for Plant Lighting Applications

A diverse group of lighting industry professionals and academics in horticulture, primarily with engineering, lighting, and/or plant physiology expertise, is developing three documents that will serve as standards for plant lighting applications. Current lighting standards are based on photometry, which is the sensitivity of light by the human eye. Although there are similarities between light perception by humans and plants, there are also distinct differences. The need to develop plant-specific standards was established in 2014 during the American Society for Horticultural Science annual conference, when Erik Runkle of Michigan State University organized a roundtable discussion on the subject with >70 participants that included university professors and researchers, LED and lighting manufacturers, test labs representatives, and lighting application specialists. The standards for plant lighting are being developed by three working groups of the American Society of Agricultural and Biological Engineers. The primary objectives of three standards documents being developed are:

1. To define the metrics of radiation for plant growth applications in controlled environments
2. To establish the methods of measurements given the diversity of horticultural applications and the metrics defined above
3. To develop recommended energy-efficiency and performance metrics for plant lighting applications to supplement or replace metrics currently based on luminous efficacy used for general illumination

In 2015, the Japan Plant Factory Association (a nonprofit organization) established a committee on LED lighting for plant factories, chaired by Eiji Goto of Chiba University. The committee proposed the tentative standards of terminology and measurement methods of LED properties. The proposal was presented in September 2016 during the annual meeting of the Japanese Society of Agricultural, Biological and Environmental Engineers and Scientists.

17.5 Efficiency and Efficacy

Academics, lighting manufacturers, and users are already embracing some of the metrics being defined for plant applications. For example, an increasing number of lighting companies are now reporting the photon efficacy (or photosynthetic photon efficacy) of lamps, which refers to the efficacy of a lamp at converting electric energy into photosynthetic photons (Nelson and Bugbee 2014). Efficiency (or efficacy) is output divided by input. The output of the fixture is µmol s^{-1} of photosynthetic photons and the input to the fixture is watts of energy, which is equal to 1 joule (J) per second, so the efficacy metric becomes µmol/J. A higher number represents a more efficient lamp for plant applications in terms of photosynthesis.

Academic studies published in 2014 and 2016 determined that a 400 W high-pressure sodium (HPS) lamp with a magnetic ballast had an efficacy value (based only on output within the photosynthetically active waveband of 400–700 nm) of 0.94 µmol/J, while double-ended, 1000 W HPS lamps with electronic ballasts had efficacy values of 1.6 or 1.7 µmol/J (Nelson and Bugbee 2014; Wallace and Both 2016). Efficacy values for LED fixtures ranged substantially from 0.8 to 1.7 µmol/J. However, some LED manufacturers have recently reported efficacy values >2.0 µmol/J, and values between 2.0 and 2.1 have been validated based on reader comments by coauthor Bruce Bugbee (Nelson and Bugbee 2014). Efficacy values for LEDs will continue to increase for at least the next several years. Although the efficacy value of horticultural lighting fixtures is an important characteristic, other parameters such as durability, longevity, emission spectrum, cost, etc. should not be overlooked.

References

Currey CJ, Lopez RG, Runkle ES (2016) Proceedings of the VIII international symposium on light in horticulture (Acta Hortic. 1134). ISHS, Leuven, Belgium

Fisher P, Runkle E (2004) Lighting up profits: understanding greenhouse lighting. Meister Media Worldwide, Willoughby, Ohio, USA

Nelson JA, Bugbee B (2014) Economic analysis of greenhouse lighting: light emitting diodes vs. high intensity discharge fixtures. PLoS One 9(6):e99010

Wallace C, Both AJ (2016) Evaluating operating characteristics of light sources for horticultural applications. Acta Hortic 1134:435–444

Part V
Light-Quality Effects on Plant Physiology and Morphology

Chapter 18
Effect of Light Quality on Secondary Metabolite Production in Leafy Greens and Seedlings

Hiroshi Shimizu

Abstract It has been experimentally found that light stimulation significantly enhances the production of specific biomolecules in plants. The biosynthetic products mentioned herein are secondary metabolites, which are organic compounds that do not participate directly in cell growth or division; these include antioxidants, vitamins, sugars, and pigments. Several studies have investigated antioxidants in a variety of plants such as leaf lettuce, microgreen, sprout, and cherry tomatoes, and they have reported that the biosynthesis of chlorogenic acid, caffeic acid, chicory acid, flavonoids, rosmarinic acid, ascorbic acid, and carotenoids, along with antioxidant activity and DPPH radical scavenging capacity, is promoted by stimulating with light of a specific wavelength. It has been also reported that ascorbic acid, carotenoids, α-tocopherol, and ergosterol for antioxidants; sucrose, fructose, and glucose for sugars; and anthocyanin and chlorophyll for pigments are enhanced by light stimulus. Although the mechanism underlying the increase in the synthesis of these molecules by light stimulus remains unclear, the use of this technique is possible if reproducibility is ensured.

Keywords Antioxidant ability • Color development • Eating quality improvement • Light-emitting diode • Plant factory • Vitamins

18.1 Introduction

It is common knowledge that plant growth can be greatly influenced by environmental conditions. Light is a fundamental environmental factor affecting the growth of plants via processes including light morphogenesis and photosynthesis. While it is difficult to effectively control the light environment in greenhouses, as

---

H. Shimizu (✉)
Graduate School of Agriculture, Kyoto University, Oiwakecho, Kitashirakawa, Sakyo-ku, Kyoto 606-8502, Japan
e-mail: hshimizu@kais.kyoto-u.ac.jp

we cannot regulate solar light, plant factories with artificial light (PFAL) are able to control the entire suite of environmental factors, including the light environment.

A major purpose of PFAL has been year-round, pesticide-free production of vegetables. Recently, research has also focused on the production of vegetables containing certain useful compounds that cannot be cultivated in the open field. In addition, rather than edible vegetables, raw materials for the production of pharmaceutical products have been put to practical use. Plant secondary metabolites in plants are often a substance having a physiological activity such as chlorogenic acid with anticarcinogenic effect, and flavonoids increase the hormonal activity, and they have attracted attention as an effective component of many functional vegetables. In this article, recent research on the effects of the light environment on plant secondary metabolites will be introduced.

18.2 Antioxidant Ability

Total phenolic compounds are functional components that exhibit antioxidant activity. Many reports on the effects of light quality and intensity on the total phenolic compounds in plants have been published. These include several studies on the effects of blue and red lights, as provided by light-emitting diodes (LEDs), on functional compounds, with fewer studies examining the effects of green and yellow lights. To date, this research has examined the sprout, baby leaf, microgreen, and leafy lettuce of plants.

In a study of baby leaf, red leaf "Multired 4," green leaf "Multigreen 3," and light green leaf "Multiblond 2," baby leaf lettuce were grown under solar light and high-pressure sodium lamps (HPS) with supplementary LEDs (Samuoliene et al. 2012). Supplementary LEDs provided 30 µmol m^{-2} s^{-1} of photosynthetic photon flux density (PPFD) at 455 nm (blue), 470 nm (blue), 505 nm (green), and 530 nm (green), with 170 µmol m^{-2} s^{-1} PPFD provided by HPS. In total, HPS and LEDs provided 200 µmol m^{-2} s^{-1} with a 16-h day^{-1} light: 8-h day^{-1} dark photoperiod. PPFD of the reference control plot was also set at 200 µmol m^{-2} s^{-1}, provided solely by HPS lamps.

The antioxidant activity of baby leaf lettuce was dependent on the variety of lettuce and the light quality (Table 18.1). Total phenol concentration in red leaf "Multired 4" and green leaf "Multigreen 3" baby leaf lettuce significantly decreased under 590-nm and 470-nm LED light, respectively, whereas total phenol concentration increased in light green leaf "Multiblond 2" baby leaf lettuce under 590 nm. Although 2,2-diphenyl-1-picrylhydrazyl (DPPH) free-radical scavenging capacity was higher in all varieties of baby leaf lettuce than in control conditions, different wavelengths of LED light caused different effects per variety. The DPPH free-radical scavenging capacity significantly increased in red leaf baby lettuce with supplement of 470 nm LED light and in green and light green baby lettuce under supplementary 455 nm and 505 nm.

Table 18.1 Variation in antioxidant system under supplementary blue and green LED light

Treatment	Total phenols (mg g^{-1}, FM)	DPPH (μmol g^{-1}, FM)	Total anthocyanins (mg g^{-1}, FM)
Red leaf "Multired 4"			
HPS	1.23*	3.67**	31.9
HPS + 455 nm	1.14	4.06	42.6*
HPS + 470 nm	1.27*	5.43*	65.1*
HPS + 505 nm	1.15	4.31	37.4
HPS + 590 nm	0.99**	4.54	33.9
Green leaf "Multigreen 3"			
HPS	0.86**	7.78**	27.84*
HPS + 455 nm	1.16*	10.41*	7.21*
HPS + 470 nm	0.93**	9.05	12.27*
HPS + 505 nm	1.15*	9.90*	3.07**
HPS + 590 nm	1.00	9.29	1.83**
Light green leaf "Multiblond 2"			
HPS	0.63**	9.14**	ND
HPS + 455 nm	0.66	9.94*	ND
HPS + 470 nm	0.60**	9.19**	ND
HPS + 505 nm	0.62**	9.41*	ND
HPS + 590 nm	0.85[a]	8.80**	ND

Modified from Samuoliene et al. (2012)
FM fresh mass, *ND* not detected
Significant differences are denoted by asterisk – * under $P \leq 0.05$, ** above $P \leq 0.05$

The concentration of anthocyanin was significantly promoted with supplementary 455-nm and 470-nm LED light in red baby leaf lettuce, though it decreased with 505-nm and 550-nm LED supplementary lighting.

Effects of the spectrum of light on compounds produced by plants have been studied in the red leaf lettuce "Banchu Red Fire." The effect of raising seedlings under different light quality treatments was determined by Johkan et al. (2010). After transplant, seedlings were cultivated under solar light with supplemental fluorescent lamps (FL), and the influence of light quality on the raising stage of the harvest plant was studied. Lettuce seeds were germinated and grown under FL for 10 days. At 10 days after sowing (DAS), seedlings were transplanted into either

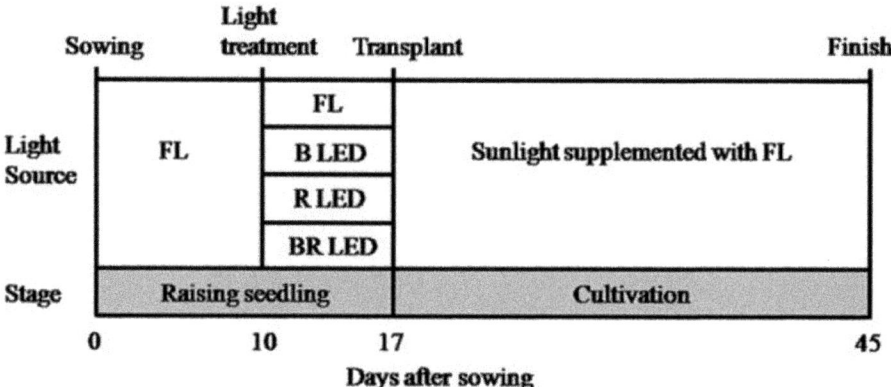

Fig. 18.1 Schematic of light treatment applied. *FL* fluorescent lamp, *B* blue, *R* red, *BR* blue + red (Adapted from Johkan et al. 2010)

conditions of FL, blue (468 nm), red (660 nm), and blue plus red light (467 and 655 nm) for the nursery stage. At 17 DAS, seedlings were transplanted to a greenhouse under solar light with supplementary FL from 1600 to 1900 h (Fig. 18.1).

Significant differences in total polyphenols, chlorogenic acid, and TAS (total antioxidant system) were apparent in relation to light quality at 17 DAS, the final day of the nursery stage under variable light quality treatments (Table 18.2). However, differences were not maintained post seedling stage.

The effects of different blue/red light ratios across the entire growth period on morphological changes, growth characteristics, and accumulation of antioxidant phenolic compounds in two lettuce cultivars have been reported (Son and Oh 2013). Seeds of "Sunmang" (red leaf) and "Grand Rapid TBR" (green leaf) were germinated and grown under a combination of FL and HPS for 18 days. Seedlings were transplanted into six light quality treatments with varying proportions of blue and red LEDs (0B/100R, 13B/87R, 26B/74R, 35B/65R, 47B/53R, and 59B/41R) at 171-µmol m^{-2} s^{-1} PPFD, and grown for 4 weeks before measurement of antioxidant activity.

Total phenolic concentration significantly increased in "Sunmang" lettuce under 47B/53R conditions to levels 1.4 and 2.4 times those under control conditions (0B/100R). The total phenolic concentration in "Grand Rapid TBR" under 26B and 59B was 2.2- and 2.7-fold that observed in the control treatment, respectively (Fig. 18.2). Thus, the LED ratio of blue/red light was found to significantly alter the total phenolic concentration in these cultivars, with enhanced phenol accumulation apparent with increases in blue wavelengths of light.

The response of antioxidant activity to light quality treatments was comparable to that of total phenol concentrations. In "Sunmang" lettuce, antioxidant production was promoted by an increase in the ratio of blue/red LEDs and became maximal under 47B/53R. "Grand Rapid TBR" also showed a high antioxidant capacity under blue-rich light environments. However, no significant difference was observed in

Table 18.2 Effects of LED light quality treatments on phenols, chlorogenic acid, and TAS concentrations in red leaf lettuce

DAS[z]	Spectrum[y]	Total phenols[x] (nmol mg^{-1} DMW)	Chlorogenic acid (nmol mg^{-1} DMW)	TAS[w] (nmol mg^{-1} DMW)
10	FL[v]	55.4	2.5	278
17	FL	75.7^c	6.2^b	332^c
	Blue (470 nm)	118.7^b	15.0^a	547^b
	Red (660 nm)	47.0^d	1.9^c	194^d
	Blue + red (470 + 660 nm)	161.6^a	14.7^a	749^a
45	FL	138.8^a	8.1^a	376^a
	Blue (470 nm)	126.8^a	6.7^a	424^a
	Red (660 nm)	139.9^a	8.7^a	421^a
	Blue + red (470 + 660 nm)	138.3^a	9.1^a	481^a

Modified from Johkan et al. (2010)
Different letters indicate significant difference
$P < 0.05$; Tukey's multiple range test, $n = 6$
[z]Days after sowing. 17 DAS = raising seedlings; 45 DAS = final cultivation
[y]Photosynthetic photon flux 100 ± 10 μmol m^{-2} s^{-1}
[x]Chlorogenic acid equivalent
[w]Total antioxidant system, Trolox equivalent
[v]White fluorescent lamp

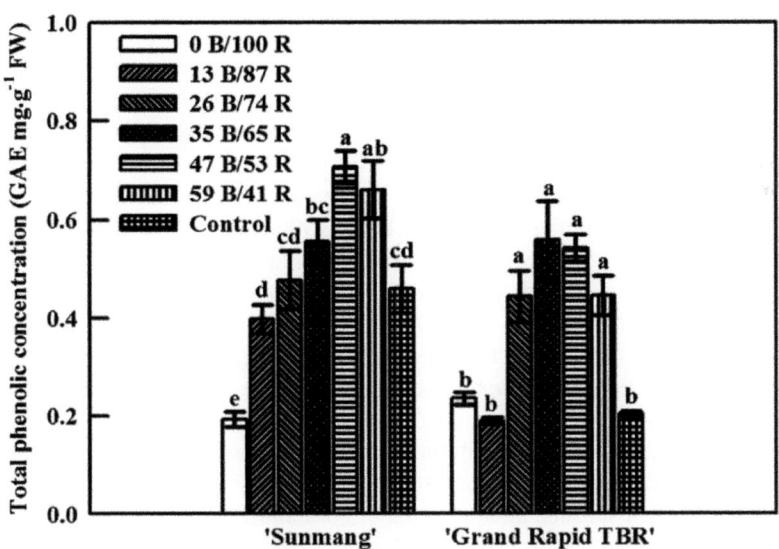

Fig. 18.2 Total phenolic concentration of lettuce grown under various combinations of blue and red LED light treatments after 4 weeks of growth (Adapted from Son and Oh 2013)

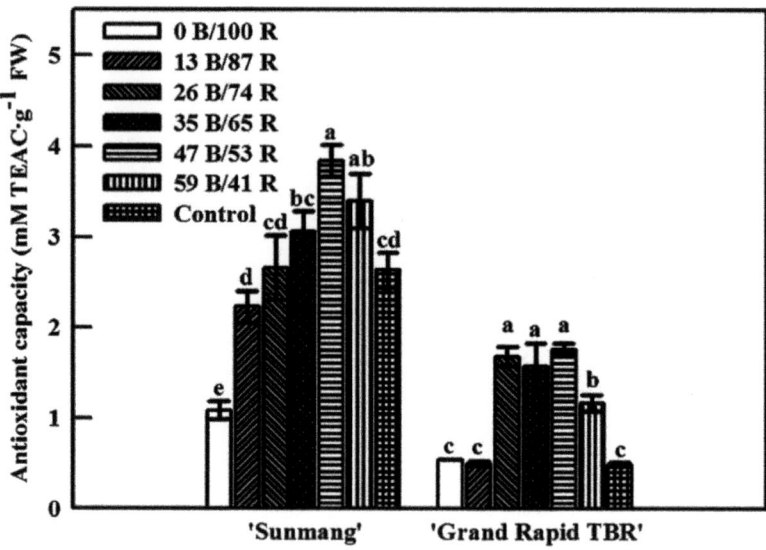

Fig. 18.3 Antioxidant capacity of lettuce grown under various combinations of blue and red LED light treatments after 4 weeks of growth (Adapted from Son and Oh 2013)

the antioxidant capacity of "Grand Rapid TBR" grown under light quality treatments with greater than 26 % blue light (Fig. 18.3). Comparing the two lettuce cultivars, "Sunmang" lettuce had greater antioxidant concentrations than "Grand Rapid TBR" under all light quality treatments, indicating species-specific responses.

In addition to stimulation by blue and red wavelengths, there are also reports of stimulation by yellow wavelengths (596 nm). For example, Urbonaviciute et al. (2009) investigated the total synthesis of an antioxidant in a sprout grown under a combination of HPS and flashing yellow LEDs (596 nm). Yellow LEDs pulsed at a frequency of 2.9 Hz (250-ms "on"; 100-ms "off") at a PPFD of 35 µmol m^{-2} s^{-1}. The intensity of the pulsed light was approximately 50 % of the total PPFD provided, with an 18-h photoperiod (Urbonaviciute et al. 2009).

Radical-binding activity increased approximately 1.5-fold due to the effect of flashing yellow LEDs on radish sprouts (Fig. 18.4). Similar trends were also apparent for the phenol component. Urbonaviciute et al. (2009) assumed such reactions were the result of photoinduced stress, i.e., the stimulation of the antioxidant activity of natural defense mechanisms mainly against damage by light oxidation. In radish sprouts, an approximate 30 % increase in phenolic compounds was observed under conditions of supplemental flashing yellow light. It is most likely that the formation of new pigments in the photosynthetic apparatus of young leaves, and the rapid adaptive response, including the synthesis of phenolic compounds, occurred as a reaction to the light quality treatment conditions (Fig. 18.5).

Studies on the biosynthesis of antioxidants in relation to light quality in plants other than leafy vegetables have also been reported. For example, the seedlings of

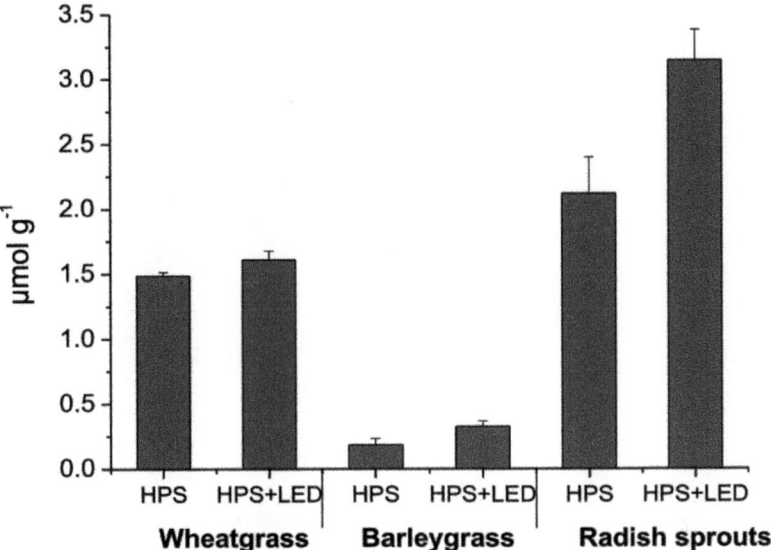

Fig. 18.4 Antioxidant activity of the extracts, as the ability to bind DPPH free radicals, in green sprouts grown solely under HPS lamps and with supplementation by flashing amber LED light (Adapted from Urbonaviciute et al. 2009)

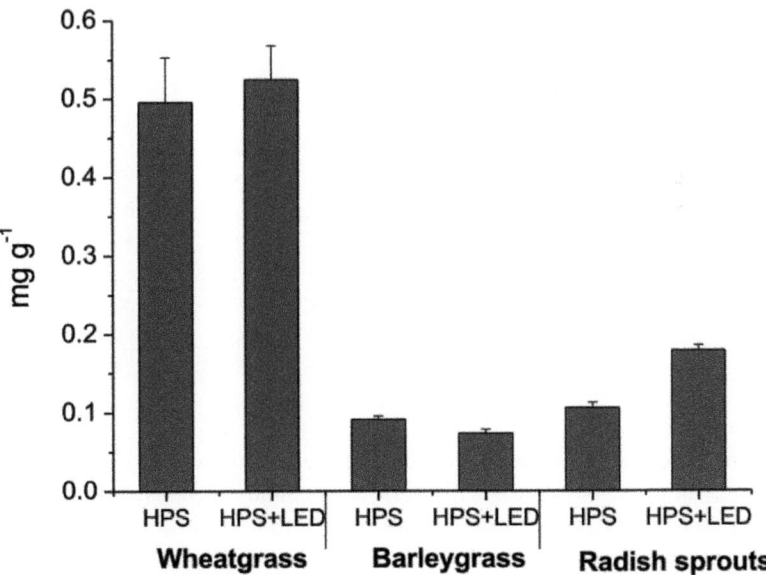

Fig. 18.5 Concentration of phenolic compounds in the fresh matter of green sprouts grown solely under HPS lamps and with supplementation by flashing amber LED light (Adapted from Urbonaviciute et al. 2009)

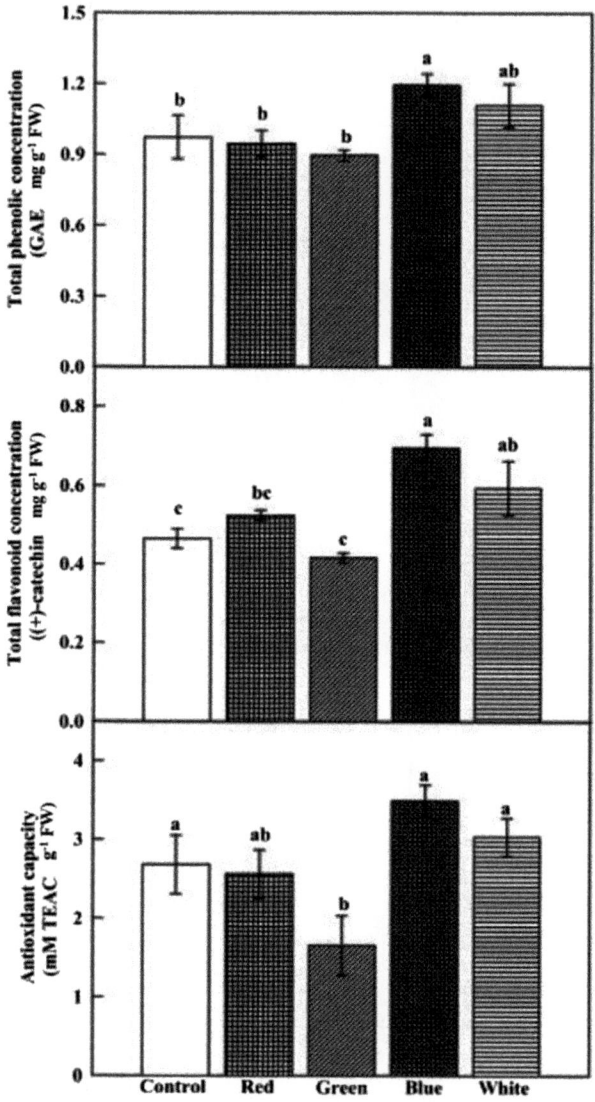

Fig. 18.6 Total phenolic concentration, total flavonoid concentration, and antioxidant capacity of cherry tomato seedlings (Adapted from Kim et al. 2014)

cherry tomatoes were grown under red, green, blue, and white LEDs and a fluorescent lamp, and the antioxidant phenolic compounds were analyzed (Kim et al. 2014). Total phenol concentration, total flavonoid concentration, and antioxidant capacity in seedlings grown under blue LEDs were significantly higher compared with those grown under other light quality treatments (red, green, and FL). In addition, red LEDs induced increases in these parameters in seedlings as compared to those grown under green LEDs and control conditions (FL) (Fig. 18.6).

In addition, Kim et al. (2013) analyzed the activity of the total phenolic concentration and antioxidant enzymes to examine the effect of light quality (white,

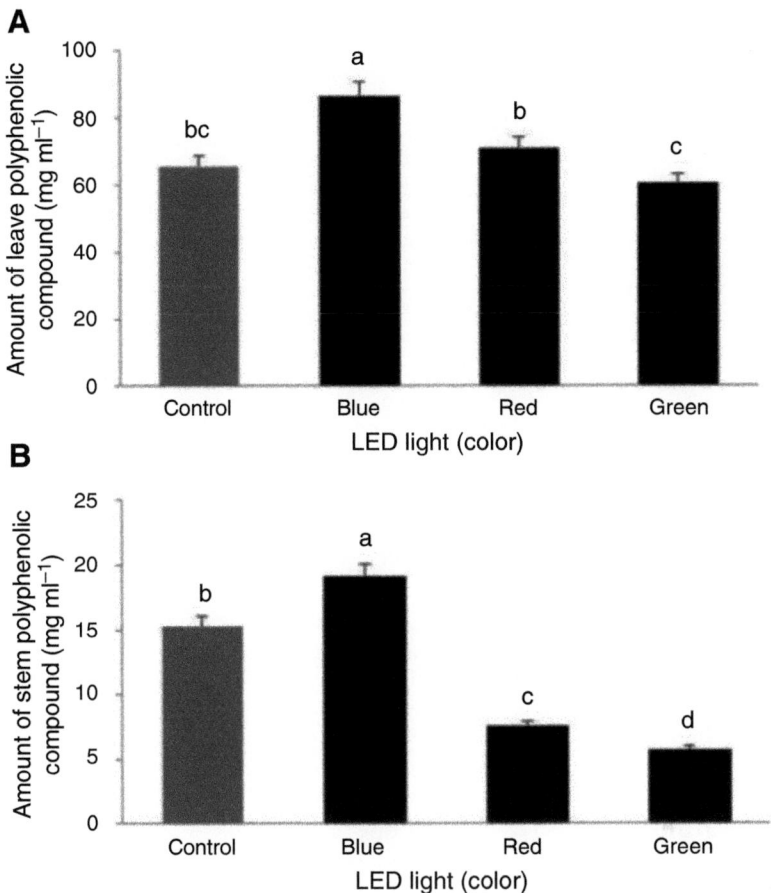

Fig. 18.7 The concentration of total phenolic compounds in the leaves and stems of tomatoes grown for 21 days under LED light treatments (Adapted from Kim et al. 2013)

blue, red, and green LEDs) on tomatoes (cv. Toy-mini tomato). When tomato seedlings were grown under blue LEDs, the concentration of total phenolic compounds was significantly in the leaves (1.3-fold) and the stem (1.2-fold) than in those of tomatoes grown under white LEDs (Fig. 18.7). In contrast, the concentration of total phenolic compounds in the leaves of tomatoes grown under red and green LEDs showed no difference to those grown under white LEDs. These findings are in line with the results of Kim et al. (2014).

Table 18.3 Variation in vitamin C under supplementary blue and green LED light quality treatments

Treatment	Vitamin C (mg g^{-1}, FM)
Red leaf "Multired 4"	
HPS	2.91*
HPS + 455 nm	1.45**
HPS + 470 nm	1.31**
HPS + 505 nm	1.83
HPS + 590 nm	1.95
Green leaf "Multigreen 3"	
HPS	0.29*
HPS + 455 nm	0.27
HPS + 470 nm	0.23**
HPS + 505 nm	0.29*
HPS + 590 nm	0.29*
Light green leaf "Multiblond 2"	
HPS	0.23
HPS + 455 nm	0.22
HPS + 470 nm	0.23
HPS + 505 nm	0.24
HPS + 590 nm	0.24

Modified from Samuoliene et al. (2012)
FM fresh mass, *ND* not detected
Significant differences are denoted by asterisk – * under $P \leq 0.05$; ** above $P \leq 0.05$

18.3 Vitamins

With respect to vitamins, the effects of light quality have been reported for ascorbic acid, carotenoids, α-tocopherol, and ergosterol. Experiments on baby leaf lettuce (red leaf "Multired 4," green leaf "Multigreen 3," light green leaf "Multiblond 2" baby leaf lettuce) were performed by Samuoliene et al. (2012), using supplementary LEDs (455, 470, 505, and 530 nm, 30-μmol m^{-2} s^{-1} PPFD) and HPS (170-μmol m^{-2} s^{-1} PPFD). The ascorbic acid concentration of red leaf lettuce was reduced under blue and green supplementary LED treatments approximately 2-fold and 1.5-fold, respectively (Table 18.3).

The effects of light stimulation on vitamins in microgreens have also been investigated (Samuoliene et al. 2013). Microgreen lettuces (kohlrabi [*Brassica oleracea* var. *gongylodes*, "Delicacy Purple"], mustard [*Brassica juncea* L., "Red Lion"], red pak choi [*Brassica chinensis*, "Rubi F1"], and tatsoi [*Brassica rapa* var. *rosularis*]) were grown under five light intensities (545-, 440-, 330-, 220-, and 110-μmol m^{-2} s^{-1} PPFD) with a mixture of light qualities using four different LEDs (455 nm [blue], 638 nm [red], 660 nm [red], and 735 nm [far red]). Ascorbic acid in the red pak choi and tatsoi showed high values under the low PPFD of 110-μmol m^{-2} s^{-1} PPFD, which was 3.8 and 3.5 times those under normal 220-μmol m^{-2} s^{-1} PPFD, respectively (Table 18.4). Given that the experiments

Table 18.4 Combinations of LED photosynthetic photon flux density

%	Total PPFD ($\mu mol\ m^{-2}\ s^{-1}$)	Blue 445 nm	Red 638 nm	Red 665 nm	Far red 735 nm
100	545	41	225	275	4
80	440	34	181	221.5	3.5
60	330	25	136	166	3
40	220	17	90	111	2
20	110	8	45	55	1

Modified from Samuoliene et al. (2013)

with baby leaf used LEDs as a supplemental light additional to HPS, it is not possible to compare the results directly; the higher investigated PPFD level had uneven effect on ascorbic acid accumulation. Lower PPFD promoted the biosynthesis of ascorbic acid during this experiment.

In addition, α-tocopherol concentration was varied by PPFD, and significant accumulation of α-tocopherol was observed under minimum PPFD in mustard, red pak choi, and kohlrabi, and significant increase was occurred in tatsoi under 220- and 110-$\mu mol\ m^{-2}\ s^{-1}$ PPFD. The accumulation of α-tocopherol and ascorbic acid under high PPFD was not consistent. In mustard plants grown under 545-$\mu mol\ m^{-2}\ s^{-1}$ PPFD, α-tocopherol concentration increased to 1.6 times values observed in plants grown under 220-$\mu mol\ m^{-2}\ s^{-1}$ PPFD. In contrast, α-tocopherol concentration in the red pak choi and tatsoi grown under 545-$\mu mol\ m^{-2}\ s^{-1}$ PPFD was significantly lower as compared to plants grown under normal PPFD of 220 $\mu mol\ m^{-2}\ s^{-1}$ (Fig. 18.8).

Li and Kubota (2009) examined the influence of supplementary UV-A, blue, green, red, and far-red light, with FL as the main light source, on phytochemicals (total anthocyanins, carotenoids, chlorophyll, total phenolic compounds, and ascorbic acid) and growth (biomass, the stem length varies the leaves under the amount of auxiliary light, leaf length, and leaf width) in leafy lettuce (*Lactuca sativa* L. "Red Cross") (Table 18.5).

The concentrations of carotenoids in lettuce leaves were affected by light quality, with xanthophylls and β-carotene increasing by 6–8 % in blue auxiliary light, though they reduced by 12–16 % under far-red light (Table 18.6). Xanthophylls have an absorption peak at 446 nm in the visible region and are produced in order to protect plants growing under high-energy blue wavelengths of light. At present, however, the mechanism of β-carotene increase remains unknown. Carotenoid and chlorophyll concentrations reduced by 12–16 % with supplementary far-red light.

Jang et al. (2013) investigated the effect of light quality (darkness, FL, blue, green, yellow, and red LEDs) on growth of the mushroom (Bunashimeji). Results demonstrated an increase in ergosterol concentration to a maximum under blue LEDs, with a minimum under red LEDs, indicating that both PPFD and quality (spectrum) had an effect on the biosynthesis of vitamins (Fig. 18.9).

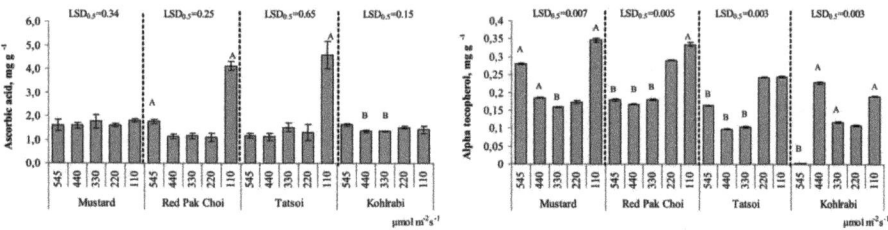

Fig. 18.8 Changes of ascorbic acid and α-tocopherol of microgreens grown under different light quality and intensity treatments (Modified from Samuoliene et al. 2013)

18.4 Eating Quality Improvement

Sugar concentration is closely related to the taste of vegetables, with several studies reporting that the light environment also influences sugar concentration. Nutritional values of herbs (white mustard [*Sinapis alba* "Yellow mustard"], spinach [*Spinacia oleracea* "Geant d'hiver"], rocket [*Eruca sativa* "Rucola"], dill [*Anethum graveolens* "Mammoth"], parsley [*Petroselinum crispum* "Plain Leaved"], green onion [*Allium cepa* "White Lisbon"]) grown using red LEDs as supplementary light were evaluated (Bliznikas and Zulauskas 2012). Natural light (average 300 μmol m^{-2} s^{-1} PPFD) was the main light source in this experiment. When natural PPFD was low, supplementary light was provided by HPS with a photoperiod of 12-h day^{-1} and intensity of 130-μmol m^{-2} s^{-1} PPFD. At the preharvest stage of 3 days, plants were provided with supplementary illumination by red (638 nm) LEDs from 0500 h to 1000 h and 1700 h to 0000 h.

The response to red LED light differed in each variety of plant. Marked increase in ascorbic acid concentration was observed in vegetables under red LED treatment. Although there was a significant accumulation of fructose and glucose with supplementary red LED treatment, the accumulation of sucrose showed a different effect. In dill and parsley grown under supplementary red light, a significant increase in monosaccharides (especially fructose and glucose) and a reduction of nitric acid concentration were observed (Table 18.7). No significant increase in carbohydrates was observed in vegetables, while nitrate concentrations either increased (mustard, rocket, and onion leaves) or remained unchanged (spinach).

Samuoliene et al. (2012) investigated the effects of PPFD on sucrose concentration in microgreens. A mixture of spectral irradiance was provided by LEDs with different peak wavelengths (455 [blue], 638 [red], 660 [red], and 735 nm [far-red]), and experiments were conducted using different light intensities (545-, 440-, 330-, 220-, and 110-μmol m^{-2} s^{-1} PPFD), as previously described.

Sucrose accumulation in microgreen leaves was shown to be related to PPFD, with a tendency for decreasing sucrose concentration with decreasing PPFD observed (Table 18.8). However, the PPFD resulting in maximal sucrose concentrations differed between microgreen varieties. Sucrose synthesis in mustard plants decreased with a decrease in PPFD, and the sucrose concentration of kohlrabi was

Table 18.5 Spectral qualities tested for the white (W), the white light with supplementary UV-A (WUA), the white light with supplementary blue (WB), the white light with supplementary green (WG), the white light with supplementary red (WR), and the white light with supplementary far-red (WFR) light

Parameter	Treatment					
	W	WUV	WB	WG	WR	WFR
Photon flux ($\mu mol\ m^{-2}\ s^{-1}$)[a]						
UV-A (350–400 nm)	4.6 ± 0.3	20.9 ± 1.1	2.5 ± 0.4	2.1 ± 0.3	2.2 ± 0.3	4.4 ± 0.3
Blue (400–500 nm)	70.5 ± 3.0 (23.2 %)	70.3 ± 3.1 (23.1 %)	166.9 ± 16.0 (55.1 %)	48.6 ± 3.2 (16.3 %)	38.5 ± 3.5 (12.6 %)	69.5 ± 3.1 (22.8 %)
Green (500–600 nm)	159.1 ± 7.0 (52.3 %)	157.5 ± 6.8 (51.9 %)	95.2 ± 6.7 (31.4 %)	211.1 ± 13.0 (70.1 %)	89.5 ± 7.2 (29.5 %)	156.5 ± 7.1 (51.4 %)
Red (600–700 nm)	75.9 ± 5.6 (24.5 %)	76.8 ± 4.6 (24.9)	41.5 ± 3.7 (13.5 %)	41.5 ± 3.8 (13.6 %)	177.2 ± 12.8 (57.9 %)	79.7 ± 5.3 (25.8 %)
Far red (700–800 nm)	6.6 ± 0.4	6.7 ± 0.3	4.1 ± 0.2	3.9 ± 0.2	4.6 ± 0.3	160.4 ± 8.8
PPF (400–700 nm)	305.4 ± 10.0 (100 %)	304.6 ± 8.2 (100 %)	303.5 ± 7.4 (100 %)	301 ± 8.3 (100 %)	305.2 ± 8.8 (100 %)	305.7 ± 9.3 (100 %)
Ratios						
Red/far red	11.5	11.4	10.2	10.7	38.5	0.5
P_{fr}/P_{total}[b]	0.84	0.83	0.81	0.84	0.87	0.56

Modified from Li and Kubota (2009)

[a]Total photon flux of the background white fluorescent lamp and the supplemental LEDs

[b]P_{fr}/P_{total}: phytochrome photostationary state

Table 18.6 Xanthophyll and β-carotene and ascorbic acid concentration in lettuce under various light quality treatments

Treatment	Xanthophylls (mg g^{-1} DW)	β-Carotene (mg g^{-1} DW)	Ascorbic acid (mg g^{-1} DW)
Experiment 1			
W1	0.49 a	0.25 a	2.32 a
WUV	0.50 a	0.25 a	2.42 a
WR	0.47 ab	0.23 ab	2.36 a
WFR	0.43 b	0.21 b	2.27 a
Experiment 2			
W2	0.52 b	0.26 b	2.19 a
WB	0.55 a	0.28 a	2.34 a
WG	0.51 b	0.26 b	2.07 a

Modified from Li and Kubota (2009)
p-value from one-way ANOVA

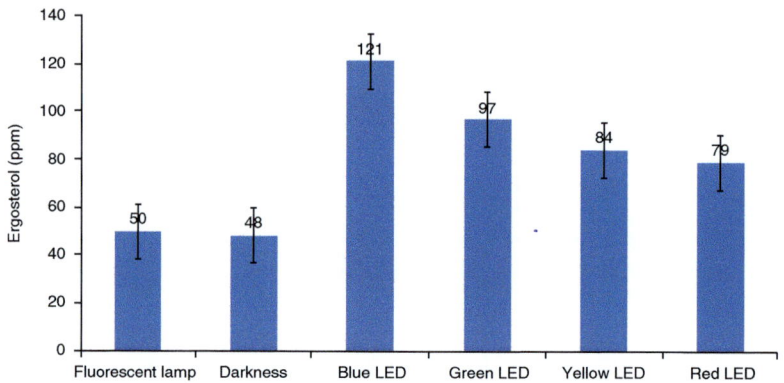

Fig. 18.9 Ergosterol concentration of *Hypsizygus marmoreus* under different LED lights (Adapted from Jang et al. 2013)

maximal at 545-µmol m^{-2} s^{-1} PPFD. In contrast, sucrose concentration in red pak choi peaked at 440-µmol m^{-2} s^{-1} PPFD and that in Tatsoi was maximal at 330-µmol m^{-2} s^{-1} PPFD.

In studies concerning the effects of light quality on growth and synthesis of important compounds by plants, several experiments with blue and red light, the spectra preferentially absorbed by chlorophyll and photoreceptors, have been conducted. An experiment conducted with yellow spectra and sprout species (green sprouts of wheatgrass [*Triticum aestivum* L. "Sirvinta"], barley grass [*Hordeum vulgare* L. "Luoke"], and leafy radish [*Raphanus sativus* L. "Tamina"]) has also been reported. Urbonaviciute et al. (2009) analyzed sugar concentrations of sprouts grown under yellow flashing light (596 nm, 2.9 Hz

Table 18.7 Nutritional properties of green vegetables after 3 days of red LED treatment before harvesting

	Vitamin C (mg g^{-1} FW)		Nitrate (mg kg^{-1} FW)	
	HPS	HPS + LED	HPS	HPS + LED
Mustard	0.46 ± 0.011	0.58 ± 0.057	3011 ± 160.6	5361 ± 363.6
Spinach	0.34 ± 0.017	0.41 ± 0.044	2989 ± 127.3	2783 ± 102.0
Rocket	0.35 ± 0.032	0.39 ± 0.007	4120 ± 188.8	4491 ± 80.4
Dill	0.52 ± 0.008	0.87 ± 0.016	4879 ± 80.9	4313 ± 154.5
Parsley	0.53 ± 0.021	0.43 ± 0.015	4486 ± 408.5	2675 ± 35.3
Onion leaves	0.33 ± 0.012	0.35 ± 0.017	2142 ± 26.1	2650 ± 33.2
	Carbohydrates					
	Fructose (mg g^{-1} FW)		Glucose (mg g^{-1} FW)		Sucrose (mg g^{-1} FW)	
	HPS	HPS + LED	HPS	HPS + LED	HPS	HPS + LED
Mustard	–	0.33 ± 0.027	–	0.22 ± 0.011	0.03 ± 0.07	0.58 ± 0.068
Spinach	–	0.92 ± 0.047	–	0.98 ± 0.003	0.06 ± 0.53	–
Rocket	0.26 ± 0.034	1.61 ± 0.316	0.61 ± 0.016	3.76 ± 0.084	–	0.95 ± 0.002
Dill	4.11 ± 0.360	10.37 ± 0.141	2.48 ± 0.610	13.82 ± 2.177	–	9.16 ± 0.092
Parsley	3.25 ± 0.103	3.32 ± 0.150	3.34 ± 0.375	7.30 ± 0.266	4.04 ± 0.113	16.12 ± 0.050
Onion leaves	1.65 ± 0.262	3.06 ± 0.608	1.59 ± 0.019	2.86 ± 0.101	2.51 ± 0.033	2.36 ± 0.236

Modified from Bliznikas et al. (2012)

Table 18.8 Sucrose concentration of microgreens grown under different irradiation levels

	Sucrose (mg g^{-1})			
PPFD (μmol m^{-2} s^{-1})	Mustard	Red Pak Choi	Tatsoi	Kohlrabi
545	0.83 ± 0.04	0.62 ± 0.05	0.92 ± 0.15	2.62 ± 0.33*
440	0.81 ± 0.41	3.47 ± 0.29*	0.91 ± 0.30	0.77 ± 0.07*
330	0.65 ± 0.07	0.94 ± 0.25*	5.71 ± 0.28*	0.35 ± 0.07
220	0.53 ± 0.19	0.58 ± 0.06	0.60 ± 0.17	0.23 ± 0.10
110	0.34 ± 0.09	0.61 ± 0.26	0.71 ± 0.11	0.19 ± 0.05

Modified from Samuoliene et al. (2013)
*Values are significantly ($P \leq 0.05$) higher than normal 220 μmol m^{-2} s^{-1} irradiance level

[250-ms "on" and 100-ms "off"]) supplementary to the main HPS light source. An 18-h photoperiod was employed, and the intensity of the pulsed light was approximately 50 % of the total PPFD, which was approximately 35-μmol m^{-2} s^{-1} PPFD.

Glucose concentrations in barley young leaves and radish sprouts were 2–2.5 times greater than concentrations recorded in control plots, and a slight decrease in the maltose (malt sugar) concentration in young barley leaves was observed. Although yellow light is said to inhibit the formation of chlorophyll and chloroplasts, given the monosaccharide concentration of sprouts recorded during this experiment, such an inhibitory effect was not apparent (Fig. 18.10).

The proline concentration of tomatoes (cv. Toy-mini tomato) grown under five different light sources (white LED [420–680 nm] as control, blue LED [460 nm], red LED [635 nm], and green LED [520 nm]) has been investigated. Light quality

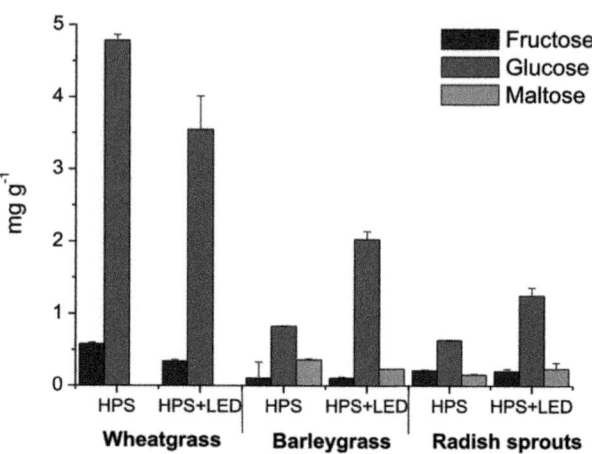

Fig. 18.10 Saccharide concentration in green sprouts grown under solely HPS lamps and with supplementation by flashing amber LED light (Adapted from Urbonaviciute et al. 2009)

had a significant effect on the proline concentration of tomato seedlings, with proline concentration 296 % in leaves and 127 % in stems grown under blue LEDs as compared to white LEDs (Kim et al. 2013).

Chicory acid related to eating quality since has a bitter taste. Ouzounis et al. (2015) investigated this secondary metabolite in two varieties of leafy lettuce ("Batavia" and "Lollo Rossa"). Lettuces were grown under supplementary blue light (45-µmol m^{-2} s^{-1} PPFD, with a variety of irradiation durations), with the main light source provided by a combination of natural light and HPS. Chicory acid concentration significantly increased with an irradiation regime of blue light from 0600 to 0800 h and 2100 to 0800 h, suggesting that an effective time period of irradiation may exist.

Sirtautas et al. (2014) examined changes in the saccharide concentrations of baby leaf lettuce "Multigreen 3." During experiments, the main light source was provided by a combination of approximately 80–120-µmol m^{-2} s^{-1} PPFD solar light and 90-µmol m^{-2} s^{-1} PPFD HPS, with approximately 15-µmol m^{-2} s^{-1} PPFD of supplementary light provided by blue (455 and 470 nm), cyan (505 nm), and green (535 nm) LEDs. The concentration of water-soluble monosaccharides in lettuce grown under blue light (455 nm) was higher than in other plots, with sucrose, glucose, mannose, and fructose concentrations 6, 2.9, 3.5, and 1.5 times those of control plots, respectively (Table 18.9). Despite 470 nm of blue wavelengths close to the 455 nm, the noticeable effect of 455 nm was not observed. The effects of sucrose, glucose, and mannose were 1.8, 2.3, and 3.3 times, respectively. In other words, a shift of the little peak wavelength in the same blue caused remarkable difference of the effect on the biosynthesis of sucrose, glucose, mannose, and fructose. Green light (535 nm) had no effect on sucrose biosynthesis. However, the concentration of glucose and mannose was higher under green (535 nm) than blue light (455 nm). In contrast, only a small effect of cyan light (505 nm) was apparent on any monosaccharide concentration.

Table 18.9 Soluble saccharide concentrations (mg g^{-1} fresh mass) in baby leaf lettuces grown under HPS and supplementary LED light quality treatments

	Soluble saccharides			
	Sucrose	Glucose	Mannose	Fructose
HPS	0.69 ± 0.03	2.49 ± 0.04	1.20 ± 0.03	0.23 ± 0.04
HPS + 455 nm	4.18 ± 0.04*	7.32 ± 0.02*	4.33 ± 0.04*	0.36 ± 0.03*
HPS + 470 nm	1.22 ± 0.05*	5.70 ± 0.03*	3.96 ± 0.03*	0.28 ± 0.01*
HPS + 505 nm	1.08 ± 0.04*	3.47 ± 0.04*	2.93 ± 0.05*	0.28 ± 0.03*
HPS + 535 nm	0.77 ± 0.03	7.50 ± 0.04*	5.96 ± 0.05*	0.29 ± 0.04*

Modified from Sirtautas et al. (2014)
*Values in each column are significantly different ($P \leq 0.05$) from HPS by Fisher's LSD test

18.5 Color Development

The color of vegetables is related to their nutritional value, and research has demonstrated that color is also controlled by light quality. Seeds of leafy lettuce were germinated and grown under FL for 10 days, after which seedlings were raised under different light qualities until 17 DAS. Following the raising stage, seedlings were transplanted to a greenhouse under solar light with supplementary FL and grown until 45 DAS (Johkan et al. 2010).

Anthocyanin concentration at 17 DAS in lettuce seedlings treated with blue light was higher than in seedlings at 10 DAS cultivated under FL (Table 18.10). Blue-containing LED lights significantly increased the anthocyanin concentration in lettuce seedlings, and maximal anthocyanin concentration was observed in seedlings raised under a combination of blue and red light. However, the anthocyanin concentration in lettuce at 45 DAS was lower than in seedlings at 17 DAS. The anthocyanin concentration at 17 DAS in lettuce seedlings grown under red light was significantly lower than in those raised under FL. However, the concentration at 45 DAS reduced by half. Overall, no significant difference was observed in the anthocyanin concentration of seedlings grown under any light quality treatment at 45 DAS relative to FL controls, indicating that the effects of light quality observed at 17 DAS were not retained.

Li and Kubota (2009) examined the color responses of red leafy lettuce (*Lactuca sativa* L. cv. "Red Cross") grown with FL as the main light source and supplementary UV-A (373 nm), blue (476 nm), green (526 nm), red (658 nm), and far-red (734 nm) light provided by LEDs, across the entire period of growth. Anthocyanin concentration of lettuce grown with supplementary UV-A and blue light increased by 11 % and 13 %, respectively, though decreased by 40 % under far-red illumination. Carotenoids (xanthophylls and β-carotene) increased 6–8 % with supplementary blue light, but were reduced by 12–16 % under far-red light.

Chlorophyll concentration was reduced by 12 % in lettuce grown under supplementary far-red light, and phenol concentration increased by 6 % in the red light quality treatment compared to controls (Table 18.11). Previously, Ninu et al. (1999) demonstrated that light quality had a strong influence on anthocyanin concentration, with blue light indicated as one of the most effective wavelengths for

Table 18.10 Effects of LED light quality treatments on the concentration of anthocyanin in red leaf lettuce

DAS[z]	Spectrum	Anthocyanin (OD530 mg^{-1} DMW)
10	FL	0.07
17	FL	0.08 c
	Blue (470 nm)	0.15 b
	Red (660 nm)	0.06 d
	Blue + red (470 + 660 nm)	0.27 a
45	FL	0.11 a
	Blue (470 nm)	0.14 a
	Red (660 nm)	0.12 a
	Blue + red (470 + 660 nm)	0.13 a

Modified from Johkan et al. (2010)
Different letters indicate significant difference
$P \leq 0.05$; Tukey's multiple range test, $n = 6$
[z]Days after sowing. 17 DAS = rising seedlings, 45 DAS = final cultivation
[v]White fluorescent lamp
[y]Photosynthetic photon flux was 100 ± 10 μmol m^{-2} s^{-1} for all light treatments

Table 18.11 Total anthocyanin and chlorophyll concentration in red leaf lettuce (For light treatments in the table, see Table 18.5)

Treatment	Anthocyanins (mg g^{-1} DW)	Chlorophyll (mg g^{-1} DW)
Experiment 1		
W1	3.31b	0.51a
WUV	3.68a	0.53a
WR	3.47ab	0.47a
WFR	1.97c	0.45a
Experiment 2		
W2	3.20b	0.50a
WB	4.18a	0.53a
WG	2.95b	0.54a

Modified from Li and Kubota (2009)
p-value from one-way ANOVA

controlling anthocyanin biosynthesis in tomatoes due to the response of cryptochrome blue/UV light receptors.

In addition to research conducted with red leaf vegetables, studies have also performed experiments using green cabbages. The pigmentation in green and red varieties of cabbage (*Brassica oleracea* var. *capitata* L. "Kinshun" [green leaf type] and "Red Rookie" [red leaf]) was examined under different monochromic LEDs (blue [470 nm], blue green [500 nm], green [525 nm], and red [660 nm]) (Mizuno et al. 2011).

In Red Rookie, leaf anthocyanin concentration was greater than or equal to 0.6 g m^{-2} leaf in all light quality treatments, with the greatest concentrations observed in cabbage grown in the red light quality treatment (Fig. 18.11).

In general, chalcone synthase, a precursor of anthocyanin biosynthesis, was reduced by blue light or UV-B radiation. In contrast, there is little evidence to suggest that red light has the ability to promote anthocyanin production. Miura and Iwata (1981) reported that the anthocyanin concentration of *Polygonum hydropiper* was higher when grown in red light than when grown at other wavelengths. Phytochrome is a red light receptor that may have a role in the anthocyanin biosynthesis of Red Rookie cabbage.

In a study of fruit and vegetable seedlings, the chlorophyll concentration of the tomato hybrid "Magnus" F1, sweet pepper variety "Reda," and cucumber hybrid "Mirabelle" F1 was evaluated under supplementary LED light at 455, 470, 505, and 530 nm, with solar light and HPS as the main light source. PPFD in experimental plots was daylight plus 15-µmol m^{-2} s^{-1} PPFD LEDs and 90-µmol m^{-2} s^{-1} PPFD HPS, with a photoperiod of 18 h. The control plot was provided with daylight and HPS (110-µmol m^{-2} s^{-1} PPFD) (Samuoliene et al. 2012).

Chlorophyll, a concentration in the leaves of cucumber seedlings, was significantly increased under supplementary light of 470-nm and 530-nm wavelengths. In addition, although chlorophyll, a concentration, increased in all supplementary light quality treatments, supplementary LED light had no effect on chlorophyll a:b ratios. In the tomato seedling, chlorophyll a and chlorophyll b concentration increased significantly under the 470-nm light quality treatment, and 470-nm supplementary light significantly increased the photosynthetic pigment concentration in pepper leaves. In addition, chlorophyll a and carotenoid concentration significantly increased in the 455-nm light quality treatment (Table 18.12).

Similarly, the chlorophyll concentration in baby leaf "Multigreen 3" was investigated under solar light (~80–120-µmol m^{-2} s^{-1} PPFD) plus HPS (~90-µmol m^{-2} s^{-1} PPFD) as a main light source with supplemental LEDs (~15-µmol m^{-2} s^{-1} PPFD): blue (455 and 470 nm), cyan (505 nm), and green (535 nm) (Sirtautas et al. 2014). Generally, photosynthetic capacity enhances with the increase of the ratio of chlorophyll a to b. However, the chlorophyll a:b ratio was shown to decrease under blue light of 455-nm wavelength, whereby monosaccharide concentration was increased (Table 18.13). In contrast, the chlorophyll a:b ratio was significantly increased under blue light of 470-nm wavelength. However, this was due to a decrease in chlorophyll b concentration as opposed to an increase in chlorophyll a. Treatment with cyan light of 505-nm wavelength resulted in a significant effect on photosynthetic pigments, with chlorophylls a and b and carotenoid concentrations increasing up to 1.2 times relative to controls.

18.6 Concluding Summary

In this article, recent studies describing the effect of light environment on plant secondary metabolites are discussed. It is known that biosynthesis of a particular molecule is promoted by stimulating with light of a specific wavelength. Since the light stimuli used in these studies were different with respect to wavelength, PPFD,

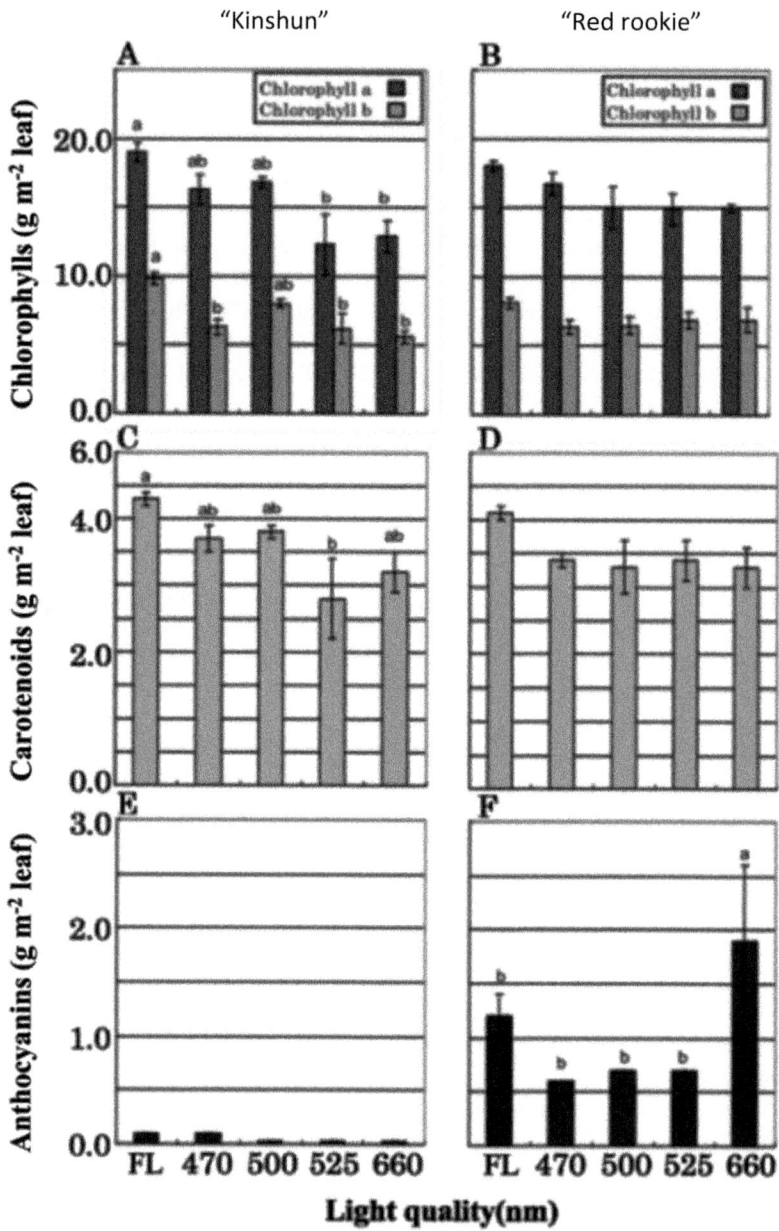

Fig. 18.11 Effects of light quality on the concentrations of chlorophylls, carotenoids, and anthocyanins in cabbage seedlings (Adapted from Mizuno et al. 2011)

Table 18.12 The pigment concentration and chlorophyll a to b ratio of vegetable transplants grown under different light quality treatments

	Chlorophyll a	Chlorophyll b	Carotenoids	Chlorophyll a and b ratio
	Cucumber "Mirabelle" F_1			
HPS	0.63 ± 0.038	0.23 ± 0.014	0.25 ± 0.015	2.73 ± 0.051
HPS + 455 nm	0.73 ± 0.029	0.27 ± 0.006*	0.26 ± 0.010	2.65 ± 0.074
HPS + 470 nm	0.74 ± 0.026*	0.27 ± 0.006*	0.27 ± 0.017	2.72 ± 0.081
HPS + 505 nm	0.86 ± 0.041**	0.32 ± 0.012**	0.30 ± 0.016**	2.69 ± 0.031
HPS + 535 nm	0.76 ± 0.031*	0.28 ± 0.011*	0.27 ± 0.012	2.77 ± 0.052
	Tomato "Magnus" F_1			
HPS	0.54 ± 0.053	0.18 ± 0.018	0.18 ± 0.019	3.07 ± 0.049
HPS + 455 nm	0.76 ± 0.047*	0.25 ± 0.019*	0.24 ± 0.014*	3.00 ± 0.085
HPS + 470 nm	0.79 ± 0.073*	0.27 ± 0.029*	0.25 ± 0.022	2.98 ± 0.050
HPS + 505 nm	0.77 ± 0.074*	0.26 ± 0.018*	0.26 ± 0.029	3.05 ± 0.085
HPS + 535 nm	0.60 ± 0.046	0.20 ± 0.017	0.19 ± 0.012	2.95 ± 0.059
	Sweet pepper "Reda"			
HPS	0.63 ± 0.031	0.25 ± 0.014	0.23 ± 0.009	2.49 ± 0.077
HPS + 455 nm	0.83 ± 0.041**	0.29 ± 0.011	0.30 ± 0.015*	2.89 ± 0.045**
HPS + 470 nm	1.00 ± 0.37**	0.34 ± 0.018*	0.35 ± 0.019**	2.99 ± 0.059**
HPS + 505 nm	0.79 ± 0.059	0.28 ± 0.022	0.28 ± 0.024	2.85 ± 0.031**
HPS + 535 nm	0.62 ± 0.036	0.21 ± 0.016	0.22 ± 0.012	2.88 ± 0.057**

Modified from Samuoliene et al. (2012)
Mean significantly (*$P < 0.05$; **$P < 0.01$) different from control (HPS) plants as determined by paired *t*-test

Table 18.13 The concentration of photosynthetic pigments in baby leaf lettuce under HPS and supplemental LED light quality treatments

	Photosynthetic pigments			
	Chlorophyll a	Chlorophyll b	Carotenoids	Chlorophyll a and b ratio
HPS	0.71 ± 0.01	0.27 ± 0.01	0.24 ± 0.02	2.63 ± 0.01
HPS + 455 nm	0.67 ± 0.04	0.29 ± 0.03	0.24 ± 0.03	2.31 ± 0.03
HPS + 470 nm	0.70 ± 0.02	0.24 ± 0.02	0.24 ± 0.01	2.91 ± 0.02*
HPS + 505 nm	0.88 ± 0.04*	0.32 ± 0.01*	0.29 ± 0.03*	2.75 ± 0.03
HPS + 535 nm	0.60 ± 0.03	0.24 ± 0.03	0.20 ± 0.03*	2.50 ± 0.03

Modified from Sirtautas et al. (2014)
Mean significantly (*$P < 0.05$) different from control (HPS) plants as determined by paired *t*-test

and photoperiod, the comparison cannot be straightforward. Even when the same stimulus is used, different responses are observed depending on the plant species. However, if reproducibility can be ensured, this can be utilized as a highly efficient technology. Future studies should elucidate the mechanisms underlying these differences and should demonstrate efficient and effective methods of providing light stimuli; after these improvements in methodology, high-value-added vegetables that cannot be grown in the open field can be produced in plant factories.

References

Bliznikas Z, Zulauskas A (2012) Effect of supplementary pre-harvest LED lighting on the antioxidant and nutritional properties of green vegetables. Acta Hortic 939:85–92

Bliznikas Z, Žukauskas A, Samuolienė G, Viršilė V, Brazaitytė A, Jankauskienė J (2012) Effect of supplementary pre-harvest LED lighting on the antioxidant and nutritional properties of green vegetables. Acta Hortic 939:85–91

Jang M, Lee Y, Ju Y et al (2013) Effect of color of light emitting diode on development of fruit body in *Hypsizygus marmoreus*. Mycobiology 41(1):63–66

Johkan M, Shoji K, Goto F et al (2010) Blue light-emitting diode light irradiation of seedlings improves seedling quality and growth after transplanting in red leaf lettuce. Hortscience 45 (12):1809–1814

Kim K, Kook H, Jang Y et al (2013) The effect of blue-light-emitting diodes on antioxidant properties and resistance of *Botrytis cinerea* in tomato. J Plant Pathol Microbiol 4:203. doi:10. 4172/2157-7471.1000203

Kim E, Park S, Park B et al (2014) Growth and antioxidant phenolic compounds in cherry tomato seedlings grown under monochromic light-emitting diodes. Hortic Environ Biotechnol 55(6):506–513

Li Q, Kubota C (2009) Effects of supplemental light quality on growth and phytochemicals of baby leaf lettuce. Environ Exp Bot 67:59–64

Miura H, Iwata M (1981) Chlorophylls and carotenoids: pigments of photosynthetic biomembranes. Methods Enzymol 148:350–383

Mizuno T, Amaki W, Watanabe H (2011) Effects of monochromatic light irradiation by LED on the growth and anthocyanin concentrations in leaves of cabbage seedlings. Acta Hortic 907:179–184

Ninu L, Ahmad M, Miarelli C et al (1999) Cryptochrome 1 controls tomato development in response to blue light. Plant J 18:551–556

Ouzounis T, Parjikolaei B R, Frette X, Rosenqvist E, Ottosen C (2015) Predawn and high intensity application of supplemental blue light decreases the quantum yield of PSII and enhances the amount of phenolic acids, flavonoids, and pigments in Lactuca sativa. Front. Plant Sci 26: http://dx.doi.org/10.3389/fpls.2015.00019

Samuoliene A, Brazaityte A, Sitrautas R, Novickovas A, Duchovskis (2012) The effect of supplementary LED lighting on the antioxidant and nutritional properties of lettuce. Acta Hortic 952:835–842

Samuoliene G, Brazaityte A, Jankauskiene J et al (2013) LED irradiance level affects growth and nutritional quality of Brassica microgreens. Cent Eur J Biol 8(12):1241–1249

Sirtautas R, Virsile A, Samulioene G et al (2014) Growing of leaf lettuce (*Lactuca sativa* L.) under high-pressure sodium lamps with supplemental blue, cyan and green LEDs. Zemdirbyste-Agriculture 101:75–78. doi:10.13080/z-a.2014.101.010

Son K, Oh M (2013) Leaf shape, growth and antioxidant phenolic compounds of two lettuce cultivars grown under various combinations of blue and red light-emitting diodes. Hortscience 48(8):988–995

Urbonaviciute A, Samuoliene G, Sakalauskiene S et al (2009) Effect of flashing amber light in the nutritional quality of green sprouts. Agron Res 7:761–767

Chapter 19
Induction of Plant Disease Resistance and Other Physiological Responses by Green Light Illumination

Rika Kudo and Keiji Yamamoto

Abstract Research on lighting as part of integrated pest management (IPM) has been attracting attention recently. This chapter presents the authors' research regarding the effects of green light on defense reaction in plants against plant pathogens, such as the induction of disease resistance in plants, endogenous production of antibacterial substances, and reinforcement of cell walls through elevation of the levels of lignin which is a main component of cell walls. This chapter also shows how green light, which had generally been regarded as having little use in plants, is effective not only in disease control but also in spider mite suppression, promotion of plant growth, and increase in functional substances. It describes how the authors developed green LED light sources for disease resistance induction and conducted experiments on major horticultural crops such as strawberries, tomatoes, perillas, and garlic chives, considering their economic benefits and practicality.

Keywords Green light • Disease resistance induction • Defense reaction • Elicitor • Strawberry anthracnose (*Glomerella cingulata*) • Spider mite control • Quality improvement • Dormancy suppression • Green LED

19.1 Introduction

As a growing number of people seek food safety and security, the need for cultivation methods that use no or reduced amounts of agricultural chemicals is rising. On the other hand, disease and pest damage should be controlled for stable production at cultivation sites. This has given rise to a significant need for new disease control techniques that can replace agricultural chemicals.

To satisfy those requirements, the use of integrated pest management (IPM) is spreading, combining agronomic, biological, and physical controls while reducing

R. Kudo (✉) • K. Yamamoto
Shikoku Research Institute Inc., 2109-8, Yashima-nishimachi, Takamatsu, Kagawa 761-0192, Japan
e-mail: rkudou@ssken.co.jp

environmental impact by limiting spraying of agricultural chemicals. Yellow or green moth repellent lights are examples of IPM used to suppress the activities of owlet moths.

Plants activate various defense reactions when exposed to stresses, such as disease damage, insect damage, and temperature- and water-related stresses. Substances that induce disease resistance reaction in plants are called elicitors, which can be divided into biological elicitors such as molecules in the cell wall of pathogenic microbes and nonbiological elicitors such as heavy metals and surface-active agents (Robatzek and Somssich 2001). It has been reported that red light and ultraviolet light act as nonbiological elicitors (Islam et al. 1998; Arase et al. 2000; Kobayashi et al. 2013).

This chapter focuses on green light as a nonbiological elicitor while considering its safety in greenhouses and describes the authors' findings on the induction of disease resistance and other physiological responses of plants by green light illumination, which has mostly escaped attention until now in studies on the use of visible light on plants.

19.2 Induction of Disease Resistance by Green Light Illumination

19.2.1 Effects of Light Quality on Gene Expression Related to Disease Resistance

Light-emitting diodes (LEDs) were used to illuminate (or irradiate) various wavelengths of the visible light spectrum on tomato seedlings, and the expression of various genes related to disease resistance was examined (Table 19.1). As a result, the expression of allene oxide synthase (AOS) genes, which are necessary for biosynthesis of jasmonic acid, a type of plant hormone that is considered to be related to disease resistance, was specifically observed only when the seedlings were illuminated with green light (Kudo et al. 2009) (Fig. 19.1).

Increased expression of lipoxygenase (LOX) genes, which, similarly to the AOS genes, are related to biosynthesis of jasmonic acid, and induction of gene expression of chitinase and various other pathogenesis-related (PR) proteins were also observed with illumination of green light (Kudo et al. 2009) (Figs. 19.2 and 19.3).

Furthermore, genes that increased their expression as a result of green light illumination on rice and strawberries were examined using the deoxyribonucleic acid (DNA) microarray. The results showed that green light illumination increased expression of heat shock proteins and other stress-related genes as well as osmotin-like proteins and various genes related to disease resistance.

These results showed that green light illumination, which had mostly been disregarded due to the perception that it contributes little to plant photosynthesis, has an effect on the expression of genes related to plant disease resistance. It is

Table 19.1 Wavelength range and peak wavelengths of LED light sources

LED light	Blue	Green	Yellow	Red
Wavelength range (nm)	400–500	480–590	560–650	560–790
Peak wavelength (nm)	470	520	590	660

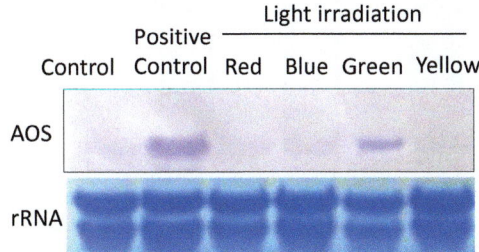

Control: Left in dark place, Positive Control: Wounding,
Light irradiation : LED illumination

Fig. 19.1 Effect of light quality on induction of tomato AOS gene expression by light illumination (Kudo et al. 2009). Cultivar, "Momotaro 8" (Takii & Co., Ltd.); air temperature, 25 °C for 2 weeks to raise the seedlings; LED light sources, blue, green, yellow, and red. Illumination: 2 h from 1 cm above the top of the plant body of the tomato seedlings. The tomato leaves were collected immediately after the illumination and chilled using liquid nitrogen and ground to powder. RNA was extracted from the powder, and analysis was carried out on allene oxide synthase (AOS) genes by northern blotting

assumed that green light illumination causes moderate stress in plants, which prompts induction of disease resistance (Kudo et al. 2009).

Based on the above findings, a system was created in which diseases are caused artificially at the probability of 100 % to assess, at a laboratory level, the effects of green light illumination in controlling diseases in major horticultural crops. The results showed that green light illumination was effective in controlling some of the main diseases in horticultural crops, namely, strawberry anthracnose caused by *Glomerella cingulata*; gray mold caused by *Botrytis cinerea* in tomatoes (*Lycopersicon esculentum* Mill.), peppers (*Capsicum annuum* L.), and eggplants (*Solanum melongena* L.); damping-off of spinach caused by *Pythium ultimum*; and corynespora leaf spot disease in perillas caused by *Corynespora cassiicola* (Kudo et al. 2010).

As the next step, an experiment was conducted, with the cooperation of greenhouse growers, on the crops for which green light illumination was observed to be effective at the laboratory level to verify its effect in controlling diseases in horticulture. Incidentally, the authors are not sure which term is more appropriate, illumination (effects of photons with particular wave numbers) or irradiation (effects of energy with particular wavelength).

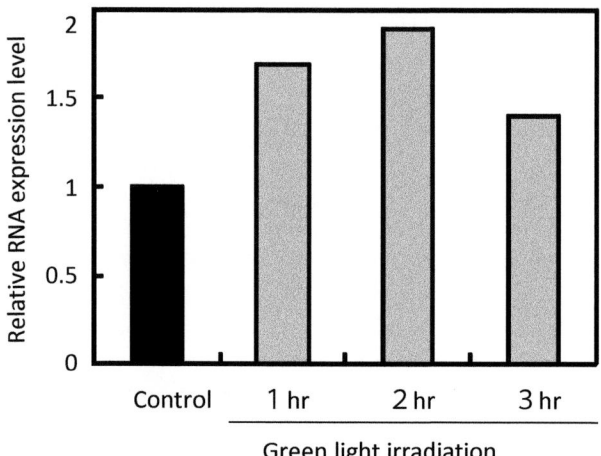

Fig. 19.2 Effect of green light illumination period on induction of tomato LOX gene expression (Kudo et al. 2009). Cultivar: "Momotaro 8" (Takii & Co., Ltd.). The tomato seedlings were raised in a nursery room at a temperature of 25 °C under fluorescent lamps for 2 weeks. Green light illumination: 80 µmol m^{-2} s^{-1} for 1, 2 or 3 h at 1 cm above the top of the tomato seedlings. The tomato leaves were collected immediately after the illumination and chilled using liquid nitrogen and ground to powder. Ribonucleic acid (RNA) was extracted from the powder and analysis was carried out on lipoxygenase (LOX) genes by real-time polymerase chain reaction (PCR)

19.2.2 Effects of Green Light on Strawberry Anthracnose

There is a substantial need for controlling strawberry anthracnose at cultivation sites, and so the effects of green light illumination on strawberry anthracnose were examined. Strawberry anthracnose occurs particularly during a period of high temperature and high relative humidity while raising strawberry seedlings or after the seedlings are planted in the field, causing blighting of strawberry roots. It is highly infectious and develops rapidly. New types of strawberry anthracnose resistant to agricultural chemicals have emerged, and the risk of strawberry anthracnose is rising with the recent global warming.

After irradiating various colors of LED on strawberry tissue culture seedlings for 2 h, the leaf surface of the seedling was inoculated with a conidial suspension of strawberry anthracnose pathogen (*Glomerella cingulata*) and cultivated in a growth chamber under given conditions. After 2 weeks, the leaf surface was examined for strawberry anthracnose lesions. The results showed that whereas illumination of blue, red, or yellow light was not effective in controlling the development of lesions, only green light illumination was effective in significantly controlling the development of lesions (Fig. 19.4). The optimal conditions of green light illumination for disease control have so far been found to be photon flux density of 80 µmol m^{-2} s^{-1}, illumination time of 2 h at night, and illumination interval of once every 3 days (Kudo et al. 2011).

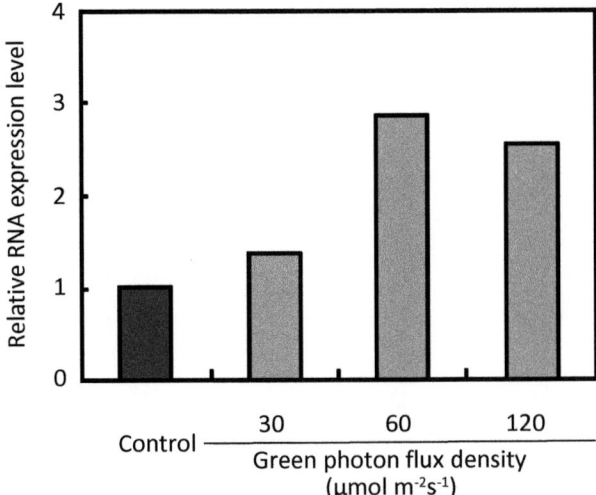

Fig. 19.3 Effect of green light illumination on induction of cucumber chitinase gene expression (Kudo et al. 2009). Cultivar: "Alpha Fushinari" (Kurume Vegetable Breeding Co., Ltd.). The seedlings were raised in a nursery room at a temperature of 25 °C under fluorescent lamps for 2 weeks. Green light illumination for 2 h: 30, 60 and 120 μmol m^{-2} s^{-1} at 1 cm above the top of cucumber seedlings. The cucumber leaves were collected immediately after the illumination and chilled using liquid nitrogen and ground to powder. RNA was extracted from the powder and analysis was carried out on chitinase genes by real-time PCR

Next, to ascertain the practicality of this technique, green light was irradiated to strawberries using LED lamps, fluorescent lamps, and metal halide lamps at the above-described illumination conditions during the raising of strawberry seedlings and cultivation, and a comparison was made with strawberries in plots without illumination in the incidence of strawberry anthracnose. The experiments were conducted in several locations in each of the four prefectures of Shikoku island, Japan. The results showed that the incidence of strawberry anthracnose declined at each test site by a third to two thirds (Kudo et al. 2011). Also, a simple diagnosis by ethanol immersion (Ishikawa 2013), a method for diagnosing latent infection by strawberry anthracnose, was used to examine changes in the incidence of latent infection by strawberry anthracnose. The results showed that green light illumination tended to control latent infection and that it was effective in reducing the incidence risk.

These findings suggest that the use of green light illumination during management and growth of the parent plants and during the raising of the seedlings may reduce the subsequent incidence of the disease.

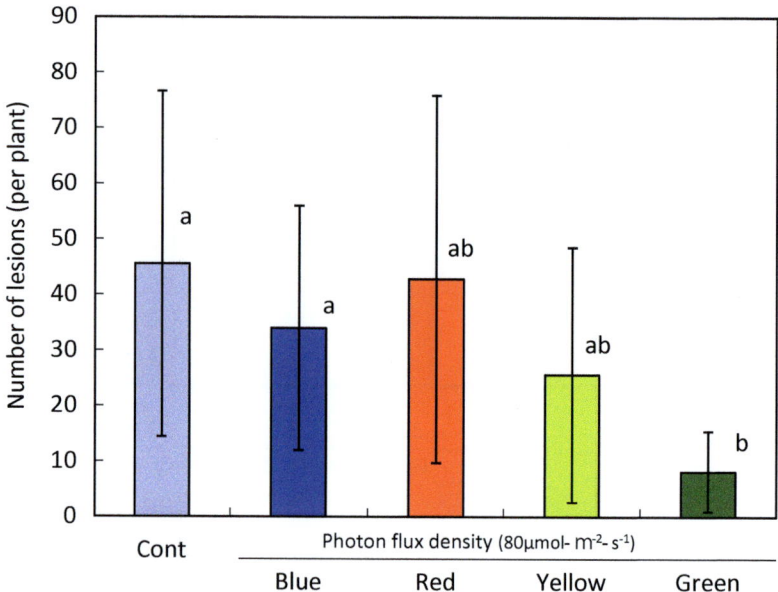

Fig. 19.4 Effect of light quality on the incidence of strawberry anthracnose (Kudo et al. 2010). Cultivar: "Smile Heart" (Shikoku Research Institute Inc.). LED light sources: blue, green, yellow, and red. Photon flux density: 80 μmol m^{-2} s^{-1}. The tissue-cultured transplants were illuminated for 2 h in an incubator set at 25 °C. A suspension of strawberry anthracnose pathogen was inoculated on the seedlings immediately by dispersion after illumination. The seedlings were then cultivated at 25 °C in an incubator for 2 weeks, after which the seedlings were assessed for incidence of strawberry anthracnose. *Based on Tukey's multiple comparison procedure, there are significant differences at the 5 % level between the English letters ($n = 5$)

19.2.3 Effects of Green Light on Corynespora Leaf Spot Disease

Perilla (*Perilla frutescens* L.), also known as Japanese herb in Japan, is a food ingredient in demand throughout the year for use in sashimi (raw fish). When symptoms of corynespora leaf spot disease (*Corynespora cassiicola*) appear in perilla leaves, the leaves lose entire commercial value and are discarded, causing a substantial decrease in yield. The paucity of registered agricultural chemicals for controlling disease in perilla makes it very difficult to control corynespora leaf spot disease in perilla.

Against this backdrop, perilla was cultivated using green LED light instead of the customary incandescent lamp as a source of artificial light for controlling plant growth in greenhouses. The results showed that green light illumination had a similar effect as incandescent lamps in suppressing bolting and that it was also effective in suppressing corynespora leaf spot disease (Kudo and Yamamoto 2013) (Fig. 19.5).

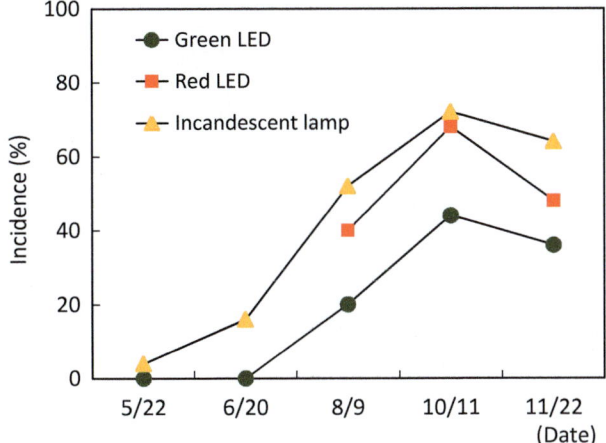

Fig. 19.5 Effect of light quality on the incidence of corynespora leaf spot disease (Kudo and Yamamoto 2013). Date of starting illumination, April 24, 2012; ending date of experiment, November 27, 2012. Perilla seedlings were planted in the plastic greenhouse for soil cultivation on March 20, 2012, and were subjected to the test using artificial lighting to control plant growth. Starting on April 24, perillas were illuminated from 22:00 to 23:30 every day using incandescent lamps, red LEDs, and green LEDs. The LEDs were placed 2 m apart. The photon flux density at the growth point under 1 m from LEDs was 0.5–1.0 µmol m^{-2} s^{-1}

Illuminating perilla with green LED light reduced the loss of perilla leaves due to disease and increased the yield and also reduced the frequency of agrichemicals application by half.

19.3 Various Effects of Green Light

While examining the effects of green light on controlling diseases in strawberry, perilla, and other crops in greenhouses, it became evident that green light had other effects on plants as well. Although green light had been regarded as having little effect on plants, it became clear that with the right illumination conditions, green light influenced plants' morphogenesis and had benefits on greenhouse horticulture. Specifically, in demonstration tests conducted in commercial greenhouses, it was discovered that green light not only was effective in controlling plant diseases but was also effective in controlling spider mites, promoting growth (fruit enlargement), improving quality, and controlling plant growth as an artificial light source.

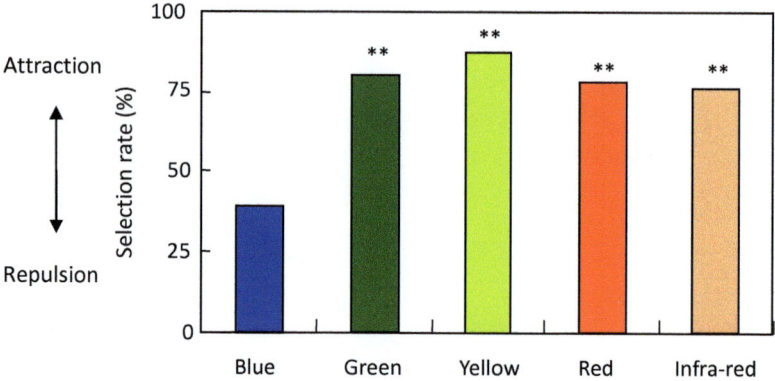

Fig. 19.6 Effect of light quality on the behavior of the predatory mite *Amblyseius* (*Neoseiulus*) *californicus* ($n = 30$). No significant tendencies were observed in the experiment on the effect of light illumination on phototaxis of spider mites. Therefore, another study was conducted on the effect of light illumination on phototaxis of the predatory mite (*Amblyseius* (*Neoseiulus*) *californicus*), which is used as a natural enemy of spider mites, using LEDs (blue, green, yellow, and red), at the photon flux density of 40 µmol m^{-2} s^{-1}. ** Significantly different at the 1 % level

19.3.1 Spider Mite Control

The phototaxis of phytoseiid mite (*Phytoseiidae*), which is a natural enemy of the spider mite, was examined. The results showed that phytoseiid mites move toward green light and light with longer wavelengths (Fig. 19.6). This finding suggests that when phytoseiid mites are released as a natural-enemy pesticide against spider mites, green light promotes settlement of the phytoseiid mites into the plants irradiated with green light and more intensive preying on spider mites.

19.3.2 Growth Promotion

It was found that illumination of green light during strawberry cultivation induced growth-related gene expression of cellulose synthase 2 and extensin (Fig. 19.7). Green light illumination was also observed to enlarge the leaf area during cultivation. TTC analysis at the end of the cultivation also revealed increased root activity from green light illumination (Figs. 19.8 and 19.9).

While it was believed that green light was not easily absorbed by plants, research in recent years has reported absorption of green light in plants and the use of green light for CO_2 fixation in mesophyll and for photosynthesis (Terashima et al. 2009; Sun et al. 2009). In addition to promotion of photosynthesis where green light acts as a signal for growth-related genes, there is a potential for use of green light to more directly promote photosynthesis. It is assumed that increased root activity also

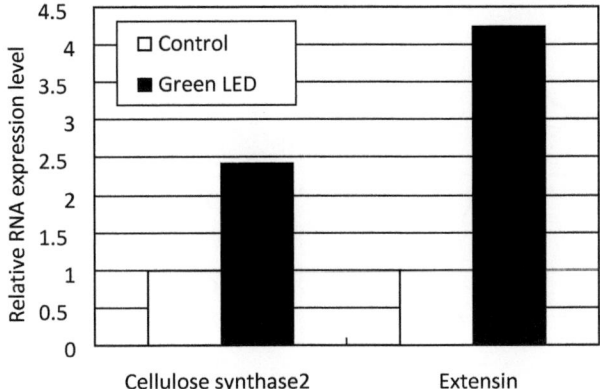

Fig. 19.7 Effect of green light illumination on expression of genes related to strawberry growth. Date of starting illumination, October 24, 2012; ending date of test, May 20, 2012. Illumination: 2 h at night, three times a week. The RNA samples extracted from the third leaf from the top of the strawberry *Sachinoka*, which was cultivated in the plastic greenhouse with illumination of green LEDs, were used to analyze induction of gene expression by real-time PCR. The samples were collected in February

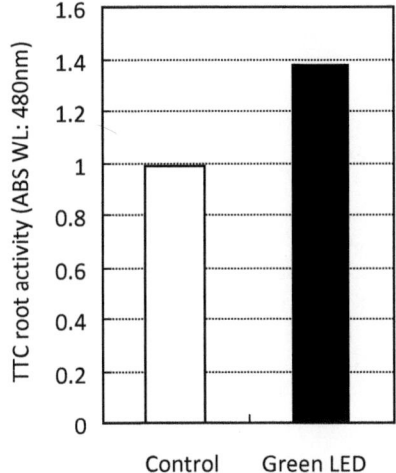

Fig. 19.8 Improvement in TTC root activity of strawberries by green light illumination. Date of starting illumination, October 24, 2012; ending date of test, May 20, 2012. Illumination: 2 h at night, three times a week. The root samples of the strawberry *Sachinoka*, which was cultivated in the plastic greenhouse with illumination of green LEDs, were used at the end of the cultivation to analyze root activity by triphenyl tetrazolium chloride (TTC) staining. The samples were collected in May

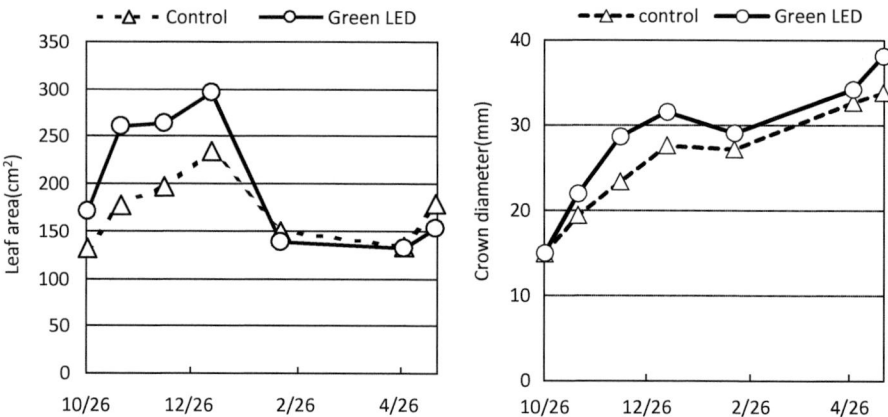

Fig. 19.9 Effect of green LED light illumination on strawberry leaf area and crown diameter. Date of starting illumination, October 24, 2012; ending date of test, May 20, 2012. Illumination: 2 h at night, three times a week. The data show changes in the leaf area of the third leaf from the top and the crown diameter of the strawberry "Sachinoka" during cultivation. "Sachinoka" was cultivated in the plastic greenhouse with illumination of green LEDs

resulted in increased nutrient absorption, the combined effect of which led to promotion of photosynthesis and growth.

19.3.3 *Increase in Functional Substances and Sugar Content*

Improvement in the sugar-acid ratio, which is an indicator of the quality of strawberries, and increase in such functional substances as total polyphenol and vitamin C in leaf vegetables were also observed. The total polyphenol content increased in perillas and leaf lettuces that were irradiated with green light. The increase in vitamin C in spinach and lycopene in tomatoes was observed after green light illumination.

Polyphenols are produced by biosynthesis as a secondary metabolite of the shikimic acid pathway. Phenylalanine ammonia-lyase (PAL) is an important enzyme that catalyzes the branching reaction from the primary metabolism to the secondary metabolism in the shikimic acid pathway of plants. It became clear that green light illumination stimulated PAL activity in leaf lettuces and increased the total polyphenol content (Figs. 19.10 and 19.11)

As for strawberries, green light illumination increased sugar content in winter. It is assumed that the rise in sugar content during the low-temperature season was due to the activation of a defense reaction to increase resistance against stress. The green light may have acted as a stressful stimulus to help plants accumulate functional substances (polyphenols, vitamin C, lycopene) and sugar as part of their resistance reaction.

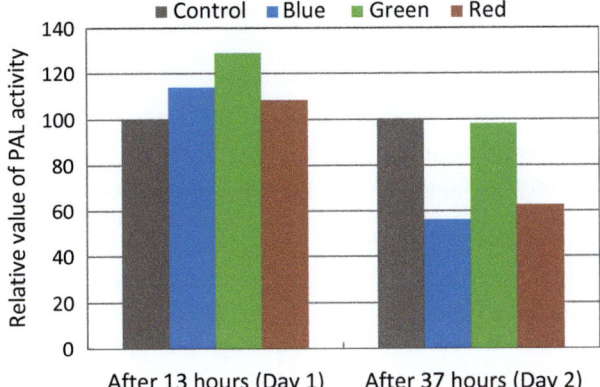

Fig. 19.10 Effect of light quality on PAL activity of leaf lettuces. After the seedlings of the leaf lettuce "Banchu Red Fire" were raised for 20 days, the seedlings were planted on a hydroponic styrofoam panel and cultivated in a nutrient flow technique (NFT) system in a plastic greenhouse. Fluorescent lamps (blue, green, and red) were used as the light source to illuminate at the photon flux density of 65 µmol m^{-2} s^{-1} for 2 h from 20:00 to 22:00 at the interval of once every 3 days. The PAL activity was measured 28 days after planting based on the Koukol-Conn method ($n = 6$)

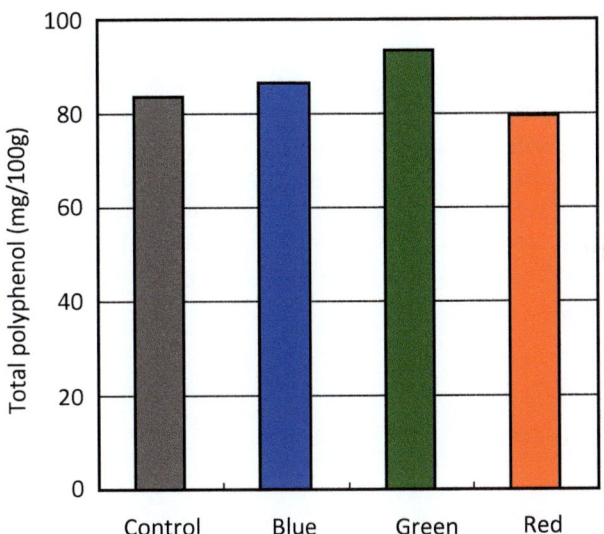

Fig. 19.11 Effect of light quality on polyphenol content of leaf lettuces. After the seedlings of the leaf lettuce "Banchu Red Fire" were raised for 20 days, the seedlings were planted on a hydroponic styrofoam panel and cultivated in a NFT system in the plastic greenhouse. Fluorescent lamps (blue, green, and red) were used as the light source to irradiate light at the photon flux density of 65 µmol m^{-2} s^{-1} for 2 h from 20:00 to 22:00 at the interval of once every 3 days. The total polyphenol content was measured ($n-3$) 45 days after planting based on the Folin-Denis method

19.3.4 Dormancy Suppression, Flower Bud Differentiation, and Bolting

Green LED light was used in place of incandescent lamps and fluorescent lamps, which are the conventional artificial light sources used for dormancy suppression in strawberries and bolting suppression in perillas. The results showed that green light had similar effects as the conventional light sources and induced photoperiodic responses in plants (Sun et al. 2009).

Perilla is a short-day plant, forming a raceme in summer and autumn on a stem that has differentiated from the leaf axil. Artificial lighting is used in yearlong cultivation of perillas to suppress bolting through photoperiodic adjustment and to enable continuous harvesting of the perilla leaves.

Green LED light was used instead of the customary incandescent lamp to study its effect on suppressing bolting of perillas. The results showed that green light had a similar effect in suppressing bolting as the incandescent lamp (Kudo and Yamamoto 2011, 2013).

In the winter, the growth of garlic chives is suppressed due to dormancy under the conditions of low temperature and short day. It was found that the illumination of green LED light on garlic chives that had not been subject to the use of artificial light to control their growth substantially promoted growth and increased yield.

On the other hand, when garlic chives are irradiated with incandescent lamps and other light sources that contain a large amount of red wavelengths to create a long-day condition, flower buds develop in early spring. It was found that green light suppressed flower bud differentiation and development. As this is perceived as having advantages in increasing yield during the winter and controlling diseases in early spring and after, the use of green LED lighting has been widely introduced recently in major regions in Japan where garlic chives are produced.

As green light can adjust photoperiodic flower bud differentiation and dormancy and suppress diseases, green LED light is gaining attention as a new source of artificial lighting.

19.4 Conclusion

In an effort to promote the use of light in agriculture, we have been developing "pest management by lighting" techniques in integrated pest management (IPM). As a result, it was found that the illumination of green light on plants enhanced plants' intrinsic disease resistance and prevented infection and incidence of diseases. It also became clear that green light illumination on horticultural products was not only effective in disease control but was also effective in spider mite control, promotion of growth (fruit enlargement), and quality improvement. Green light illumination does not require registration such as agricultural chemicals do, and there are no issues related to safety. For these reasons, green light illumination is considered to have substantial value as a new type of technique using light.

At present, we are working on commercializing green LED lighting sources (product name: Midorikikuzo®) and on establishing horticultural production technology with reduced environmental impact by combining the use of natural enemies, which are key components of IPM, microbial pesticides, and organic materials, based on green light illumination techniques.

We hope to develop pest management technology based on LED lighting, broaden its application to many crops, and encourage the use of LEDs in horticulture and plant factories as an eco-friendly light technology.

Acknowledgment The authors are thankful to Dr. Toyoki Kozai for his advice and critical reading of the manuscript.

References

Arase S, Fujita K, Uehara T, Honda Y, Isota J (2000) Light-enhanced resistance to *Magnaporthe grisea* infection in the rice Sekiguchi lesion mutants. J Phytopathol 148:197–203

Ishikawa S (2013) Studies on identification of pathogen, epidemiology and integrated disease management of strawberry anthracnose. Bull Tochigi Agric Exp Stn 54:1–172

Islam S, Honda Y, Arase S (1998) Light-induced resistance of broad bean against *Botrytis cinerea*. J Phytopathol 146:479–485

Kobayashi M, Kanto T, Fujikawa T, Yamada M, Ishikawa M, Satou M, Hisamatsu T (2013) Supplemental UV radiation controls rose powdery mildew disease under the greenhouse condition. Environ Control Biol 51:157–163

Kudo R, Yamamoto K (2011) Studies on light culture by green light irradiation of perilla. Hortic Res 10(2):11–155 (in Japanese)

Kudo R, Yamamoto K (2013) Effects of green light irradiation on corynespora leaf spot disease in perilla. Hortic Res 12(2):13–157 (in Japanese)

Kudo R, Yamamoto K, Suekane A, Ishida Y (2009) Development of green light pest control systems in plants. I. Studies on effects of green light irradiation on induction of disease resistance. SRI Res Rep 93:31–35 (in Japanese)

Kudo R, Yamamoto K, Ishida Y (2010) Development of green light pest control systems in plants. II. Studies on effects of green light irradiation on strawberry anthracnose (*Glomerella cingulata*) and cucumber anthracnose (*Colletotrichum orbiculare*). SRI Res Rep 94:33–39 (in Japanese)

Kudo R, Ishida Y, Yamamoto K (2011) Effects of green light irradiation on induction of disease resistance in plants. Acta Hortic ISHS 907:251–254

Robatzek S, Somssich IE (2001) A new member of the *Arabidopsis* WRKY transcription factor family, AtWRKY6, is associated with both senescence- and defence-related processes. Plant J 28(2):123–133

Sun J, Nishino JN, Vogelmann C (2009) Green light drives CO_2 fixation deep within leaves. Plant Cell Physiol 39:1020–1026

Terashima I, Fujita T, Inoue T, Chow WS, Oguchi R (2009) Green light drives leaf photosynthesis more efficiently than red light in strong white light: revisiting the enigmatic question of why leaves are green. Plant Cell Physiol 50:684–697

Chapter 20
Light Quality Effects on Intumescence (Oedema) on Plant Leaves

Kimberly A. Williams, Chad T. Miller, and Joshua K. Craver

Abstract Intumescence is a physiological disorder that is characterized by abnormal outgrowths of epidermal and/or palisade parenchyma cells on the leaf, petiole or stem surfaces of affected plants. Intumescences are a different disorder than oedema based on anatomy of affected cells and causal agents. This disorder is most often observed on crops produced in controlled environments and has been reported on a wide range of plant species, including ornamental sweet potato (*Ipomoea batatas*), cuphea (*Cuphea* spp.) and solanaceous crops of tomato (*Solanum lycopersicum*) and potato (*Solanum tuberosum*). When susceptible crops are grown under ultraviolet-deficient environments, such as light-emitting diode (LED) sole-source lighting that supplies only red wavelengths, intumescences are most severe. However, the incidence of the disorder can be diminished or prevented if crops are grown in environments providing ample blue or ultraviolet wavelengths of light. End-of-day far-red lighting has also shown some promise in mitigating the disorder.

Keywords Blue light • Oedema • Far-red light • LED • Tomato • Ultraviolet light

20.1 Introduction

Intumescence is a physiological disorder that is characterized by abnormal outgrowths of cells on leaf, petiole and stem surfaces (Morrow and Tibbits 1988; Wetzstein and Frett 1984). The disorder has long been observed on many crops produced in controlled environments and, more recently, on those grown under LEDs lacking wavelengths in the blue spectrum. The first description of the disorder in the literature was by Sorauer in 1899, who reported that intumescences developed on many different plant species, including mono- and dicotyledonous

---

K.A. Williams (✉) • C.T. Miller
Kansas State University, Manhattan, KS, USA
e-mail: kwilliam@ksu.edu

J.K. Craver
Department of Horticulture and Landscape Agriculture, Purdue University, 625 Agriculture Mall Drive, West Lafayette, IN 47907-2010, USA

© Springer Science+Business Media Singapore 2016
T. Kozai et al. (eds.), *LED Lighting for Urban Agriculture*,
DOI 10.1007/978-981-10-1848-0_20

angiosperms, gymnosperms and ferns (La Rue 1932). While there are many disorders with similar symptoms to intumescence, the terms used to name them are often used interchangeably and include intumescence, edema or oedema, excrescence, neoplasm, gall, genetic tumour, leaf lesion and enation (Pinkard et al. 2006). The body of literature with reports of these disorders includes a wide variety of plant species, but development on solanaceous crops such as tomato (*Solanum lycopersicum*) and potato (*Solanum tuberosum*) is the most commonly documented.

In addition to the long lists of terminology and plant species subject to these disorders, the list of environmental factors that have been reported to contribute to their development is also extensive. Plant water status; temperature; hormones, including ethylene; plant genetics; pest damage, including injury by insects and fungal infection; mechanical and chemical injury; nutrient status; and, of course, light quality and availability have all been reported as agents that contribute to foliar intumescence in various plant species (Pinkard et al. 2006). Thus, the literature is muddled by lack of clarity regarding the nomenclature and proper identification of these disorders, as it is unlikely that they are all the same.

In particular, the terms intumescence and oedema are frequently used interchangeably to describe such disorders on several plant species produced in controlled environments. However, Lang and Tibbits (1983) suggested that these terms should instead refer to completely different disorders. They defined oedema as 'a 'watery swelling of plant organs or parts,' resulting from water congestion in plant tissue'. On the other hand, they indicated that the disorder observed in their research with tomatoes should be called intumescence because plants showed symptoms when the relative humidity 'was low and there was no water congestion in the tissue'. Morrow and Tibbitts (1988) added to this differentiation by stating that 'oedema' typically forms under conditions where excess water and high relative humidity prevent sufficient transpiration by the plant, whereas their research showed that UV radiation aided in preventing intumescence on solanaceous species. Further, in research of Rangarajan and Tibbitts (1994) where far-red light failed to inhibit oedema injury on ivy geranium, they suggested that the causative factors and physiological systems that regulate oedema on *Pelargonium* spp. and intumescence on *Solanum* spp. were different. Williams et al. (2015) used light and scanning electron microscopy to study lesions as they developed on five plant species and concluded that the anatomy of intumescence in tomato (*Solanum lycopersicum* 'Maxifort'), ornamental sweet potato (*Ipomoea batatas* 'Blackie'), and cuphea (*Cuphea llavea* 'Tiny Mice') appeared different than oedema in interspecific (*Pelargonium* x 'Caliente Coral') and ivy geraniums (*Pelargonium peltatum* 'Amethyst 96'). Therefore, because environmental causes and anatomy of intumescence and oedema have been found to be different, the two terms do not seem to refer to the same physiological disorder. Interactions may exist between absence of ultraviolet light and other environmental factors, such as plant water status, but this remains to be studied with future research. The focus of this chapter is narrowed to light quality effects on the occurrence of what we believe to be intumescence, though the disorder may have been reported as other abiotic leaf damage.

20.2 Description and Impact of Intumescences

20.2.1 Anatomy and Morphology

Intumescence begins with a group of epidermal cells or palisade parenchyma cells just below the epidermal layer, enlarging and undergoing hypertrophy. As the disorder progresses, the cells become increasingly elongated and translucent (Figs. 20.1 and 20.2), often coalescing into lesions comprised of groups of cells. Depending on plant species, lesions may begin on the adaxial or abaxial leaf surface or both. The affected area expands as lesions develop, though under some conditions, a single intumescence will senesce and abscise from the leaf surface with no further expansion. Typically, though, as lesions senesce, the apex blackens (ornamental sweet potato, Fig. 20.2), turns brown and collapses (tomato) and/or coalesces into larger, necrotic spots (e.g. cuphea, Fig. 20.3; Craver et al. 2014a).

20.2.2 Genetics

The formation of intumescences has been shown to be under genetic control for the variety of plant species on which the disorder has been studied, including potato, *Solanum tuberosum* (Petitte and Ormrod 1986; Seabrook and Douglass 1998); tomato, *Solanum lycopersicum* (e.g. Sagi and Rylski 1978); cuphea, *Cuphea* sp. (Jaworski et al. 1988); and ornamental sweet potato, *Ipomoea batatas* (Craver et al. 2014b). For example, in Petitte and Ormrod's (1986) work with potato, they found that two early to midseason cultivars ('Norchip' and 'Superior') were resistant to the disorder, while two late-maturing cultivars ('Kennebec' and 'Russet Burbank') were susceptible. In Sagi and Rylski's work with tomato, the variety 'Hosen Eilon' developed the disorder, while 'Viresto' did not. In Jaworski's work on *Cuphea* sp., it was noted that intumescences could be avoided in *C. wrightii* by selecting accession numbers of plants that were not susceptible to the disorder (Jaworski et al. 1988). In a large screening trial of ornamental sweet potato varieties, only 8 out of 36 varieties were considered highly symptomatic with 20 % or more leaves affected by intumescence (Craver et al. 2014c). Rud (2009) and Craver et al. (2014c) suggested that perhaps one of the best options for control of intumescence was the selection and production of resistant cultivars.

20.2.3 Photosynthesis and Yield

One of the most detrimental impacts of severe intumescence development is the impairment of photosynthesis (Lang et al. 1983; Pinkard et al. 2006). The negative impacts of intumescence on plant tissue can include chlorosis, senescence

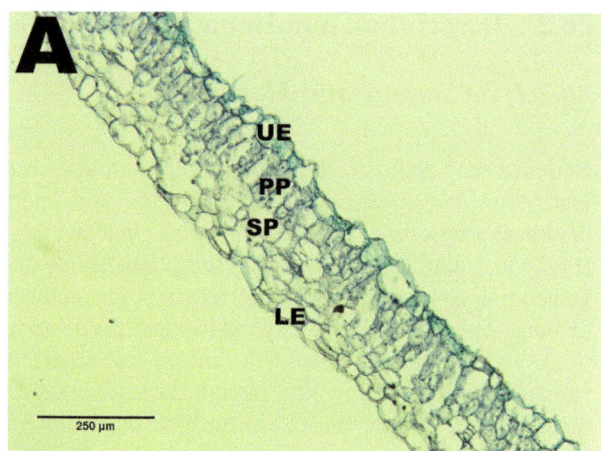
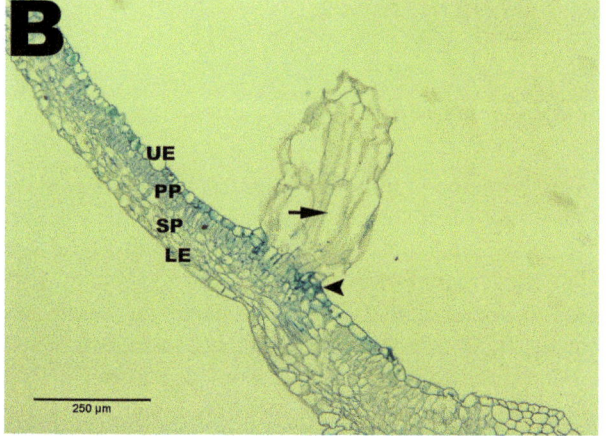

Fig. 20.1 Light microscopy cross sections of asymptomatic ornamental sweet potato leaf (**a**) and symptomatic intumescent leaf (**b**) stained with Toluidine Blue O. Upper epidermis (*UE*); palisade parenchyma (*PP*), spongy parenchyma (*SP*) and lower epidermis (*LE*). Hypertrophic PP cells are shown (*arrow*) protruding above the UE, pushing aside UE cells (point). (Images by Craver)

(Wetzstein and Frett 1984), leaf abscission (Rud 2009) and the downward curling of leaves (Figs. 20.2, 20.3 and 20.4). Lang et al. (1983) reported that hypertrophied palisade and chlorenchyma cells of tomato (*S. lycopersicum*) had few or no chloroplasts. Pinkard et al. (2006) furthered the idea of impaired photosynthesis by stating that intumescences may reduce the amount of leaf surface area available for light absorption (Fig. 20.5). Eventual plant death can occur if the disorder is left unchecked (Rud 2009).

Craver et al. (2014b) found that ultraviolet light prevented intumescence development on the susceptible cultivar of ornamental sweet potato, 'Ace of Spades'. In this study, plant growth as measured by fresh and dry weights and width at harvest was similar across treatments with and without ultraviolet light. These results indicate that if intumescence development is not severe, growth is not measurably affected.

Fig. 20.2 Ornamental sweet potato (*Ipomoea batatas*) showing early to mid-stage intumescence development on the adaxial leaf surface (*top*) and late stages of intumescence development with necrotic lesions (*bottom*). Note the downward curling of the leaf blade (*bottom*). (Images by Craver)

20.2.4 Aesthetic and Economic Impact

Some of the most intumescent or symptomatic species grown for commercial use (e.g. cuphea, ornamental sweet potato) are produced solely for ornamental purposes. Intumescences not only impair the physiological processes of the plant but also negatively affect the overall aesthetic quality and, ultimately, marketability of the crop. Severely intumescent plants may appear to be diseased which decreases salability and product value (Figs. 20.2, 20.3 and 20.4).

20.3 Light Quality Affects Intumescence

Light quality has long been viewed a strong candidate for a causative factor or potential preventative measure for intumescence development because of observations that the disorder occurs predominantly in controlled environments.

Fig. 20.3 *Cuphea llavea* 'Tiny Mice' leaves showing early to mid-stages (*top*) and late stages with necrotic regions (*bottom*) of intumescence development on leaves. Note the downward curling of the leaf blade (*bottom*). (Images by Craver)

20.3.1 Ultraviolet Light

Ultraviolet (UV) radiation includes UVA (315–400 nm), UVB (280–315 nm) and UVC (100–280 nm). This range of light quality is thought to be related to intumescence development because many greenhouse glazing materials block UV wavelengths, and intumescences have frequently been observed on crops grown under cover, but not on the same crops growing outside of the protected environment. Greenhouse plastic, or polyethylene, for example, includes a UV-block additive to slow degradation under solar radiation and extend its useful life.

Recent research has linked specific benefits of UV light to plants, including regulation of plant metabolism and morphology (Jansen and Bornman 2012; Jenkins 2009). Bridgen (2015) and Franz et al. (2012) have suggested that UV light could be used to manage plant growth during crop production. Reductions in leaf growth have commonly been observed in response to UV radiation. Wargent et al. (2009a) studied leaf response across a range of organizational scales to evaluate underlying mechanisms of effect.

Fig. 20.4 An ornamental sweet potato (*Ipomoea batatas* 'Ace of Spades') leaf showing severe intumescence development, primarily along the major leaf veins, and *downward* leaf curling. (Image by Craver)

Fig. 20.5 Tomato (*Solanum lycopersicum* 'Maxifort') leaves showing severe intumescence damage, with large, necrotic areas that result in reduced photosynthetic area for the plant. (Image by Craver)

20.3.1.1 Prevention of Intumescence

The benefit of UV light related to intumescence is the consistent findings that UVB radiation effectively prevents the development of this disorder on susceptible plant species. At low enough doses, intumescence is prevented by UVB light and negative impacts on growth from these wavelengths do not occur. While discussing the many physiological disorders that occur during crop production for life support systems in space, Wheeler (2010) indicated that near UV light (300–400 nm) can effectively prevent intumescence from occurring.

Lang and Tibbitts (1983) placed tomato plants in exposure boxes constructed of Plexiglas G (blocked wavelengths below 330 nm) and Plexiglas G II-UVT (transmitted wavelengths above 230 nm). They found that the tomato plants grown in the Plexiglas G-II-UVT boxes were free from injury, while those plants grown in the Plexiglas G boxes displayed severe intumescence. Continuing with this line of research, Morrow and Tibbitts (1987) looked at intumescence development on tomato leaf disks. Leaf disks were induced with intumescences by blocking the UV radiation emitted from cool-white fluorescent lamps with UV-absorbing Plexiglas.

Rud (2009) found similar results in that UVB light greatly reduced the occurrence of intumescences on tomato (*Solanum lycopersicum* 'Maxifort'). She hypothesized that there may be a threshold mechanism involved with this response. That is, 'if a plant susceptible to the development of intumescences receives a certain amount or intensity of UV light, intumescence development may be prevented'.

In greenhouse experiments, Craver et al. (2014b) established a similar relationship between lack of UVB and intumescence development on ornamental sweet potato 'Ace of Spades', a susceptible variety. The addition of UVB radiation significantly reduced the number of leaves affected with intumescences when compared to plants grown under other light treatments lacking these wavelengths.

20.3.1.2 Molecular Mechanisms

UV RESISTANCE LOCUS8 (UVR8) has been shown to regulate UV-protective gene expression responses and is involved in controlling aspects of leaf growth and morphogenesis, though inhibition of epidermal cell division in response to UVB is largely independent of UVR8 (Wargent et al. 2009b). Interestingly, while recent studies have shown that UV radiation reduces epidermal cell expansion (e.g. Jacques et al. 2011; Wargent et al. 2009a), other studies provide evidence that intumescences—which occur in the absence of UVB—are hypertrophied epidermal and/or palisade parenchyma cells (e.g. Craver et al. 2014a).

There is no research that has evaluated gene regulation or the genetic mechanism of intumescence occurrence. One study subjected leaf tissue with and without intumescences from a susceptible tomato variety, *Solanum lycopersicum* 'Maxifort', to microarray analysis for comparison of gene expression profiles

(Williams et al. 2011). A number of genes that could be induced by UVB treatment but that were suppressed in the leaves with intumescences were identified. Among them, 3-beta-hydroxysteroid dehydrogenase (*3β-HSD*) was identified as potentially playing a key role in UVB inhibiting intumescence development.

As Robson et al. (2015) indicate, future research will need to disentangle the complex interactions that occur at the threshold UV dose where metabolic regulation and stress-induced morphogenesis overlap.

20.3.2 Blue and Green Light

Blue and green light have been found to have an inhibitory effect on intumescence development. In very early work, Dale (1901) observed that intumescence would not develop on hibiscus (*Hibiscus vitifolius*) under blue, yellow or green 'glasses'. Similarly, Morrow and Tibbitts (1988) found that intumescences were not induced when tomato leaf disks were exposed to blue wavelengths of light providing a photosynthetic photon flux density (*PPFD*) of 25 µmol m^{-2} s^{-1}, and only 3 % of leaf disk area had intumescences when exposed to green wavelengths at the same *PPFD*.

Seabrook and Douglass (1998) reported the only exception to this result with blue-green light. They observed that intumescences were reduced in potato (*Solanum tuberosum* 'AC Brador' and 'Shepody') plantlets grown in vitro under a yellow filter that eliminated the blue-green (380–525 nm) portion of the spectrum. This result suggested that blue-green light is putatively involved in the occurrence of intumescence.

In early research with LED lamps, Massa et al. (2008) reported that when the percentage of blue light (440 nm peak) was maintained at less than 10–15 % of total photosynthetic photons, resulting in a red light-dominant (660 nm peak) environment, cowpea (*Vigna unguiculata*) developed intumescences on older leaves. 'Triton' pepper (*Capsicum annuum*) plants grown with either intracanopy or overhead red (R) + blue (B) LED lighting also developed severe intumescences, though for this species, symptoms were not mitigated by using higher percentages of blue light as occurred for cowpea.

In recent work with LED lamps, Wollaeger and Runkle (2013) observed intumescence on tomato 'Early Girl' in all treatments grown under only 10 % each of B and green (G) LEDs with the remaining light from orange (O; peak = 596 nm) to R (peak = 634 nm) to hyper-red (HR; peak = 664 nm) wavelengths; a *PPFD* of 160 µmol m^{-2} s^{-1} was provided. In a follow-up study, Wollaeger and Runkle (2014) grew tomato 'Early Girl' under combinations of B, green (G) and R LEDs or under fluorescent lamps, again providing a *PPFD* of 160 µmol m^{-2} s^{-1}. The resulting LED treatment percentages were $B_{25}+G_{25}+R_{50}$, $B_{50}+G_{50}$, $B_{50}+R_{50}$, $G_{50}+R_{50}$, R_{100} and B_{100}. Tomato plants developed the most leaflets with intumescence when grown under solely red light, while the disorder was absent or nearly so when grown under at least 50 % blue light or fluorescent light. Additionally, the blue light (31 µmol m^{-2} s^{-1}) and small quantity of UVA (<5 µmol m^{-2} s^{-1}) light emitted from the fluorescent lamp were almost sufficient to prevent intumescence development.

Next, Wollaeger and Runkle (2015) found that tomato 'Early Girl' grown under $B_{50}+G_{50}$ or fluorescent lighting resulted in plants with no or nearly no intumescences (Wollaeger and Runkle 2015). Intumescences developed on 40 % fewer leaflets for tomatoes grown under $G_{50} + R_{50}$ compared to R_{100}, but $G_{50} + R_{50}$ treated plants had about 70 % more leaflets with intumescences than those grown under $B_{25}+G_{25}+R_{50}$. This work supports the mitigation of the disorder with increasing blue wavelengths.

As more research with LED lights occurs and the greenhouse industry adopts their use, observations that are consistent with the above trends are being reported for other crops. For example, Yelton et al. (2014) observed that the occurrence of 'oedema' on basil was reduced under higher levels of blue light; treatments were 0, 8, 16, 24 and 32 % B with a *PPFD* of 250 µmol m^{-2} s^{-1} under a 14 h day^{-1} photoperiod.

20.3.3 Red and Far Red

UV and blue light have not been the sole focus of studies involving the relationship of light and intumescence development. Research by Morrow and Tibbitts (1988) looked at the potential involvement of phytochrome in intumescence development on leaf disks of a wild tomato, *Solanum lycopersicum* var. *hirsutum*. Various photon spectra were created by using different lamps and filters. They found that solely red light resulted in intumescence on 63 % of the sample leaf area of the disks and that increased photon flux density of red light resulted in greater occurrence of the disorder. However, when leaf disks were also subjected to far-red light, the effects of the red light on intumescence development seemed to diminish, such that if ample amounts of far-red light were made available, intumescence injury was effectively inhibited. The authors speculated that this inhibitory action by far-red wavelengths suggested the involvement of phytochrome in this disorder. They expanded this hypothesis by proposing that there were two photosynthetic photon density responses involved—a prolonged red response and a reversible red/far-red response. In this case, the prolonged red response would control induction of this disorder on the plant, while the red/far-red response would directly control expression (Morrow and Tibbitts 1988).

Wollaeger and Runkle (2014) have routinely observed the greatest incidence of intumescence on tomato when produced under only red wavelengths of light. As discussed above, plants grown under the $G_{50} + R_{50}$ treatment developed 40 % fewer leaflets with intumescences than those under R_{100}.

In recent research, Eguchi et al. (2015) reported on using end-of-day far-red lighting from LEDs as a cost-effective alternative to UV to inhibit intumescences in tomato seedlings. Tomato rootstock 'Beaufort' were grown under LED light ratios of $B_{10} + R_{90}$ or $B_{75} + R_{25}$ providing a *PPFD* of 100 µmol m^{-2} s^{-1} under an 18 h day^{-1} photoperiod. Varied doses (1, 2, 4, 9 or 74 mmol m^{-2} d^{-1}) of end-of-day far-red lighting were applied to the seedlings, and intumescence development was compared to a control that did not receive any additional far-red light. Evaluation for intumescence occurred 16 days after seeding. For the seedlings grown under the $B_{10} + R_{90}$ treatment, the 1 mmol m^{-2} d^{-1} end-of-day far-red treatment (5.2 µmol m^{-2} s^{-1} for 3.3 min) reduced the incidence of stem intumescence from 63 % to none and the incidence of leaf intumescence from 62 % to 42 % compared to control plants. However, the efficacy of the end-of-day far-red lighting was the same regardless of dose. The authors concluded that, in combination with production under blue wavelengths, the subjection of tomato seedlings to end-of-day far-red lighting was useful in mitigating the occurrence of intumescences.

20.3.4 Concluding Summary

Intumescence appears to be a physiological disorder that occurs at the threshold between beneficial and harmful UV dose. Research is rapidly enhancing our understanding of the role of light quality in various aspects of plant metabolism. At the same time, LEDs are increasingly popular for many reasons, including their ability to expose crops to specific wavelengths of light. With the development of a mechanistic understanding of the underlying causes of intumescence, LED spectra can be modified in commercial applications to optimize plant growth and development while minimizing physiological disorders such as intumescence.

References

Bridgen M (2015) Using ultraviolet-C light as a plant growth regulator. Acta Hortic 1085:167–169

Craver JC, Miller CT, Williams KA et al (2014a) Characterization and comparison of lesions on ornamental sweetpotato 'Blackie', tomato 'Maxifort', interspecific geranium 'Caliente Coral', and bat-faced cuphea 'Tiny Mice'. J Am Soc HortScience 139(5):603–615

Craver JC, Miller CT, Williams KA et al (2014b) Ultraviolet radiation affects intumescence development in ornamental sweetpotato (*Ipomoea batatas*). HortScience 49(10):1277–1283

Craver JC, Miller MG, Cruz et al (2014c) Intumescences: further investigations into an elusive physiological disorder. Greenhouse Product News (GPN) 9:32–40

Dale E (1901) Investigations on the abnormal outgrowths or intumescences on *Hibiscus vitifolius* Linn.—a study in experimental plant pathology. Phil Trans R Soc Lond B 194:163–182

Eguchi T, Hernandez R, Kubota C (2015) End-of-day far-red lighting to mitigate intumescences on tomato seedlings grown under LEDs. HortScience 50(9):S219–S220, Abstr

Frantz J, Heckathorn SA, Rud N et al (2012) Short-term UV light exposure can lead to long-term plant growth regulation. HortScience 47(9):210–211, Abstr

Jacques E, Hectors K, Guisez Y et al (2011) UV radiation reduces epidermal cell expansion in *Arabidopsis thaliana* leaves without altering cellular microtubule organization. Plant Signal Behav 6(1):83–85

Jansen M, Bornman J (2012) UV-B radiation: from generic stressor to specific regulator. Physiol Plant 145(4):501–504

Jaworski C, Bass MH, Phatak SC et al (1988) Differences in leaf intumescences between Cuphea species. HortScience 23:908–909

Jenkins G (2009) Signal transduction in responses to UV-B radiation. Ann Rev Plant Biol 60:407–431

La Rue C (1932) Intumescences on poplar leaves. I. Structure and development. Am J Bot 20(1):1–17

Lang S, Tibbitts T (1983) Factors controlling intumescence development on tomato plants. J Am Soc HortScience 108(1):93–98

Lang S, Struckmeyer BE, Tibbitts TW (1983) Morphology and anatomy of intumescence development on tomato plants. J Am Soc HortScience 108(2):266–271

Massa G, Kim H, Wheeler RM et al (2008) Plant productivity in response to LED lighting. HortScience 43(7):1951–1956

Morrow R, Tibbitts T (1987) Induction of intumescence injury on leaf disks. J Am Soc HortScience 112(2):304–306

Morrow R, Tibbitts T (1988) Evidence for involvement of phytochrome in tumor development on plants. Plant Physiol 88:1110–1114

Petitte J, Ormrod D (1986) Factors affecting intumescence development on potato leaves. HortScience 21(3):493–495

Pinkard E, Gill W, Mohammed C (2006) Physiology and anatomy of lenticel-like structures on leaves of *Eucalyptus nitens* and *Eucalyptus globulus* seedlings. Tree Physiol 26:989–999

Rangarajan A, Tibbitts T (1994) Exposure with far-red radiation for control of oedema injury on 'Yale' ivy geranium. HortScience 29(1):38–40

Robson TM, Klem K, Urban O et al (2015) Re-interpreting plant morphological responses to UV-B radiation. Plant Cell Environ 38(5):856–866

Rud, N (2009) Environmental factors influencing the physiological disorders of edema on ivy geranium (*Pelargonium peltatum*) and intumescences on tomato (*Solanum lycopersicum*). Master's thesis. Kansas State University, Manhattan, KS

Sagi A, Rylski I (1978) Differences in susceptibility to oedema in two tomato cultivars growing under various light intensities. Phytoparasitica 6(3):151–153

Seabrook J, Douglass L (1998) Prevention of stem growth inhibition and alleviation of intumescence formation in potato plantlets in vitro by yellow filters. Am J Potato Res 75:219–224

Wargent JJ, Moore JP, Ennos AR et al (2009a) Ultraviolet radiation as a limiting factor in leaf expansion and development. Photochem Photobiol 85:279–286

Wargent JJ, Gegas VC, Jenkins GI et al (2009b) UVR8 in *Arabidopsis thaliana* regulates multiple aspects of cellular differentiation during leaf development in response to ultraviolet B radiation. New Phytol 183:315–326

Wetzstein H, Frett J (1984) Light and scanning electron microscopy of intumescences on tissue-cultured, sweet potato leaves. J Am Soc HortScience 109:280–283

Wheeler R (2010) Physiological disorders in closed environment-grown crops for space life support. 38th COSPAR scientific assembly

Williams KA, Wu Q, Park S et al (2011) Understanding the mechanisms regulating the development of intumescences in tomato through genomic analyses. HortScience 46(9):S267–S268, Abstr

Williams KA, Craver JK, Miller CT et al (2015) Differences between the physiological disorders of intumescences and edemata. Acta Hortic XXIX International Horticultural Congress on Horticulture: sustaining lives, livelihoods and landscapes (IHC2014): 1104, pp 401–406

Wollaeger H, Runkle E (2013) Growth responses of ornamental annual seedlings under different wavelengths of red light provided by light-emitting diodes. HortScience 48(12):1478–1483

Wollaeger H, Runkle E (2014) Growth of impatiens, petunia, salvia, and tomato seedlings under blue, green, and red light-emitting diodes. HortScience 49(6):734–340

Wollaeger H, Runkle E (2015) Growth and acclimation of impatiens, salvia, petunia, and tomato seedlings to blue and red light. HortScience 50(4):522–529

Yelton M, Byrtus J, Chan G (2014) Better tasting basil grown with LED lighting technology. LumiGrow, Inc., Research Brief. http://www.lumigrow.com/download-the-basil-steering-study/

Part VI
Current Status of Commercial Plant Factories with LED Lighting

Chapter 21
Business Models for Plant Factory With Artificial Lighting (PFAL) in Taiwan

Wei Fang

Abstract Plant factory with artificial lighting (PFAL) is an emerging industry in Taiwan and attracting much attention. The number of companies involved in PFAL operations has increased from fewer than 10 to more than 100 in the past 5 years. Unlike traditional agricultural systems, PFALs can eliminate almost all the risks in production, although profitability cannot be guaranteed unless the business model is properly designed and implemented. This chapter focuses on introducing business models currently utilized in Taiwan.

Keywords PFAL • Business model • Taiwan

21.1 Introduction

Plant factory with artificial lighting (PFAL) is an emerging industry in Taiwan, which has been booming since 2010. In 2010, only 1 University and 13 companies were involved in PFALs, as loosely defined at that time. The activities of at least 6 out of the 13 companies that were related to hydroponics in greenhouses are not strictly PFALs. By the end of 2015, 14 Universities and research institutes were undertaking PFAL research and 98 private companies were operating PFAL business in Taiwan.

PFAL is an attractive technology package. It is both new and relevant to consumers' environmental, health, and food safety concerns. Thanks to advances in technology, production risks can be significantly reduced, although profitability cannot always be guaranteed. In Japan, only 25–30 % of PFALs are making profit (Kozai 2014) and in Taiwan is less than that for most of the PFALs in Taiwan are rather "fresh" in the industry. Currently, the number of companies involved is increasing rapidly in Taiwan, making the industry appear promising. However, without a proper business model, setting up a PFAL could still be a possible cash strap.

W. Fang (✉)
Department of Bio-Industrial Mechatronics Engineering, National Taiwan University, Taiwan, Republic of China
e-mail: weifang@ntu.edu.tw

21.2 Business Models

Various business models of PFAL are being utilized in Taiwan. The products can be plants themselves in the form of whole plants, loose leaves, or baby leaves. Product presentation is very important, including packaging in sealed soft plastic bags, soft plastic bags with air vents, or sealed hard plastic boxes. Different types of packaging convey different messages to consumers, for example, "this product does not need to be washed before eating" or "is just as safe as an item grown in a greenhouse or open field." It is also important to emphasize vegetables that are locally grown, not imported. If dressings are provided with salad greens, they must be exceptionally tasty. Thoughtful design of packaging bags and boxes can make the purchasing experience more pleasant compared with that for traditional agricultural products.

Sales channels can be membership based, through websites, within individual companies, or in local communities. Selled by the third-party should be limited. It is clear that selling the products through supermarket chains (owned by others) can only be a temporary measure, since shelf-charge rates are generally high. Therefore, the business-to-consumer (B2C) is more advantageous than the business-to-business (B2B) model for PFAL products.

If a PFAL operation cannot sell all its fresh products, there are other options. One company in Taiwan developed more than ten types of processed items such as ice cream, egg rolls, bread, noodles, face masks, skin cleansers, etc. with vegetable additives. Nutritional supplements in juice, powder, and tablet forms are alternative products that may be priced differently depending on the ingredient added. For example, noodles supplemented with lettuce and with ice plant vary in cost.

Ice plants are a good example of PFAL innovation. They have a salty fruit taste (similar to wax apples). The saltiness can be adjusted via nutrient solutions. Products containing ice plant are currently sold at premium prices due to their novelty, and creative entrepreneurs have invented attractive recipes utilizing it, including ice cream desserts, juices, cocktails, duct breast, and wrapped ground shrimp. It can be marinated in liquor or red wine or given a smoked-pine flavor.

One construction company incorporated the PFAL concept in its plan for a community development to promote a green lifestyle. Each home will be equipped with an appliance to grow vegetables, and the community will have a service division providing seeds, seedlings, stock nutrient solutions, etc. to the residents.

Several companies focus on the development and sale of home appliance-style plant production units and indoor green walls for residential use. One produces aquaponics units for hobby gardeners and homeowners. PFALs with attached restaurants or stands selling organic products are another popular business model in Taiwan. Such shops are normally chain operations located within cities.

Most enterprises constructing PFALs for others have demonstration facilities for the education of potential customers. The most successful demonstration sites are on a scale of daily production of at least 100 plants, have been in stable operation for several months or longer, and have established sales channels. Unfortunately, only a few companies have achieved these goals so far, which makes it difficult to convince the public to start their own PFALs.

Many PFAL businesses are providers of related hardware, such as light-emitting diodes (LEDs), clean rooms, air-conditioning and hydroponic systems, power supplies, and thermal insulation material. These companies generally start by building PFAL demonstration sites and learning to grow plants. Becoming a PFAL turnkey provider is a common goal.

To summarize, there are several distinct PFAL business models utilized in Taiwan (Fang 2014, 2015):

1. Some PFALs grow leafy greens for their own use, for example, restaurant owners and corporations providing fresh, safe, nutritious vegetables to their own customers and employees.
2. PFALs produce leafy greens for Internet customers and/or a membership. Some companies are flexible, even exchanging memberships with other health-related organizations such as yoga clubs. Members can shop at specific PFAL companies and dine in their chain restaurants at discounted prices.
3. PFALs both grow and sell vegetables and offer processed items containing them such as egg rolls, ice cream, bread, noodles, facial and skin care items, and nutritional supplements.
4. PFALs sell their leafy greens through organic chain stores.
5. PFALs entrepreneurs devise creative recipes showcasing the unique flavors of their leafy vegetables such as ice plants, rucola, and basil, mainly for direct sale to restaurants.
6. In a joint venture with construction companies, a service company was established to supply nutrient solutions, seeds, and seedlings to residents in newly built housing developments.
7. Appliance-style PFAL module providers work in collaboration with the construction industry.
8. Appliance-style PFAL module providers operate demonstration facilities.
9. Related hardware providers build PFAL demonstration rooms to sell their products, with the goal of becoming turnkey providers.
10. Related hardware providers build cargo container-style PFALs for transplant production.
11. PFAL turnkey builders and consultants may operate with or without PFAL demonstration rooms.
12. PFALs-related educational services are also viable businesses. In Japan and Taiwan, some consulting firms and universities offer short (half-day to 5-day) and long-term (50-day) workshops. National Taiwan University has offered 30-h courses twice yearly for five consecutive years, in which more than 500 people have been trained.
13. Some PFAL-related firms/foundations/associations organize technical tours domestically and/or internationally.
14. One PFAL turnkey provider works with a rental company that regards production facilities as equipment and therefore can be rental items. Companies

involved in plant production need less initial investment, and monthly rental fees are tax deductible, resulting in significantly more financial flexibility.
15. Some solar panel manufacturing companies rent the roofs of PFALs to install solar panels. The electricity generated is then sold to power companies.

Although PFALs in Taiwan are small in scale, they are flexible and innovative in adopting various business models. At present, some hold great promise, while others have failed. As in other emerging industries, even enterprises following the same business model may succeed or fail depending on a host of factors.

Hydroponics is a well-accepted technology academically and commercially. However, in Taiwan, it is still difficult to convince people that the technology is more than just safe. Books related to plant factory for academic usage (Fang 2011b, 2012) and for general public (Fang and Chen 2014) were published. Answers to 24 frequently asked questions on hydroponics and artificial light grown vegetables were published (Fang 2011a), but some continue to doubt the nutritive value of vegetables grown hydroponically and will not accept those not grown in the ground and/or not using solar light.

The hydroponic industry in Taiwan was booming from 1985 to 1990 for leafy green vegetable production using the nutrient-flow technique (NFT) or deep-flow technique (DFT) and for tomato and cucumber production using drip irrigation with rock wool. However, due to the low dissolved oxygen concentration and bacterial outbreaks in greenhouses during the hot season, the industry gradually faded from 1990 to 2010. With the current PFAL boom, companies involved in hydroponic-related businesses are coming back to life.

21.3 Conclusion

PFALs are booming worldwide. In Taiwan, with no financial and policy supports from the government, private companies are viewing this new industry with great interest. PFAL-related nonprofit organizations have been established to take advantage of horizontal and vertical connections, foster cooperation among companies, and promote business opportunities.

At present, no private agricultural organizations are involved in PFALs in Taiwan. Several farmers' associations considered converting their unused warehouses into PFALs, but finally dropped the idea. The high initial cost is the first concern for potential PFAL operators; difficulties in finding a sufficient number of quality workers and managers are another, at least currently. In addition to academic training in undergraduate and graduate schools, National Taiwan University also offers a 30-hr workshop twice annually. Although more than 550 have taken the workshops, fewer than 10 % actually became involved in the PFAL business. Training of qualified managers and workers for the PFAL industry is an issue that must be dealt with to provide worldwide business opportunities.

To solve the high initial cost problem in the PFAL industry, working with rental companies is one possible solution. However, the design of production systems including multi-layer cultural benches, lighting, hydroponics, controls, air circulation, etc. must be integrated, similar to equipment used in manufacturing. The equipment itself can serve as collateral for start-up costs.

Many companies become involved in PFALs with the ultimate aim of taking advantage of the business opportunities of turnkey projects. However, some were unable to demonstrate that their systems could grow higher-quality plants efficiently. It is unfortunate that some enterprises consider PFALs an opportunity to generate quick profits, which has led to lawsuits and public confusion. The approximately ten international turnkey PFAL projects constructed by Taiwanese enterprises are all in PR China.

Some consumers still question the use of artificial lights and hydroponics to grow food. "Unnatural" chemicals have been a frequently cited complaint in the past 3–4 years. Greater public awareness of how PFALs can lessen food safety and environmental problems, and more frequent media coverage of their benefits, will help consumers learn about and appreciate the technologies involved, as well as increase their willingness to pay premium prices for PFAL products. Important tasks remaining are lowering costs, increasing value, and expanding the vegetable varieties grown. PFALs are expected to coexist with organic and traditional agriculture and will undoubtedly play a key role in urban agriculture in smart/intelligent cities.

References

Fang W (2011a) Some remarks regarding plant factory. Agriculture extension booklet number 67. College of Bioresource and Agriculture, National Taiwan University (in traditional Chinese)

Fang W (2011a) Total controlled plant factory. Harvest farm magazine, Taiwan (in traditional Chinese)

Fang W (2012) Plant factory with artificial light. Harvest Farm Magazine, Taiwan (in traditional Chinese)

Fang W (2014) Industrialization of plant factory in Taiwan. Invited lecture in the Proceedings of the Greenhouse Horticulture & Plant Factory Exhibition/Conference (GPEC), pp 131–181. Japan Protected Horticulture Association (in Japanese)

Fang W (2015) Business models of plant factory in Taiwan. Invited lecture in the Proceedings of Plant Factory 2015: International Symposium on Role and Contribution of Plant Factory Technology and Indoor Vertical Farming to Urban Agriculture and Future Life. Institute of Plant Science and Biotechnology, Research Institute of Agriculture and Life Science, Seoul National University

Fang W, Chen GS (2014) Plant factory—a new thought for the future. Grand Times Publisher, Taiwan (in traditional Chinese)

Kozai T (2014) Topic and future perspectives of plant factory. Invited lecture in the Proceedings of Greenhouse Horticulture & Plant Factory Exhibition/Conference (GPEC), pp 63–96. Japan Protected Horticulture Association (in Japanese)

Chapter 22
Current Status of Commercial Plant Factories with LED Lighting Market in Asia, Europe, and Other Regions

Eri Hayashi

Abstract Plant factories with artificial lightings (PFALs) have been drawing tremendous attention worldwide particularly after 2010. Possibilities of PFALs for the contribution to solve environmental, agricultural, and other social challenges that our globe has been facing have been vigorously discussed globally. Each region or country has their own backgrounds and motivations on the PFALs movements. Various terms to describe this field other than PFAL are used in each region, such as vertical farm, city farm, indoor agriculture, and controlled environment agriculture, and each term has slightly different definitions. This chapter focuses on PFAL with LED lightings. The LED grow lights are one of the key factors of PFALs to economically and technically move the industry forward. PFALs with LED lighting market in Asia, Europe, and other regions are discussed in this chapter.

Keywords Plant factory • PFAL • LED lighting • Market trends • Japanese market • Taiwan • Asia • Europe • MIRAI • SPREAD • Toshiba • PlantLab

22.1 Introduction

PFALs have been drawing tremendous attention worldwide particularly after 2010. Possibilities of PFALs for the contribution to solve environmental, agricultural, and other social challenges that our globe has been facing have been vigorously discussed globally.

Each region or country has their own backgrounds and motivations on the PFALs movements. Moreover, there are indeed various terms to describe this field other than PFAL: vertical farm, city farm, indoor agriculture, and controlled environment agriculture, and each term has slightly different definitions.

E. Hayashi (✉)
Japan Plant Factory Association, 6-2-1 Kashiwanoha, Kashiwa, Chiba 277-0882, Japan
e-mail: ehayashi@npoplantfactory.org

The LED grow light is one of the key factors of PFALs to economically and technically move the industry forward. There have been numerous global lighting companies entering into this field, which has been gradually expanding the LED PFAL market.

22.2 Current Market Status in Japan

22.2.1 Background on the Japanese PFAL Industry

After the 1980s and 1990s, Japan has been experiencing a commercial PFAL boom since 2009.

The first commercial PFAL was established in Japan in the early 1980s. This facility, Miura Nouen, grew lettuce with one-layer cultivation bed and triangular form panel system under HID lamps (Takatsuji 2007). In 1985, another PFAL "Bio Farm" was installed at a vegetable section of a shopping mall in Chiba. Around that time till the mid-1990s, HID lamps were the common light source for commercial PFALs. In 1990s, a number of commercial PFALs equipped with HID lamps were built and started operating. One of the major farms was TS Farm Shirakawa, operated by food company, Kewpie. Today, this farm grows lettuce 4,500 heads per day using aeroponics system, and the cultivation floor area is approximately 2,000 m^2. Many of the farms established around the 1990s are still in production and commercially growing vegetables to date.

Since the late 1990s, commercial PFALs with fluorescent lamps (FLs) and multi-layered cultivation system began to be apparent. While the number of PFALs had been increased gradually each year in the 2000s, the year 2009 was the start of the third PFAL boom in Japan. Besides government-funded projects, mainly by the Ministry of Economy, Trade, and Industry (METI) and Ministry of Agriculture, Forestry, and Fisheries (MAFF), a large number of Japanese companies with various backgrounds started to enter the PFAL industry as growers and/or system manufacturers.

22.2.2 Current Trends of Japanese PFAL Industry

One prominent current trend among the Japanese commercial PFALs industry is the quantities and varieties of industry players. Not only entrepreneurs, food-related companies, and agriculture-related companies, but multiple firms from a wide variety of industries such as electronics, chemical, energy, engineering, materials, transportation, IT, and so forth, including major players of those sectors, have entered the PFAL market. It could be noted that most of them took a step into the industry in the process of searching for new business opportunities. Some have

Fig. 22.1 Nisshinbo strawberry PFAL: cultivation area (*left*) and strawberries (*right*) (Source: Nisshinbo)

become commercial farmers simultaneously being a system supplier or having a vision of being a turnkey provider at some stage.

As of 2015, around 200 "commercial" PFALs in Japan are operating, actually growing, and distributing vegetables. This is the largest number worldwide so far. Besides the fact that the number of commercial PFALs has been increasing every year, the average scale of commercial farms has been increased since 2010. Most of the farms today grow leafy greens, baby greens, or culinary herbs. In addition to those plants, some companies including Nisshinbo have been operating a large-scale strawberry PFAL and distributing strawberries to their multiple clients. Nisshinbo's strawberry PFAL can be seen in Fig. 22.1.

We can see polarization in Japanese PFAL market patterns: (1) aiming for cost reduction of leafy greens in mass production and (2) developing high value-added plant market – i.e., highly nutritious plants applying to health foods or pharmaceuticals, for instance. In regard to business models in the Japanese PFAL market, not only being a farm to grow and distribute vegetables but also offering rental farm spaces for urban farming, being social welfare facilities, and so forth have been getting popular since around 2012. Among many social welfare PFALs, Seiden Kogyo is one such company. Along with being a supplier of their original PFAL systems, called "Social Kitchen," Seiden Kogyo operates their social welfare PFAL "Social House" nearby the headquarters in Takasaki in Gunma prefecture.

Although PFAL market has been gradually expanding, it could be said that the market size is still too limited considering a number of companies scrambling limited opportunities. Particularly for large companies, the industry needs to grow significantly in order for the large companies to maintain and develop their PFALs entities in their firms.

22.2.3 Current Japanese LED PFAL and LED Lighting Market Trends

The majority of PFALs operating in Japan have been equipped with fluorescent lights since the 1990s. However, there was indeed commercial PFAL with LED lighting since the early 2000s. For instance, KupidoFare, a social welfare facility, has been operated since 2003 in Northern Japan.

Since 2011, LED PFALs have been built gradually. Among PFALs newly built in the year of 2012 and 2013, nearly half of them were equipped with LED lightings. Approximately 20 % of all the commercial PFALs operating in 2013 were using LED grow lights. After 2014, large-scale commercial LED PFALs, such as MIRAI or Green Clocks at Osaka Prefecture University, started growing vegetables. Tables 22.1, 22.2, and 22.3 show backgrounds, descriptions of farms, LED used in farms and visions, etc. of MIRAI, SPREAD, and Toshiba, respectively. Also, a couple of pictures of SPREAD's PFAL can be seen in Figs. 22.2 and 22.3. More PFALs, including PFALs equipped with FLs in Japan, are also shown in another book entitled *Plant Factory: An Indoor Vertical Farming System for Efficient Quality Food Production* (Kozai et al. 2015).

Although it is now common that new facilities are equipped with LEDs, there are still more than 100 PFALs estimated to use fluorescent lights in Japan. Those FL PFALs have been examining multiple LED lights to compare lighting options in order to choose the most suitable lightings for their targeted plants. This means there is a largest replacement demand for LED lightings from FL in the Japanese PFAL market today.

As Japan has been facing a transition from FLs to LED lightings, there is a challenge to solve once PFALs replace their light source from FLs to LEDs: besides finding their suited grow lights and lighting environment, how can you maintain the quality of the plant products comparable to the existing products grown with FLs? For those who already have commercially growing vegetables, different light sources give them some obstacles to replicate FL-grown vegetables as LED-grown vegetables including texture, color, and shape of plants, and they look for climate, nutrient, cultivation techniques, etc. suitable to their targeted vegetables.

Almost all the growers raise common questions: "which LED works best for us?", "how can we compare and choose a right one for us?", "where and how can we find practical and objective information on LED grow lights to make a decision?", etc. Some progress has been made in the industry lately to address these concerns by improving transparency of LED grow lights specifications.

As for LED lighting players, Philips and Showa Denko were the major two big players at the beginning of the large-scale LED PFALs era. However, the term "diversity" best describes the market situation especially since 2013 or 2014. In addition to those two players, more LED lighting companies have moved into a variety of lighting businesses. The lighting firms active in Japanese PFAL market include:

Table 22.1 MIRAI Co., Ltd

1.	Location of company headquarters/location of plant factory operating			
	Kashiwa City, Chiba, Japan/ (1) Kashiwa City, Chiba, (2) Miyagi			
2.	Vision for the company/view of the plant factory and vertical farming movement			
	MIRAI Co., Ltd owns two mother factories in Japan, and our core business is twofold: distribution of vegetable produced by the factory for domestic market and distribution of plant factory system itself for both domestic and overseas markets. Through trial and error taking place daily in our own factories, we have been accumulating know-how on plant factory operation and plant cultivation. As we plan to expand our client base into overseas markets, we plan to offer cultivation consulting with our accumulated know-how, and the clients will receive our support until successful vegetable production is achieved even in overseas			
3.	Description of plant factory with LED lighting			
	Location	Miyagi Fukko Park, Tagajo-shi, Miyagi, Japan		
	Start date	July 2014	Floor area	Approx. 2,300 m^2
	Cultivation rack	6~15 layers × 18 lanes		
	Production capacity	10,000 heads of leafy lettuce per day		
	LED lights	LED developed by GE specifically for plant factory, 17,500 pcs		
	Feature	Converted a former Sony Corporation semiconductor factory into the plant factory. Building design, construction, and engineering of cultivation rack system by Kajima Corporation and development and manufacturing of LEDs by GE		
4.	Crops			
	Lettuce, green leaf lettuce, frill lettuce, romaine lettuce, Korean lettuce, basil, Italian parsley, kale, mustard green, pepper mint, spear mint, dill, sage, thyme, watercress, etc.			
5.	Target market			
	Global market including North America, the Middle East, CIS countries, Caribbean countries, etc.			
6.	Current LED being used			
	The LED fixtures were developed by GE specifically designed for better plant growth. We believe the LEDs last longer, consume 40 % less power than fluorescent lights, and result in 50 % higher yield due to emit light at wavelengths optimal for plant growth			
7.	Description of the ideal LED needed for the company			
	(1). Low cost			
	One of the major problems for plant factory business is huge initial cost. As we use approximately 20,000 pcs of LEDs, the investment on LEDs accounts for a large portion of the initial cost. Therefore, cost reduction of LEDs is a key factor for cost reduction of plant factory			
	(2). Less replacement frequency and easy maintenance			
	Long-lasting LEDs with consistent illuminance and high waterproof capability suited in humid environment inside plant factories are preferable			
	(3). Easy installation			
	To save time and labor upon installation of LEDs, it would be desirable to have well-designed LED fixtures for cultivation rack and electrical cable for easy installation			
	(4). Well-suited for cultivation procedures			
	The factory operation entails a variety of procedures ranging from seeding to harvesting. Sometimes LED electrical cable may disturb harvesting work or falling of LEDs may happen due to defectiveness of LED fixtures. It would be desirable if LEDs were designed as			

(continued)

Table 22.1 (continued)

	a part of the entire cultivation equipment including cultivation rack, irrigation pipe, pump, etc., rather than a standalone piece, such that we could minimize the risk of LED problems interfering with the factory operation
8.	Thoughts or opinions on where "lighting" for vertical farming may be headed in the future
	As sensing method and IoT technology progress, MIRAI expects that the company will accumulate more data useful for better LED development and that the most optimized LED development adjustable for each cultivar will progress based on the accumulated data on wavelength, light intensity, etc. Also, as described in 7, the demand for low-cost and easy-to-use LEDs is still high and is expected to continue improving

For this publication one of the plant factories commercially operating and finding success in Japan, MIRAI (Plant Sales and Fresh Vegetables Business General Manager; Nagateru Nozawa and Plant Sales, Manager, Shohei Yoshimoto) was contacted in February, 2016

- Showa Denko
- Philips
- Nihon Advanced Agri
- Keystone Technology
- Ushio Lighting
- Toshiba
- Kyocera
- Stanley
- GE
- Shinetsu Kagaku

One of the interesting Japanese LED lighting trends is that not a few lighting players are indeed involved in commercial farming – distributing plants grown under their LED lightings. Moreover, some companies in Japan focus on distributing their LEDs as part of a package of cultivation system, rather than distributing LEDs as a sole product. On top of that, nowadays there is a new trend on LED lighting business: development of custom-made LEDs in collaboration with PFAL growers. It could be stressed out that PFAL growers or system manufacturers custom-order LED grow lights for themselves from lighting companies. Grower-driven LEDs could be another direction the industry will be moving forward. Flexibility would definitely be essential for achieving optimal lighting environment.

22.3 Current Market Status in Europe, Asia, and Other Regions

22.3.1 *Current Market Status in Europe*

As Asia, North America, or other regions, Europe has also been experiencing a PFAL boom since 2013. It seems this movement in Europe was encouraged mostly

Table 22.2 SPREAD Co. Ltd.

1.	Location of company headquarters/location of plant factory operating	
	Kyoto, Japan	
2.	Brief background/history of the company	
	SPREAD was established in 2006 under the TRADE GROUP, the Japanese market leader for the distribution of fresh produce with business activities from production to sales. Using the knowledge gained from the vegetable market, SPREAD constructed the Kameoka Plant, the world's largest plant factory in terms of production that produces 21,000 heads of lettuce per day	
	The lettuce, branded as Vegetus, is sold in approximately 2,000 supermarkets domestically and SPREAD is one of the few plant factory operators to achieve profitability. SPREAD will begin operations of their highly automated Vegetable Factory™ in 2017	
3.	Vision for the company/view of the plant factory and vertical farming movement	
	We do not think plant factories will solve all of the food problems. However, we want to be able to bring fresh vegetables and prosperity to people around the world by promoting agriculture as a whole with the cooperation of farmers and continuing to put efforts into solving food and environmental problems. We believe that each form of agriculture will play a vital role in the creation of a truly sustainable society	
4.	Description of plant factory with LED lighting	

	Kameoka plant	Vegetable Factory™
Production per day	21,000 lettuce heads	30,000 lettuce heads
Workers	50 people	25 people
Energy cost	1.75 kWh	1.20 kWh
Water usage per head	0.825 L	0.11 L
Lighting	Fluorescent	LED (developed in-house)
Automation	No	Seedling raising to harvest

5.	Cost and energy: improvement on new factory
	Labor costs will be reduced by 50 % due to full automation of the cultivation process from seedling raising to harvest. Energy costs will be reused by 30 % by using low-cost LED lighting that was developed in-house. Water recycling, filtering, and sterilization have been improved with water recycling rates of 98 %
6.	Crops
	Frilly lettuce, pleats lettuce, romaine lettuce, and green and red lettuce
7.	Target market
	Our target market is the end consumers who buy lettuce from retail chains and also businesses such as restaurants or hotels that use lettuce
8.	Current LED being used
	The LEDs that will be used in the new Vegetable Factory™ have been developed in-house
9.	Description of the ideal LED needed for the company
	The ideal LED will be low cost with diverse functionality, because depending on the plant and what kind of product you want to produce, the ideal type of LED will be different
10.	Thoughts or opinions on where "lighting" for vertical farming may be headed in the future
	LED lighting will most likely replace the current standard lights that are available, but there is still much room for improvement. We don't believe anybody has perfected the technology yet

For this publication one of the plant factories commercially operating and finding success in Japan, SPREAD Co. Ltd. (CEO, Shinji Inada), was contacted in February, 2016

Table 22.3 Toshiba corporation

1.	Location of company headquarters/location of plant factory operating
	Shibaura, Minato-ku, Tokyo/Yokosuka, Kanagawa, Japan
2.	Vision for the company/view of the plant factory and vertical farming movement
	The plant factory should be existed to provide safe, clean, long-life, special, and functional vegetables/fruits to any place in the world. Production in a complete enclosed type (artificial light type) plant factory is quite different from the conventional agriculture. Knowledge of agriculture is of course important, but vegetable production in the enclosed type plant factory has aspects of engineering manner that is based on data sensing, statistical analysis, optimization of the process parameters, production control, etc. It can be said that the production in the plant factory is one of the industries as same as semiconductors or electrical products. Thus, manufacturing companies could be the right entity to produce the vegetables in the enclosed type plant factory
3.	Description of plant factory
	We have established our plant factory "Clean room farm YOKOSUKA" in Sep. 2014. This 2,000 m^2 factory has about 8,000 FL light sources, and we are now replacing this FL to TENQOO LED lighting. It is still about 20 % replaced, but finally we are expected to be able to reduce the energy consumption to about 50 %
4.	Crops
	Frill lettuce, cos lettuce, Mizuna, Spinach, Swiss chard
	Processed vegetables (cut vegetables, in-cup mixed salad)
5.	Target market
	Supermarket, salad shop, restaurant, company cafeterias, online shop
6.	Current LED being used
	Toshiba TENQOO LED lighting
7.	Comment of where your business is headed and its view of the future of urban agriculture
	Providing plant factory solutions with keeping own plant factory business

For this publication one of the plant factories commercially operating and finding success in Japan, Toshiba Corporation (Research & Development Division, Marketing Strategic Office, Plant Factory Project Team, Noriaki Matsunaga and Yuichiro Ikeda), was contacted in February, 2016

Fig. 22.2 SPREAD plant factory (Source: SPREAD)

by the concept of urban agriculture. In Europe, it is often called "City Farm" instead of PFAL or vertical farming. The Netherlands, one of the world's dominant providers of greenhouse, has been playing a positive role on the European PFAL

Fig. 22.3 Postharvest process (Source: SPREAD)

market. A noticeable trend in recent years is that greenhouse industry has been emerging their business domain with PFAL market, i.e., entry of leading greenhouse players into PFAL industry. Outstanding example would be Philips. Besides distributing LEDs for supplemental lighting, Philips has been actively placing their grow lights in multiple large-scale PFALs in Japan, North America, and Europe. Furthermore, Philips opened their fully owned PFAL facility in Eindhoven, the Netherlands. Other European LED lighting companies seem to follow this movement as well.

One of the internationally well-known commercial PFAL company is PlantLab. PlantLab started operating their newly built facility in 2014. It is a 6,000 m^2 floor area PFAL with their patent original system, Plant Production Units (PPUs), and is equipped with LED lightings of the US lighting company, Illumitex. Besides headquartered in the Netherlands, PlantLab also has their base in the USA. Not only PFAL manufacturers but most of the major European LED grow light companies have their sales offices in the North America and seem actively involved in the US market. Background and description of PlantLab is described in Table 22.4, and their facility and inside the farm can be seen in Figs. 22.4 and 22.5.

England is another country experiencing PFAL movements. In 2014, an international conference on vertical farming and urban agriculture was held by the University of Nottingham. People often point out that there is a tremendous social demand on safe and locally grown foods particularly in a big city like London. Growing Underground is an underground PFAL in the heart of London. This PFAL grows microgreens and some leafy greens with the multi-layered hydroponic system. Another PFAL in London is GrowUp Urban Farms. This company was founded in 2013. The farm grows microgreens, baby leaf salad, some herbs, and

Table 22.4 PlantLab

1.	Location of company headquarters/location of plant factory operating
	's-Hertogenbosch, the Netherlands
2.	Background/history of the company
	Our mission is to ensure that crops can reach their full potential, so that we can live in a world where a growing population has access to a sustainable source of safe, tasty, affordable, and nutritious food
	Based on our in-depth knowledge of plant physiology, PlantLab developed a patented technology to grow crops in completely closed environment, Plant Production Units (PPUs), without daylight. By utilizing PPUs and our proprietary plant growth algorithms, state-of-the-art LED systems, and air, water, and plant nutrition control solutions, PlantLab's innovations remove the typical variables that hinder crop growth. It is now possible to control crop yields, harvest planning, nutritional content, taste, etc.
	PlantLab's method can be applied for production of vegetables, fruits, flowers, and ingredients for medicines, nutraceuticals, flavors, fragrances, and cosmetics. Besides, our solutions enable urban farming in an economic viable way. Our objective is to partner with leading players in the various industries
3.	Crops
	Full range of arable crops for among others breeding, a broad range of high value vegetable crops, etc.
4.	List of current product portfolio and a description of other areas of commercial horticulture that your company may be involved in
	Turnkey solution for indoor farming – from plant growth recipe to turnkey implementation as well as long-term service level agreements

For this publication one of the plant factories commercially operating and finding success in Europe, PlantLab (Chief Partnership Officer, Ard Reijtenbagh), was contacted in February, 2016

Fig. 22.4 PlantLab facility (Source: PlantLab)

fish, using aquaponics system. This PFAL is equipped with Philips Green Power LED Production Modules.

There are actually many countries in Europe moving forward for the PFAL market creation. Indeed, some PFALs were built and started operating in Belgium. In Russia, Japanese PFAL manufacturers/grower, MIRAI newly built a large-scale PFAL with LED lightings. Photos of MIRAI's PFAL in Russia and vegetable grown there are shown in Figs. 22.6 and 22.7.

Fig. 22.5 PlantLab inside the farm (Source: PlantLab)

Fig. 22.6 Newly built PFAL in Russia by MIRAI (Source: MIRAI)

22.3.2 Current Market Status in Asia and Other Regions

Taiwan is the second biggest country after Japan for the number of PFALs operating in the nation. As is the case with Japanese, in the Taiwanese electric industry, in particular, the LED-related companies entered into the PFAL market in order to look for a new market. It is often said that many Taiwanese LED companies relatively took initiatives and started PFAL-related business around 2010. The majority of PFALs in Taiwan are utilizing LED lightings. The number of PFALs in Taiwan has been gradually increasing since after 2010. In actuality, most of them are small- to medium-size farms, not commercial large-scale facilities.

The Taiwanese PFAL market is best described as widely applicable on their business models. While some are of course simply growing and distributing leafy greens, there is a variety of business models which enable them to apply the plants grown in their facilities into multiple products and services. One good example would be NiceGreen. This company cultivates various kinds of leafy greens and uses those plants as an ingredient in their multiple products ranging from processed foods to cosmetics. Moreover, the company runs many restaurants utilizing their

Fig. 22.7 Vegetable grown in PFAL in Russia (Source: MIRAI)

vegetables for their unique menu. For more details on business models in Taiwan, readers are encouraged to refer to the work introduced in this book by Dr. Wei Fang and another book entitled *Plant Factory: An Indoor Vertical Farming System for Efficient Quality Food Production* (Kozai et al. 2015).

South Korea has also experienced PFAL movements especially around 2010. The South Korean government has been offering research fund projects, and a couple of PFAL-related projects were carried out by universities or research institutions, many times in partnership with companies. For instance, research PFAL facility equipped with LED lightings was built in 2010 at a national research institution as part of a national project. In South Korea, in addition to common leafy greens, highly functional plants including Korean ginseng are commercially grown by PFAL companies. There are PFALs built inside hospitals cultivating medicinal plants. As for South Korean LED grow light companies, Parus is one of the firms involved in this field since the early stage of the PFAL boom. They have been developing their original lightings and cultivation system, aiming to expand their business abroad. North America is one of their target areas, and they already seem to have a collaboration with a North American company to distribute their system. Particularly since late 2015, numerous PFALs have been built in South Korea, and it seems like this country has been experiencing the second PFAL boom.

Singapore has been drawing attention as one of the suitable areas to build and manage PFALs to distribute locally grown vegetables. Since this country is importing the large amount of fresh vegetables, and since high-income households represent a significant fraction of Singaporean population, Singapore has been one of the realistic and appealing target to invest for people interested in PFAL business. Panasonic started operating its PFAL in 2014. They grow various kinds of leafy greens with soil cultivation and supply them to local restaurants, for instance. There is another company called Sustenir Agriculture who operates their flagship farm in Singapore. This farm is equipped with LED lighting of a US company, Illumitex, and grows leafy greens including several kinds of kale. As in Singapore, there are already a couple of PFALs operating in Hong Kong as well. Vegetable Marketing Organization operates PFAL introducing technology of Japanese company, Mitsubishi Chemical. This farm cultivates and distributes baby leaf.

There is a common question PFAL-related companies including LED lighting firms or investors in the world have been asking each other: Is China going to be a huge PFAL market? Like other countries, China has been experiencing a PFAL boom particularly in the R&D field. Although there is almost no commercial PFALs already operating yet, some projects are in progress in multiple cities in China. Some include projects led by other countries, such as the USA and Japan. Moreover, LED lighting industry is one of the major fields that this country as a whole has been and will be focusing on. It is predicted that we would see some improvements in China, not only in PFAL operating market but also PFAL-related businesses including LED grow light industry.

Other areas such as the Middle East, Central and South America, or Africa also have been raising a great interest on PFALs, believing in possibilities of how PFALs could contribute to solve the environmental, agricultural, or other social issues that each country has been facing. There are several projects that companies or institutions in this field have been working on for years in the Middle Eastern countries.

Particularly in Central and South America, PFALs have been more popular than ever before especially since ICCEA 2015, an international conference on PFALs and highly controlled greenhouses, held in Panama by the Foundation for the Development of Controlled Environment Agriculture (FDCEA). This organization has been organizing an educational seminar throughout Latin American countries. In Panama, there is indeed a farm, Urban Farms, already operating.

22.4 Conclusion: Predictions for PFALs with LED Lightings

LED grow lights have made an improvement after 2010. In fact, more PFAL farmers have decided to install LED lightings instead of other light sources such as FLs when building a new farm in recent years. Even though light efficiency and other specifications of LEDs have been at least partly improved, it could be said that this is merely a beginning of an era of LED lightings for PFALs market, which is a globally emerging industry.

European and the North American lighting companies particularly tend to start their grow light business from supplemental lighting. It now looks like the majority of the lighting firms have already passed into the field of PFALs or vertical farms, which makes their products commercially available. Besides research on supplemental lightings, many LED lighting companies have started researching on the effects of LEDs for plant growth and other topics such as controlling nutrient of plants. Many companies are working on the research in collaboration with universities or research institutes.

Today, LED grow light users hope to see improvements in the LED, such as cost reduction, efficiency, quality, intensity, and flexibility. Moreover, almost all the growers foremost hope to learn cultivation technique or so-called plant recipe suitable for the LED that they use. Therefore, in addition to distributing LED lighting system itself, growth recipe along with LED systems would gain much more attention. This also means that software businesses of LED grow lights would be playing a more major role than it is today, and it is predicted that lighting companies with sustainable business models would expand their businesses onto software businesses in the near future.

As an emerging worldwide industry, the PFAL system will be more diversified in each region for various reasons. Reflecting the diversity, PFAL technologies and industry itself are expected to further develop in the coming years with innovative concepts, products, utilities, transparent specifications, and user-friendly growth recipes.

References

Kozai TG, Niu M, Takagaki (eds) (2015) Plant factory, 1st edition: an indoor vertical farming system for efficient quality food production. ISBN:9780128017753, Elsevier, MA, USA, pp 351–386

Takatsuji M (2007) Plant factory. Ohmsha pub, Ohmsha, p 125 (in Japanese)

Chapter 23
Current Status of Commercial Vertical Farms with LED Lighting Market in North America

Chris Higgins

Abstract The North American indoor vertical farming industry, known as plant factories (PFALs) in Japan and as city farms in Europe, is best described as nascent. Although there has been a significant amount of discussion and enthusiasm about this industry, in actuality, very few commercial vertical farms have been built in the USA and Canada. However, there have been significant developments and improvements in electric light sources for producing plants without sunlight. This chapter will focus on the following topics: background of the North American vertical farming industry, market trends for vertical farm lighting, LED companies and predictions for lighting vertical farms as a growing industry.

Keywords Trends • Vertical farming • Indoor Ag • LED grow lights • Farmbox Greens • Hort Americas

23.1 Introduction

The North American indoor vertical farming industry, known as plant factories (PFALs) in Japan and as city farms in Europe, is best described as nascent. There has been a significant amount of discussion, hype, and enthusiasm about this method of plant production. In actuality, very few commercial vertical farms have been built in the USA and Canada.

There are still many questions about whether vertical farms can be economically sustainable. However, there have been significant developments and improvements in electric light sources for producing plants without sunlight.

C. Higgins (✉)
Hort Americas, LLC, 2801 Renee St., Bedford, Texas 76021, USA
e-mail: chiggins@hortamericas.com

© Springer Science+Business Media Singapore 2016
T. Kozai et al. (eds.), *LED Lighting for Urban Agriculture*,
DOI 10.1007/978-981-10-1848-0_23

23.2 Background on the North American Vertical Farming Industry

While the North American vertical farming industry has gained in popularity as a concept over the past 6 years, the current designs of vertical farming can be traced back to the 1980s and 1990s. Today, young plant producers are growing indoors using a wide variety of methods to produce an expanding number of crops.

Dutch and Dutch-influenced greenhouse growers have tried and sometimes have successfully used multilayer growing systems to produce commercially viable ornamental and vegetable crops. Most of these systems used sunlight with the aid of supplemental light from high-intensity discharge (HID) and fluorescent light sources.

An American farmer in Illinois created a multilayer indoor farm using HID grow lamps to cultivate lettuce as early as 1992. A wide variety of indoor farmers have pushed innovation to produce cannabis, both legally and illegally, in various hydroponic growing systems using a variety of electric light sources. Unknowingly these cannabis growers may become one of the more important influencers in the vertical farming industry. Cannabis growers use similar technologies while producing a more economically valuable crop that allows them to make more investments in innovation.

These trailblazer growers are not alone in their efforts to advance the development of new technology for vertical farming. Vocal figures like Dr. Dickson D. Despommier, professor emeritus of microbiology and public health at Columbia University and author of the book *The Vertical Farm*, key thought leaders and influencers at the University of Arizona's Controlled Environment Agriculture Center (CEAC), and other researchers at prestigious North American universities are working diligently to forward the efforts of early adopters and their visions of redefining agriculture's future in the current movement in North America.

Over the past 6 years, some key vertical farms and farming systems, for a variety of reasons, have gained media attention and have inspired investors, inventors, and entrepreneurs to create the early North American vertical farming industry. Some of the high-profile vertical farms that have garnered much this media attention include: AeroFarms (New Jersey, USA), Green Sense Farms (Indiana, USA), Green Spirit Farms (Michigan, USA), FarmedHere (Illinois, USA), iBio (formerly Caliber Biotherapeutics) (Texas, USA), Urban Barns Foods (Quebec, CA), Urban Produce (Ontario, CA), and Truleaf (Nova Scotia, CA).

There are other vertical farms that are operating and finding success. For a variety of reasons, these other vertical farms either don't want the attention or the media has not yet identified them. A few of these are Uriah's Urban Farm (Florida, USA), Greener Roots Farm (Tennessee, USA), Farmbox Greens (Washington, USA), Local by Atta (New Brunswick, CA), Ecobain Gardens (Saskatchewan, CA), and Ecopia Farms (California, USA), to name a few. There are many other operations that have not been listed. As of January 2016, this list is steadily growing. Table 23.1 shows background, description of farm, and vision, etc. of

Table 23.1 Farmbox Greens

1.	Location of company headquarters
	Seattle, Washington, USA
2.	Brief background or history of the company
	Farmbox Greens is an urban, vertical farm located in Seattle, Washington, USA. We specialize in growing a wide variety of microgreens and culinary herbs. What's unique about our farm is that we grow year-round indoors, in a climate controlled facility using aeroponic and hydroponic systems and the latest lighting technologies
	We started growing using aeroponics and LEDs in early 2011. We expanded into our current facility with about 70 m^2 (750 sq ft) of growing area in 2013. We're small, but we focus on optimization and maximizing revenue per square foot within the space we have
3.	Vision of the company
	We want to grow the business to a scale that makes a difference. Where exactly that journey will take us is somewhat unknown since the industry is so new. I also believe deeply in the triple bottom line and would like us to make a positive impact on our customers, community, and industry as a whole
4.	Crops
	We grow 15+ varieties of microgreens, shoots, and culinary herbs. This includes arugula, pak choi, mizuna, radishes, peas, sunflower, basils, sorrels, cilantro, parsley, and others
5.	Current LEDs being used
	We have 17 W white TLED and Gen 1 Dark Red/Blue Philips GreenPower Production Modules
6.	Description of the ideal LED needed for the company
	We would like to see the installation process made easier. Jumpers from one light to another would be very helpful with the installation. Also, running low voltage wire with a remote transformer would mitigate heat and potential safety concerns with electrical and water being in close proximity. It would also be nice to see the efficiency continue to improve while also having a decreasing price point
7.	Thoughts or opinions on where "lighting" for vertical farming may be headed in the future
	Less expense and more efficient

For this publication, one of the vertical farms commercially operating and finding success in the North America, Farmbox Greens (Owner, Dan Albert), was contacted in February 2016.

Farmbox Green. Pictures of their products and inside the farm with LED lightings can be seen in Figs. 23.1 and 23.2.

Unfortunately, North America has already seen its fair share of vertical farms that have gone out of business. This is not surprising as the vertical farming industry is in its infancy and is trying to find a way to exist in a highly competitive, price-sensitive market in which no proven models for success exist.

In order to be successful in today's vertical farming market, the following equation is most often true:

high plant density + low light crops + short production cycle + niche crop +niche market = success

This is why almost all North American vertical farms are producing similar crops. These crops include leafy greens, baby greens, microgreens, and culinary

Fig. 23.1 Radish grown in Farmbox Greens vertical farm (*left*) and package labeling (*right*) (Source: Farmbox Greens)

Fig. 23.2 LED lighting being used in Farmbox Greens farm (Source: Farmbox Greens)

herbs. All of these crops are high value in many North American cities at certain times of the year. This is especially true because they are locally grown and can be sold year round.

23.3 Current Vertical Farming Lighting Market Trends

Current vertical farming trends in North America vary significantly. One thing that vertical farming enthusiasts have in common is that they are each trying to determine how they should be involved in the industry. Many are also trying to "catch lightning in a bottle" as predictions of the potential size of the vertical farming market continue to grow. This applies to universities, institutes, inventors, investors, small and large businesses, farmers, want-to-be farmers, seed companies, suppliers, and retailers.

There are still a lot of unanswered questions. Some of these questions revolve around light, but many of them are focused on other variables of vertical farming that are just as important, including seed technology, climate management, water, nutrition, and labor.

There is one question industry participants are asking that is the same. How does each of these factors play an integral role in making a vertical farm both environmentally and economically sustainable?

Since the focus of this chapter is on light and more specifically LEDs, here are three simple facts about plants and grow lights.

1. Everyone (hopefully) agrees that plants require light to grow.
2. Everyone, for the most part, also agrees that plants respond to light intensity or PPFD (photosynthetic photon flux density, $\mu mols/m^2/s$), light quantity (daily light integral, $mols/m^2/day$), and light quality (400–800 nm).
3. The North American vertical farming industry has all but accepted the fact that LEDs will eventually be the lights of choice for the vertical farming industry. However, based on the current cost of LEDs, there are many growers who are still investing and installing other light fixtures including plasma, induction, and fluorescent. The reason for these alternative choices is capital costs and access to financial resources.

After these facts, current trends, thoughts, and beliefs within the North American vertical farming industry seem to go in every direction. How much light is needed? What is the best way to the deliver light? What is the most efficient way to deliver light? What light produces the best quality plants? What is the best lighting fixture? What company makes the best lighting fixture? As an industry there has been little consolidation in belief and this should not be surprising based on the current state of the industry.

One of the few barriers to entry into the vertical farming market is access to capital. Those companies limited in capital or worried about the heavy upfront cost of LED lighting are looking at alternative light fixtures. They are looking for lights that promise equal performance at a lower cost than LEDs.

Another topic of debate is the best spectrum of light to grow plants under. While there seems to be less discussion of this topic among researchers, the vertical farming industry continues to argue this topic as more products and different technologies are positioned to gain market share.

For answers to these questions, refer to the work introduced in this book and another book entitled *Plant Factory: An Indoor Vertical Farming System for Efficient Quality Food Production* (Kozai et al. 2015).

At the same time, light suppliers are quickly trying to identify the proper fixture design for vertical farms. The North American market has made this extremely difficult for lighting manufacturers as almost every vertical farm has created a unique design for a propriety growing system. These propriety designs mean that every light must be installed slightly different while performing to a fairly defined set of criteria. The industry is demanding constant customization and innovation while at the same time requiring costs be reduced.

As use of LEDs in other industries continues to increase, suppliers and manufacturers have been able to focus on efficiency (defined in $\mu mol/J$). There is currently a race between lighting manufacturers to see which one can design the LED fixture that can deliver the most light for the least amount of energy consumed. This in itself is extremely important for the development of the vertical farming industry because it allows growers to grow crops more profitably as newer fixtures produce the same amount of light with less heat. This reduction in heat allows for a decrease in heating, ventilation, and air conditioning costs. Climate management is a major cost and a big challenge for many North American vertical farms.

23.4 Current Status of the LED Lights Designed for Vertical Farms

There is no shortage of lighting companies competing for market share in the North American vertical farming industry. Lighting giants including Philips, GE, Toshiba, and Panasonic are aggressively positioning themselves as leaders of not only lighting equipment but also as plant-production knowledge companies. Relatively new lighting companies like Illumitex are promoting that they are as knowledgeable as the large companies. These new companies are also saying that because of their size, they are more responsive to the constantly changing demands of vertical farmers. The purpose of this chapter is not to determine which company offers the best products. The purpose of this chapter is only to identify which companies are currently focused on the vertical farming industry and to describe their current position and offering within the vertical farming market.

The lighting industry is changing very quickly, comments made in this section were done so in January 2016. Please check with your lighting suppliers for updated information. Lighting companies that are most active in the North American vertical farming industry include:

Philips: currently offering GreenPower LEDs designed for "City Farming" applications including leafy vegetables and ripe fruit without daylight
Illumitex: currently offering the Eclipse™ series designed for sole-source lighting in environments such as vertical farms and growth chambers
LumiGrow: currently offering LED grow lights and SmartPAR software empower growers and researchers with the ability to schedule and tune light levels and spectra

Lighting companies that have developed product lines (sometimes in other markets) and are now dedicating resources that will allow them to be more active in the North American market vertical farming market include:

GE
Heliospectra
Fluence Bioengineering
Intravision
Valoya
PL Light Systems

Companies like Hort Americas, a wholesale horticultural/agricultural supply company, are not directly involved in the manufacturing of LED grow lights but are influential in both the sales and educational process of using or adapting LED lighting technology to the commercial horticulture and agriculture industries.

23.5 Conclusion: Predictions for Vertical Farm Lighting

Excuse the pun, but the future is bright for those involved with vertical farming, as well as other aspects of controlled environment agriculture. As consumers continue to demand a wider variety of food crops that are grown locally, without pesticides, guaranteed safe, and available year round, there will be growers who can meet these needs. Some of these successful growers will be vertical farmers. Lighting will play a major role in these farmers' ability to service these and other markets. At this point in time, it is not possible to predict who those successful growers will be.

What is known is that lighting has and will continue to play a critical role in the vertical farming industry.

- Plants require light to grow. Plants require a specific amount of light over a given period of time. Some plants require and also grow better under specific light spectra.
- Researchers are just starting to scratch the surface of what is possible as tools are continuously developed that allow for better control of different aspects of light under different growing conditions.
- It will only be a matter of time before seed companies and plant breeders are able to hybridize seeds that express traits that are capable of taking advantage of consistent and predictable environments, including specific light spectra, created in vertical farms.

Based on the fact that light is necessary for plant growth, lighting companies will continue to design and manufacture lights that are one of or a combination of the following:

- Less expensive
- More efficient
- More intuitive
- More flexible

With the development of these new tools, it is guaranteed that the entrepreneurial spirit of the North American farmer and businessman will take over. These tools will be used to create agribusinesses for commercial crops already discussed as well as new crops that one could never imagine being commercially viable today.

Reference

Kozai T et al (2015) Plant factory, 1st edn. An indoor vertical farming system for efficient quality food production, ISBN:9780128017753, Elsevier

Chapter 24
Global LED Lighting Players, Economic Analysis, and Market Creation for PFALs

Eri Hayashi and Chris Higgins

Abstract Lighting environment with LED grow lights is one of the main components of PFALs or vertical farming technology. We see more and more PFAL growers have lately been installing LED lightings as a light source and see the increasing number of lighting companies entering into this industry worldwide. There is a question whether producing plants with light source such as LEDs without sunlight can ever be economically feasible. Moreover, actual growers are facing challenges comparing LED lighting options and deciding a right source which is economically sustainable for them. Sharing a sample of how and what to consider while comparing light options by economic analysis should be beneficial. In this chapter, we first overview global LED lighting players, discuss market analysis on growing with LED lighting, rethink the current PFAL and LED grow light business structure trends, and finally discuss a market creation for PFALs with LED lightings.

Keywords LED lighting company • Japan • Taiwan • USA • The Netherlands • Sweden • Norway • Economic analysis • Sample spreadsheet • Comparing lighting options • Innovative mind

24.1 Introduction

Lighting environment with LED grow lights is one of the main components in PFALs or vertical farming technology. More and more PFAL growers have lately been installing LED lightings as a light source, and we see the increasing number of lighting companies entering into this industry worldwide. Accordingly, it should be no surprise to see questions rising whether producing plants with light source such as LEDs without sunlight can ever be economically feasible. Moreover, actual growers are facing challenges comparing LED lighting options and deciding a right source which is economically sustainable for them. Sharing a sample of

---

E. Hayashi (✉)
Japan Plant Factory Association, 6-2-1 Kashiwanoha, Kashiwa, Chiba 277-0882, Japan
e-mail: ehayashi@npoplantfactory.org

C. Higgins
Hort Americas, LLC, 2801 Renee St., Bedford, Texas 76021, USA

how and what to consider while comparing light options by economic analysis should be beneficial. In order to truly expand the industry, examining the current structure of market players and where this industry will be headed must be meaningful.

24.2 Global Plant Factory LED Lighting Players

24.2.1 Global Trends of Plant Factory LED Lighting Players

There has been a significant increase in the number of LED grow light companies active in worldwide PFALs market. Some lighting firms in Europe or North America entered into the field of PFAL lighting particularly sometime after 2010, after being in the supplemental lighting business. In contrast, Japanese lighting companies tended to start their grow light business targeted mainly at PFALs industry from the beginning. There are multiple greenhouses equipped with HID lamps in Japan, where mainly leafy vegetables are grown. However, greenhouse supplemental lightings are not yet as common as in European countries such as the Netherlands where natural sunlight are scarce during a certain time of the year.

24.2.1.1 Lighting Companies Designing LED Grow Light for PFALs or Vertical Farms

For this publication, the lighting companies that are active in the global PFALs, vertical farming industry, including North America, Asia, and Europe were contacted. Activity was determined by each company having supplied at least one commercial PFAL or vertical farm installation. The lighting players contacted are as follows. Countries of each firm were determined by where headquarters are located.

Asia: Keystone Technology (Japan), Kyocera (Japan), Nihon Advanced Agri (Japan), Showa Denko (Japan), Stanley Electric (Japan), Toshiba (Japan), Ushio Lighting (Japan), Solidlite (Taiwan)

Europe: Heliospectra (Sweden), Intravision (Norway), Philips Lighting (the Netherlands)

North America: Current, powered by GE (USA), Illumitex (USA)

One of the salient characteristics of global business of PFAL lighting companies is that European and North American players are already active outside of their domestic markets, while Japanese are not so far. This may mean that Japanese firms are relatively slow in developing international businesses. Almost all the major European LED grow light companies already have their branch offices in North America, or North American companies are active in Europe or in some cases in Asia.

24.2.2 Plant Factory LED Lighting Players Headquartered in Asia

Asian lighting companies contacted were Keystone Technology (Japan), Kyocera (Japan), Nihon Advanced Agri (Japan), Showa Denko (Japan), Stanley Electric (Japan), Toshiba (Japan), Ushio Lighting (Japan), and Solidlite (Taiwan), and company visions, descriptions of LED lightings, future prospects, etc. of those companies are shown in Tables 24.1, 24.2, 24.3, 24.4, 24.5, 24.6, 24.7, and 24.8, respectively. Moreover, structure and descriptions of Kyocera's LED lightings can be seen in Figs. 24.1 and 24.2, structure of single tip system of Nihon Advanced Agri in Fig. 24.3, and a picture of their plant factory in Fig. 24.4. Pictures of PFAL equipped with LED grow lights of Stanley Electric are shown in Fig. 24.5.

Table 24.1 Keystone Technology

1.	Company name and location of company headquarters
	Keystone Technology Inc.
	Yokohama, Kanagawa Japan
2.	Vision for your company and/or view of the plant factory and vertical farming movement
	We believe plant factory in urban area could be a way to settle the worldwide environmental problems. As to make sure business of our clients to succeed, we provide our clients with new market for distributing vegetables and brand consulting services so that we, beyond being just a system supplier, would spread the future food production system throughout the world
3.	Description of your plant factory with LED lighting/crops/target market
	Shin-Yokohama LED vegetable garden (showroom) and Basyamichi LED vegetable garden
	Leaf lettuce, swiss chard, basil, edible flowers
	Restaurant, farm stand, EC shopping
4.	Description of your LED lightings
	We have developed our original power LED with corresponding to the absorption spectrum of chlorophyll. We also have a system to adjust red, blue, and green LEDs independently based on the growth phase of our products (controllable wavelength for the best at each growth phase (independently controlled R/G/B)). Our highly adjustable system allows us to grow our products faster than natural sunlight, and we envision using this technology to produce vegetables with higher nutrition value
5.	List of key projects
	Colowide MD (started in 2012/Yokosuka city)
	We constructed a LED-equipped plant factory on the second floor of central kitchen of restaurant chains. This allowed us to minimize logistics cost and to stably supply high-quality vegetables
	Shune365 (started in 2015/Hatano city/floor area 300 m^2)
	This company built LED farm utilizing their own unused factory aiming to enter a new market as a part of their diversification strategy. Using "Hatano Spring Water" which is certified by Ministry of the Environment as spring water, they intend to cultivate high nutritional vegetables

(continued)

Table 24.1 (continued)

6.	List of current product portfolio and a description of other areas of commercial horticulture that your company may be involved in
	We have our original products ranging from home-use LED lighting for gardening to cultivation units for commercial PFALs
7.	Comment of where your business is headed and its view of the future of urban agriculture
	We focus on offering consulting service besides distributing our systems. We put high priority on differentiating plants grown in our PFALs system from conventional vegetables. We believe our systems are suitable for growing high functional plants
8.	Thoughts or opinions on where "lighting" for vertical farming may be headed in the future
	We see LED grow lights will be improved, nicely affected as consumer LEDs greatly spread worldwide. We have devoted ourselves to develop state-of-the art original LED lighting system, believing that "green business" is a future. There will be indeed a certain amount of demands for economical white LEDs, which we believe those would be applied for cost-conscious large-scale commercial PFALs. Our RGB LED lightings with independent control systems are suitable for relatively small- to medium-scale PFALs focusing on growing high value-added plants

For this publication, one of the LED lighting companies active in the global PFALs industry, Keystone Technology (CEO Seiichi Okazaki), was contacted in February 2016

Table 24.2 Kyocera

1.	Company name and location of company headquarters
	KYOCERA Corporation
	Kyoto, Japan
2.	Vision for your company and/or backgrounds of your business
	We developed our original LED lighting suitable for growing vegetables in plant factories and launched sales in October 2015
3.	Description of your LED lightings
	Aiming for the light similar to natural sunlight, we have developed a system to combine purple LED element with red/green/blue phosphors and adjust ratio of each phosphor and wavelength. This system allows us to generate wavelength of light suitable for plants photosynthesis. We have also developed our original ceramic material specifically designed to protect LED element, and this allows the LED lamp to maintain luminosity around 100,000 h. With these technologies, we will achieve stable vegetable production
4.	List of current product portfolio and a description of other areas of commercial horticulture that your company may be involved in
	We have been developing businesses around our high-performance LEDs based on phosphor formulation using Kyocera's purple LED as an exciting light. LEDs designed for plant production is part of such active business development in Kyocera. Our main LED lighting business areas include high CRI LED (general lighting), inspection lights, lights for museum, medical lightings, and so forth
5.	Thoughts or opinions on where "lighting" for vertical farming may be headed in the future
	While LED light has been playing one of the pivotal roles in plant productions in factories, we do not simply want to rely on the existing LEDs. We believe it is important to step back and investigate what kind of lighting environment is best suited for plant productions and how LED technologies can help the productions
	We consider it is our top priority to develop product flexibly as there are more than one academic approach suited for varieties of plant

For this publication, one of the LED lighting companies active in the global PFALs industry, Kyocera (Noritake Kimoto), was contacted in February 2016

Table 24.3 Nihon Advanced Agri

1.	Company name and location of company headquarters	
	Nihon Advanced Agri CO., LTD/Shiga, Japan	
2.	Vision for your company and/or backgrounds of your business	
	We are advancing the science of health and beauty through precision agriculture! We are a LED lighting/plant factory systems company, a farmer of high nutrient plants, a health food company, and a restaurant owner. Our business lines include LED grow lights, a unique plant factory system, high nutrient plants cultivated under our original cultivation techniques, health foods with high nutrient plants grown in our farm, and restaurant business	
3.	Description of your plant factory with LED lighting/crops/target market	
	Using our 3 wavelengths wideband LED, we operate plant factory in Shiga, growing high nutrient plants, such as ice plant, rosalina, and microgreens. We also produce health foods with ice plant, which was grown, dried, and powdered in our original farm. Those fresh plants and health foods are targeted to retail shop, wholesale market, and restaurant and also delivered directly to consumers ordered through our webpage	
4.	Description of your LED lightings	
	Our LED lights are based on a patented single-chip design that emits red, blue, and green bands which is similar to natural sunlight	
	While reducing energy consumption (23 W), our LEDs output a balanced spectrum and high light intensity	
	3 types of white LED lighting: blue-white, red-white, and white	
	Both built-in power and external power source types	
	UL/CSA, IPX4	
5.	List of key projects	
	A couple of large-, medium-, or small-scale commercial plant factories in Japan and test plants in other countries including the USA	
6.	Comment of where your business is headed and its view of the future of urban agriculture	
	We promote vertically integrated plant factory business ranging from production of nutrient-rich functional vegetables to health food, utilizing our original cultivation system and techniques which can induce functional nutrients by stressing plants. In regard to our LED grow light business, besides East and Southeast Asia and other regions in the world, we are targeting the North American and European market with UL/CSA	
7.	Thoughts or opinions on where "lighting" for vertical farming may be headed in the future	
	From agriculture to life science	

For this publication, one of the LED lighting companies active in the global PFALs industry, Nihon Advanced Agri (CEO, Akinari Tsuji), was contacted in February 2016

Table 24.4 Showa Denko

1.	Company name and location of company headquarters
	SHOWA DENKO K.K.
	Minato-ku Tokyo Japan
2.	Vision for your company and/or view of the plant factory and vertical farming movement
	Plant factory, where we can grow fresh vegetables anytime anywhere without being affected by climate or soil, is a growing market, and we see this industry will greatly expand globally
	We have a total support system in our team and are developing our service based on our original high-speed cultivation methods – S methods (SHIGYO® method). Our business lines include plant factory system with LED grow lights and related systems
3.	Description of your LED lightings
	We use highly efficient red (660 nm) and blue (450 nm) LED lights
	Our tube-type LED light (length 1.2 m) contains independent lighting control system for red and blue, which enables high-speed cultivation technique (S method)
	Our system allows adjusting light intensity and red/blue ratio suited for individual plant species. This allows us to achieve stable production as well as energy saving
4.	List of key projects. Projects should be 1000 m^2 or greater and can exist anywhere in the world
	KiMiDoRi (Kawauchi-mura Fukushima prefecture, Japan)
	Kyoei Kogyo (Gifu prefecture, Japan)
	TJ Create (Hyogo prefecture, Japan)
5.	List of current product portfolio and a description of other areas of commercial horticulture that your company may be involved in
	Cultivation system for plant factory/vertical farm
	LED grow lights for natural grass
	LED supplemental grow lights for greenhouse
	Co^2 supply system
6.	Comment of where your business is headed and its view of the future of urban agriculture
	Given our high-performance LED lightings (hardware) and high-speed cultivation techniques (software), we aim to expand our highly productive agriculture technology into plant factory, greenhouse, controlled environment agriculture, nursery system, etc.
	As for overseas business, any regions unsuited for conventional farming such as Southeast Asia, Europe, etc. are our main target
7.	Thoughts or opinions on where "lighting" for vertical farming may be headed in the future
	LED lights for agriculture is expected to have multiple effects by optical response to LED wavelength besides promoting photosynthesis

For this publication, one of the LED lighting companies active in the global PFALs industry, Showa Denko (GIP, Ryoichi Takeuchi), was contacted in February 2016

Table 24.5 Stanley Electric

1.	Company name and location of company headquarters
	STANLEY ELECTRIC CO., LTD.
	Meguro-ku, Tokyo, Japan
2.	Vision for your company and/or view of the plant factory and vertical farming movement
	We believe agriculture has to be transformed from "nonprofitable" to "profitable" agriculture
	Globally competitive agriculture is needed in Japan considering TPP issue
	We need to change the way of agriculture to realize highly productive agriculture driven by the global food issues
	We need to develop a new agriculture almost independent from climate changes in order to address the food production crisis caused by global climate change
3.	Description of your LED lightings
	Appearance configuration: thin, surface light source
	Structure: light reflection and diffusion applying our original technology
	Natural light color that enables you to visibly inspect the health and growth of plants
	Uniformity of light illumination
	Specifically designed as grow lights: wavelength suitable for photosynthesis
	Sulfur resistance
4.	List of current product portfolio and a description of other areas of commercial horticulture that your company may be involved in
	Our original LED grow light business targets not only PFALs but also seedling facilities and greenhouse. Our lighting could be utilized for PFAL, supplemental lighting or tissue culture, etc.
5.	Comment of where your business is headed and its view of the future of urban agriculture
	We expect LED lighting market to expand in numerous fields such as
	Plant factory: to be profitable market
	Supplemental lighting: to achieve high yield anywhere
	Control: utilizing lights as "signal" besides energy source
	Fuel production including algae culture or artificial photochemical
	Water sterilization treatment
	Aquaculture
	High value-added agriculture including pharmaceutical plants production

For this publication, one of the LED lighting companies active in the global PFALs industry, Stanley Electric (Nobuhisa Kanemitsu), was contacted in February 2016

Table 24.6 Toshiba

1.	Company name and location of company headquarters	
	Toshiba corporation	
	Minato-ku, Tokyo, Japan	
2.	Vision for your company and/or view of the plant factory and vertical farming movement	
	The plant factory should be existed to provide safe, clean, long-life, special, and functional vegetables/fruits to any place in the world. Production in a complete enclosed-type (artificial light-type) plant factory is quite different from the conventional agriculture. Knowledge of agriculture is of course important, but vegetable production in the enclosed-type plant factory has aspects of engineering manner that is based on data sensing, statistical analysis, optimization of the process parameters, production control, etc. It can be said that the production in the plant factory is one of the industry as same as semiconductors or electrical products. Thus, manufacturing companies are the right entities to produce the vegetables in the enclosed-type plant factory	
3.	Description of your LED lightings	
	We have developed our LED lighting for plant cultivation as one of the lineups of TENQOO series which is a product of Toshiba Lighting and Technology Co., Ltd. By using common platform, we have succeeded in reduction of cost and development time. R&D section can make prototype LED, and we actually use it in our mass production plant factory and feedback to the R&D section. So we can provide reliable products	
	Specification	
	Rated voltage: AC100 V~242 V	
	Power consumption: 46.0 W (200 V)	
	Photosynthetic photon flux PPF: 78.5 $\mu mol\ s^{-1}$	
	Light source life: 40,000 h (PPF retention rate, 90 %)	
	Built-in power unit	
	Body: steel (white)	
	LED Bar: polycarbonate (milky white)	
	Ingress protection: IP23	
	Comparison with fluorescent light	

	FL (our conventional)	LED (TENQOO for plant cultivation)
PPF ($\mu mol\ s^{-1}$)	47.6	78.5
PPF 40,000 h ($\mu mol\ s^{-1}$)	33.3	70.7
Power consumption (W)	46.0	46.0
Energy efficiency ($\mu mol/J$)	1.03	1.71
Energy efficiency 40,000 h ($\mu mol/J$)	0.72	1.54
Light source life	10,000 h	40,000 h
PPF retention rate	70 %	90 %

(continued)

Table 24.6 (continued)

4.	List of current product portfolio and a description of other areas of commercial horticulture that your company may be involved in
	LED and lighting control system/UV light for sterilization/air conditioner
	Cultivation rack systems/sensor node with 920 MHz band wireless/sensing logger
	Security camera system/functional water (alkaline for washing/acid for sterilization) generator
	Factory control system/remote monitoring system/production control system
	Energy management system/portable plant factory system (book shelf type)
	Plant factory design and construction/IT tools (Note-PC, Tablet-PC)
	Merchandise labeler/visible light photocatalysis
5.	Comment of where your business is headed and its view of the future of urban agriculture
	Providing plant factory solutions with keeping own plant factory business
6.	Thoughts or opinions on where "lighting" for vertical farming may be headed in the future
	Understanding of optimum spectrum for the plants has already been matured, we believe. It should be focused on an efficiency of the light system, cost, operation, and reliability

For this publication, one of the LED lighting companies active in the global PFALs industry, Toshiba Corporation (Research and Development Division, Marketing Strategic Office, Plant Factory Project Team, Noriaki Matsunaga and Yuichiro Ikeda), was contacted in February 2016

Table 24.7 Ushio Lighting

1.	Company name and location of company headquarters
	Ushio Lighting
	Chuo-ku, Tokyo, Japan
2.	Vision for your company and/or view of the plant factory and vertical farming movement
	1. We have to differentiate ourselves from conventional vegetables by, for instance, enriching nutrient of plants or focusing more on traceability
	2. It is clearly essential to meet consumers' various needs. Otherwise, it seems almost imperative to face cost competition with conventional vegetables
3.	Description of your LED lightings
	1. Control system enables us to control red and blue individually
	2. Wireless control from our computer for lighting control, on/off
	3. Our software on our computer allows a precise adjustment of lighting conditions and managing schedule
	4. With our wireless control system, no need for wiring for lighting/on and off control. Therefore, reduction of installing cost and workload
	5. Easy grouping of cultivating area and able to create various environments suited for each group
	6. High rigidity, due to our original bar design, not fluorescent tubes
4.	List of key projects. Projects should be 1000 m^2 or greater and can exist anywhere in the world
	Nihon Yamamura Glass (Amagasaki city, Hyogo prefecture, Japan)
	In this plant factory equipped with more than 10,000 LED lightings, we grow multiple kinds of vegetables. By adjusting red/blue emission ratio or lighting schedule on the computer, we flexibly meet the needs based on specific crops or shipment time

(continued)

Table 24.7 (continued)

5.	List of current product portfolio and a description of other areas of commercial horticulture that your company may be involved in
	Our group company Ushio Inc. has invested in Hokkaido Salad Paprika and help them establish cultivation technology such as improvement of the crop yield using supplemental lighting system. We have been working with them to establish the technology by providing our HID lamps used as their supplemental lighting system
6.	Comment of where your business is headed and its view of the future of urban agriculture
	1. Firstly, we target domestic market, specifically to farms who grow high-value crops
	2. In the future, we will plan on expanding our LED grow light business abroad by collaborating with European or Asian group companies
	3. We intend to expand our business on supplemental lightings to large-scale greenhouse in addition to our current business targeted toward plant factories
7.	Thoughts or opinions on where "lighting" for vertical farming may be headed in the future
	Polarization in LED grow light market patterns: economical vs. high-performance LED grow light market. For instance, we predict there will be simple white light LEDs to grow general leafy greens, while there will also be LED grow lights with color adjustment features that can accommodate the needs to produce high-value vegetables

For this publication, one of the LED lighting companies active in the global PFALs industry, Ushio Lighting (Toru Fukushima), was contacted in February 2016

Table 24.8 Solidlite

1.	Company name and location of company headquarters
	Solidlite Corporation
	Fongshan Village, Hukou Township, Hsinchu County 303, Taiwan
2.	List of key projects. Projects should be 1000 m^2 or greater and can exist anywhere in the world
	1. Hon-Si engineering company: growing vegetables and strawberry
	2. Arwin Corporation: growing vegetables
	3. Green Seasons Corporation: growing vegetables and mushroom
3.	List of current product portfolio and a description of other areas of commercial horticulture that your company may be involved in
	Specific application:
	1. Seedling
	2. Orchid/tissue culture, etc.
	3. Leafy vegetables
	Some other LED products:
	1. LED flat panel
	2. LED tube

(continued)

Table 24.8 (continued)

	3. T10 LED tube
	4. Track light
	5. 150 w supplementary light
	Commercial horticulture product lists:
	1. Lighting for "indoor vertical green wall"
	2. Lighting for "aquaponics"
	3. Lighting for "tissue culture"
	4. Supplementary lighting for greenhouse
4.	Comment of where your business is headed and its view of the future of urban agriculture
	As mentioned above, LED grow light is essential for PlantFactory, and PlantFactory is playing a big role to overcome food security issues, since there are serious pollutions and other issues in the world for conventional farming
5.	Thoughts or opinions on where "lighting" for vertical farming may be headed in the future
	We expect future trend of urban agriculture to be leisure style, especially for family and restaurant such as "EZ Cuppa" and "Aquaponics"

For this publication, one of the LED lighting companies active in the global PFALs industry, Solidlite (Kevin Huang), was contacted in February 2016

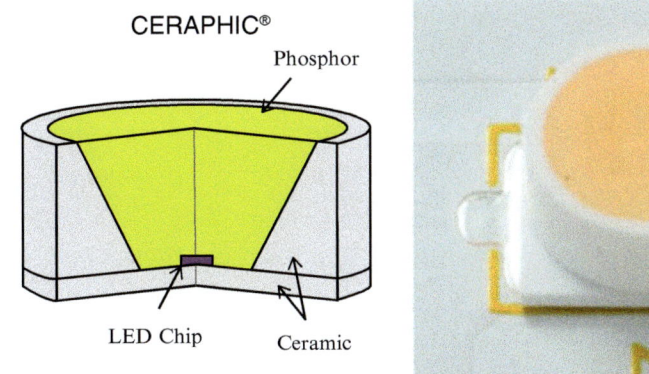

Fig. 24.1 Structure of Kyocera's LED grow lights (Source: Kyocera)

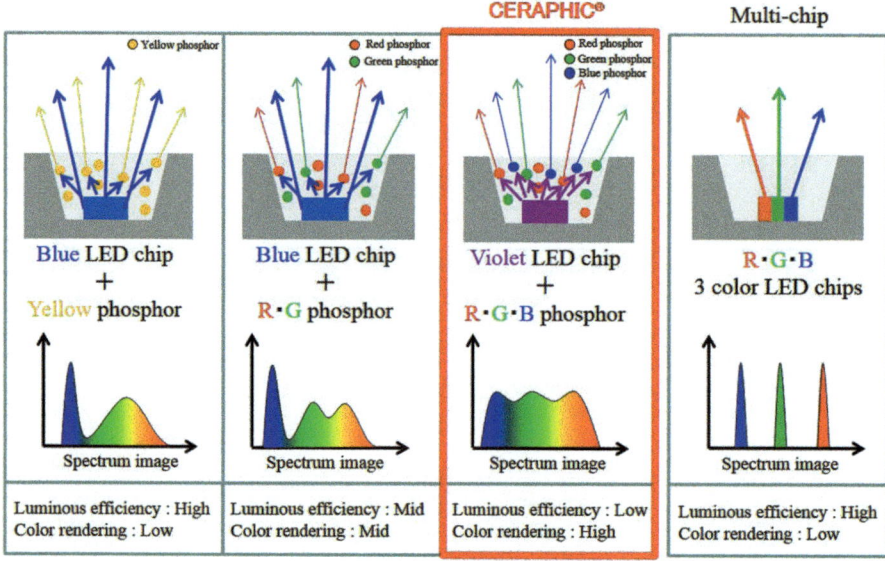

Fig. 24.2 Emission system of Kyocera's LED grow lights (Source: Kyocera)

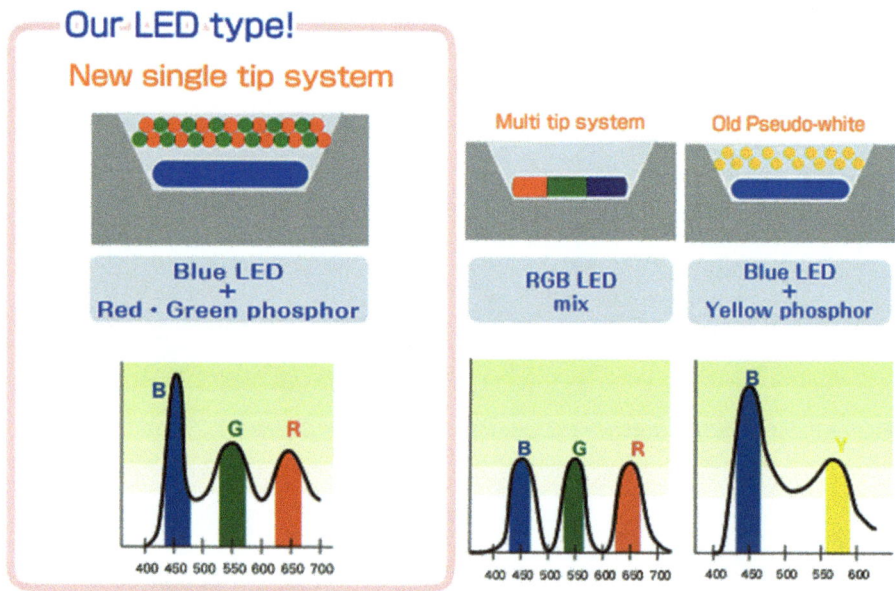

Fig. 24.3 New single tip system: Nihon Advanced Agri LED grow lights (Source: Nihon Advanced Agri)

Fig. 24.4 Plant factory: growing highly nutritious plants for health foods under grow lights (Source: Nihon Advanced Agri)

Fig. 24.5 PFAL with LED lightings of Stanley Electric at Store (Ito-ya) in Ginza (Source: Stanley Electric)

24.2.3 Plant Factory LED Lighting Players Headquartered in Europe

European players contacted were Heliospectra (Sweden), Intravision (Norway), and Philips Lighting (the Netherlands), and company visions, descriptions of their LED lightings, product portfolios, future prospects, etc. of those companies are described in Tables 24.9, 24.10, and 24.11, respectively. An overview of Heliospectra's product portfolio is shown in Fig. 24.6.

Table 24.9 Heliospectra

1.	Company name and location of company headquarters
	Heliospectra AB
	Headquarter: Göteborg Sweden
	Office: San Francisco, CA, USA
2.	Vision for your company and/or view of the plant factory and vertical farming movement
	While cannabis production grabs the headlines in North America, the most interesting story is food security. To feed a growing world population, the UN estimates that food production in the world must increase by 70 % over the next 40 years. The world is facing a growing population and stretched natural resources, pushing the agricultural industry and more specifically the commercial greenhouse market to undergo a change. Smart greenhouses are changing the way of growing, and modern farmers are adopting more efficient practices and technologies to produce our food supply, resulting in an increase in so-called urban and vertical farming. On top of this, a newfound enthusiasm on the part of consumers for quality produce grown in a sustainable fashion is spreading
	Greenhouse growers have historically relied on high-intensity discharge (HID) lights to supplement the sun in lower-light seasons. But HID lights are derived from general illumination technology, emitting a broad spectrum of light that is of little use to plants. This results in runaway energy waste and cost. For lack of a viable alternative, HID technology persisted in greenhouses even as significant advances were made in managing other growth variables including temperature, humidity, CO2, and nutrients. With the entry of LEDs, a new lighting technology was introduced to provide growers with a focused spectrum giving growers control over light. With the advent of spectrally adjustable LED lighting, such as that available from Heliospectra, growers can take it one step further and achieve consistently high-performing crops, while they keep operating costs down year-round. With the emergence of smart greenhouses and more controlled environments, we see more and more interest from growers looking for energy-efficient, smart lighting solutions
	We believe Heliospectra's technology fits well as we are one of few suppliers who offer a system solution where we integrate sensors, software, and our intelligent lights which allow us to connect with other control systems such as ventilation systems. In a not so distant future, Heliospectra's patented "biofeedback system" will enable growers to take real-time measurements on the light wavelengths that plants absorb and thereby tailor the light based on each plant's needs. This technique saves energy and produces better products for growers
3.	List of current product portfolio and a description of other areas of commercial horticulture that your company may be involved in
	Heliospectra intelligent lighting has been developed for a number of different application areas including research applications, commercial greenhouses, indoor growth facilities, and vertical farming. In conjunction with researchers and leading growers, we have developed a number of different lighting solutions to meet the needs of our customers
	Our patented LED grow lights are one of the most technologically advanced grow lights available for the horticulture market. Heliospectra's light systems allow growers to produce plants that look and taste better, have a longer shelf life, and increase the overall quality and operational yield. Our patented solution allows growers to create customized lighting spectrum recipes that may be able to affect the marketability and quality of the plants while at the same time give growers the potential to reduce the energy consumption in a grow operation by up to 50 %
	See Fig. 24.6 for an overview of our product portfolio:
	1. *Heliospectra RX30 – research applications*
	As a former research company, we believe in the scientific advantage our products offer because we use them ourselves on a daily basis in our in-house, controlled environment research facility which includes over 20 different growth chambers

(continued)

Table 24.9 (continued)

Designed by our botanists and engineers, we believe the RX30 is the world's most flexible plant science R&D tool suitable for research and laboratory use. Having our plant research lab, we understand the importance for researchers to work with tools that give them the flexibility, stability, and precision they need. The Heliospectra RX30 Series is a 400 W, programmable LED light system with nine (9) wavelength channels ranging from 380 nm (UVA) to 735 nm (far-red), designed and engineered to give researchers a scientific advantage
2. *Heliospectra LX60 and E60 – commercial growers and indoor grow facilities*
Greenhouses offer growers climate control for their crops while also promoting energy efficiency and utility savings. To meet the market's demand, our highly engineered Heliospectra light system has been designed and engineered to replace traditional lighting solutions in commercial horticulture environments
3. *Heliospectra E60*
Designed to replace a 1000 W HID. The E60 is a fixed spectrum, high-intensity LED grow light developed together with some of the world's leading greenhouse growers. The plug-and-play nature of the E60 provides instant light. Designed in Sweden to handle harsh greenhouse conditions, this fixture provides growers with high-quality light for year-round crop production. The E60 offers:
A wide, uniform PAR light distribution
Two-tiered optics
Advanced heat dissipation solution
Removable duct flange
Intense light output
4. *Heliospectra LX60*
The fully adjustable LX60 Series has been third-party verified as one of the most efficient and versatile on the market, with efficiency equal to a 1000 W HID and only half the energy consumption. It comes with our built-in control system allowing growers to create customized lighting spectrum recipes for a variety of growth regimes that improve the quality of specific plants. The LX60 comes in two different models. The LX60 offers:
A wide, uniform light distribution
Two-tiered optics
Individually controllable wavelengths
Heat dissipation solution
Removable duct flange
Software control via WiFi and Ethernet
5. *Heliospectra LightBar – vertical farming*
Heliospectra LightBar is a high-intensity, fixed spectrum LED grow light with a spectrum optimized for horticulture. The LightBar is optimized for water cooling and specifically developed for vertical farming. The LightBar offers:
A uniform light distribution
High-intensity light
Optimized light spectrum for photosynthesis and/or photomorphogenesis response
Waterproof (IP66)
Liquid cooling

(continued)

Table 24.9 (continued)

4.	Key projects	
	Spisa AB	
	Founded in 1995, Spisa is today Europe's largest provider of ecological potted herbs with operations in seven markets. In addition to growing herbs, Spisa is also a large provider of different salad greens. The environment and quality have always been a key focus, and Spisa has therefore always been seeking new technologies and improvements	
	As Sweden's northern weather did not supply the daily amount of light needed for healthy and good-looking plants, Spisa needed high-quality, energy-efficient supplemental lighting for the long winter months and was looking to replace HPS lamps	
	Heliospectra calculated the needed PAR for their crops and provided lights with a spectrum specially designed for them. Today Spisa is using Heliospectra LED lights in three locations in Sweden, including both the LX602G and E60, and in the autumn of 2015, Spisa bought another 1400 light bars for their propagation facility	
	Results: results compared to their former HPS lighting solutions includes a	
	46 % decrease in energy consumption	
	Improved plant vigor	
	Improved taste	
	Increased shelf life	
	Overall improved plant quality	
	Further, young plants' survival rate has improved and waste has been reduced	
5.	Comment of where your business is headed and its view of the future of urban agriculture	
	We see a clear trend driving the movement toward greenhouses and controlled environments agriculture (CEA) food production. With stretched natural resources and an increased need for food production, we see a need for more efficient agriculture and a trend for locally grown produce, as well as an increase in incentives for efficient production. At Heliospectra we will increase our focus in North America on commercial greenhouse growers but also see it as natural to support our customers in the emerging urban agriculture and vertical farming market segments	
	Heliospectra aims to be part of the development of these new markets and is already one of the driving forces behind future-oriented projects aiming to develop effective growing systems with minimal consumption of light and water resources. Two of these projects include a new Middle East project aiming to create a pilot facility combining energy-efficient products, thereby advancing the efficiency of plant growth. Heliospectra is also part of an international consortium which constitutes a part of the EDEN Initiative, a research program developed by the German aerospace center, DLR Institute of Space Systems (ISS). EDEN ISS' primary objective is to develop, integrate, and demonstrate various crop cultivation technologies and operating processes for safe food production on board the International Space Station and for future manned space expeditions	
	Due to their reconfigurability and considerably cooler operation vs. HID lamps, LED lights are uniquely capable of supporting crop growth in the full range of growing environments from greenhouses to converted warehouse space to purpose-built plant factories. For the same reasons, we believe LEDs are viable in every latitude and climate zone, enabling truly sustainable global urban agriculture	

For this publication, one of the LED lighting companies active in the global PFALs industry, Heliospectra (Marketing Manager, Rebecca Nordin), was contacted in February 2016

Table 24.10 Intravision

1.	Company name and location of company headquarters
	Intravision Group AS
	Headquarter: Oslo, Norway
	Offices: Canada (Intravision Light Systems) I China (Intravision Technologies)
2.	Vision for your company and/or view of the plant factory and vertical farming movement
	Short intro: Intravision Group is a bio-light and system integration company working mainly with new technologies for production of foods and plant-made pharmaceuticals. Our focus is on multiband and spectrum variable LED lights, enabling biological control and optimization of plant responses to specific wavelengths from UVA to IR light. We both initiate and partake in associated research projects. We have headquarters in Oslo, Norway, with offices in Shenzhen, China (production and sales for Asia), and in Toronto, Ontario, Canada (R&D and sales for the Americas)
	Intravision Group partnered in 2010 with the University of Guelph's Controlled Environment Systems Research Facility (CESRF). The cooperation matched Intravision's spectrum variable LED light technology with CESRF's competence and technology within controlled environment production of plants. This enabled the development of new knowledge on how to optimize light spectrum variations and photoperiods to trigger desired responses in plants and to optimize plant production practices, including: shorter generation cycles, greater yield, and improved quality of specific valuable components. Initially technology development and testing focused on high-value products like phyto-pharmaceuticals, high-value biotech products, and food security projects for extreme habitats. Presently this includes large-scale efficient indoor food production systems and supplementary light systems for greenhouse horticulture
	The controlled environment movement: Our view is based on the facts we are faced with which is a view common among our peers but increasingly in step with an advancing general awareness. The UN projects an increase in world population from the present day 7.3 billion to 10.75 billion by 2050 and describes this situation as one we are unprepared for. The available land area for farming is decreasing – due to rapidly growing urban centers and massive infrastructure developments. The availability of clean water is an even greater problem. Currently 80 % of freshwater usage is agriculture related. Compounding this problem is the fact that heavy pollution and general overuse is already creating enormous problems presently for some of the most heavily populated areas on the planet. In Shanghai the groundwater base is currently down at a depth of 300 m, with saltwater penetrating the drained basins for miles beneath the city. Further inland the Yellow River itself – the water source for some 140 million people – is drying up due to over usage. In California, the water shortage problems are proportionately similar
	Presently, we are in principle able to produce enough food for our 7.3 billion population, but we are neither able to properly distribute nor use it. According to the UN, approximately 995 million people are presently starving; one additional billion live in states of insufficient nutrition, and on the other end of the food chain, one billion struggle with obesity-related issues. In order to feed a balanced nutrition to 10.75 billion people, we will have to produce 70 % more food on the planet compared to 2015. And this will have to be done on roughly the same agricultural land area as we have today, except with substantially less usage of water
	The solution demands a multitude of innovations in areas ranging from increased productivity, GMO, new production technologies, innovations in nutrition compositions, food design, distribution, and farming. We believe the technology developed by Intravision Group, the Modular Agricultural Production System (MAPS), can be one of the important building blocks in this future providing:
	Continuous year-round production in a controlled environment
	Safe food, with no use of pesticides and other chemicals
	Optimized taste and nutrition content by controlling plant environment and optimizing nutrition and light spectrum
	Energy- and space-efficient technology using multilayer production
	Locally produced, can be distributed inside the marketplace

(continued)

Table 24.10 (continued)

3.	List of current product portfolio and key projects
	With our technical platform in spectrum adjustable LED light systems and controlled environments, Intravision delivers both turnkey systems, as well as separate LED light systems, some examples on ongoing projects/products:
	FLOW – Space-efficient automated GMP production system for GMO tobacco for Platform Cooperation (cancer medicines and vaccines). Controlled environment "flow-through" factory for large-scale systems developed by Intravision with a patent pending. The technology can be applied on all leafy greens with generation time in production of 4–5 weeks
	MAPS (Modular Agricultural Production Systems) – Intravision/CESRF – technology developed for
	1. Food security applications in Kuwait, for the Kuwait Institute of Scientific Research
	2. Controlled environment production system for flowering food plants and leafy greens
	*A high light intensity PAR1000 MAPS variation is being tailored for production of medical marijuana in cooperation with licensed Canadian company ABcann Medicinals – incorporated with water-cooled LEDs – enabling efficient heat conversion to be executed externally from the growth area
	COLUMN – Space-efficient production system tailored for strawberries in controlled environments. Strawberry plants are arranged on specially designed columns and moved around by a conveyer system. Light spectrum presently being tailored for maximizing flowering and for taste/nutrition. System tested for 2 years in a greenhouse, but presently being outfitted with a spectrum variable LED light system. Second phase project in planning to include an agricultural robot to prune plants and pick berries
	SPECTRA – Spectrum variable lamps for supplementary lights in greenhouses and for single-layer production of medical marijuana
	SPECTRA BLADES – Spectrum variable lamps tailored for applications like ZipGrow towers, in cooperation with Bright Agrotech and Modular Farms
	MODULAR FARMS – Intravision and CESRF-engineered Modular Farms, a mobile plant production system – based on a GRP Sandwich Construction – in the shape of an oversized version of a freight container. The first Modular Farm will be finalized in March 2016, and the first commercial versions will be operational in remote areas in northern Canada in the fall of 2016
4.	Comment of where your business is headed and its view of the future of urban agriculture
	Research and development will continue to be a focus led by Intravision Group in Norway; commercial sales and manufacturing will expand in both Canada and China through the Groups' sister companies. We will continue our work to understand how to optimize light spectrum variations in order to trigger desired responses in plants, aiming to shorten generation cycles in production and increase yield and quality of certain valuable components. But we will also help growers and distributors integrate light systems that consider the homogeneity of the growing environment – light, nutrition, air – necessary to achieve advanced indoor agriculture

For this publication, one of the LED lighting companies active in the global PFALs industry, Intravision (founder and CEO, Per Aage Lysaa and Business Development, Joshua D. Siteman), was contacted in February 2016

Table 24.11 Philips Lighting

1.	Company name and location of company headquarters
	Philips Lighting, Horticulture LED Solutions/Eindhoven, Netherlands
2.	Vision for your company and/or view of the plant factory and vertical farming movement
	Vertical farming will be complimentary to traditional field and greenhouse farming. The UN estimates that the world's population will grow by 80 % by 2050 and that 80 % of available land is already being used for farming. Producing with LED lighting provides another alternative to traditional methods that can help ensure there is enough food supply for the world's rapidly growing population. The moment that vertical farms will be commercially adapted depends on a lot of different factors, like the demand for local grown vegetables from consumers and retailers and the price which they want to pay for this local grown food. The moment that the systems are more common, they will be optimized, which makes them more efficient which results in lower cost prices per crop
3.	List of key projects. Projects should be 1000 m^2 or greater and can exist anywhere in the world
	Some of our projects:
	(a) GrowUp: UK
	(b) Osaka Prefecture University: Japan
	(c) Green sense farms: USA
	Our research facilities:
	(d) GrowWise: Eindhoven, the Netherlands. This facility is fully owned by Philips
	(e) BrightBox: Venlo, the Netherlands
4.	List of current product portfolio and a description of other areas of commercial horticulture that your company may be involved in
	We see the Philips GreenPower production module is most used in vertical farms. This system is based on long history and experience in lighting in the horticulture industry. We believe that it's of outstanding quality and best performing in its range. Although a sufficient lighting system is inevitable to grow indoors, we rather speak about the total growth system since its essential to integrate all ingredients of a farm. Only when the lighting, climate, CO2, fertilizers, etc. are optimally used and integrated will you get maximum return on investment, e.g., highest kilograms per mole, or best taste, or best color, etc.

(continued)

Table 24.11 (continued)

	Comment of where your business is headed and its view of the future of urban agriculture
5.	As this is a young and emerging market, not a lot of formal analysis is available on this yet. This makes it hard to provide numbers. We do see a healthy growth in this segment. At this moment, we see that there are several commercial projects successfully operational. We see that the market is more and more ready for this new technique
	Next to this, we are researching promising areas such as using light to influence factors beyond growth: things like a plant's disease resistance and the nutritional value of fruit and vegetables. So we are sure that there is a bright future ahead of us which will be reality much sooner than expected

For this publication, one of the LED lighting companies active in the global PFALs industry, Philips Lighting (Global Director City Farming, Gus van der Feltz and Global Marketing Manager City Farm, Marjan Welvaarts), was contacted in February 2016

Fig. 24.6 An overview of Heliospectra's product portfolio (Source: Heliospectra)

24.2.4 Plant Factory LED Lighting Players Headquartered in North America

North American players contacted were Current, powered by GE (USA) and Illumitex (USA), and their company visions, descriptions of their LED lightings, product portfolios, future prospects, etc. can be seen in Tables 24.12 and 24.13. Pictures of Illumitex LED lightings are shown in Figs. 24.7 and 24.8.

Table 24.12 Current, powered by GE

1.	Company name and location of company headquarters	
	Current, powered by GE	
	Boston, MA, USA	
2.	Vision for your company and/or view of the plant factory and vertical farming movement	
	The UN reports the urban population projected to be at 6.3 billion by 2050 and the food demand to increase by 60 %. The world doesn't have enough land mass to cover the food demand. Vertical farming/plant factories are going to become a way to feed the world. Indoor farms operate continuously and produce plants using a fraction of the acreage of traditional farms. With LED lighting, farms can layer plants such as lettuce or greens often as high as 9 m. Artificial lighting is a must where it enables producers to grow indoors. GE LED horticultural lamps are designed to help grow through the different stages of plant growth	
3.	List of key projects. Projects should be 1000 m^2 or greater and can exist anywhere in the world	
	Current, powered by GE with GE Japan Corp, developed customized LED lighting for the Mirai lettuce farm in the city of Tagajo that is located in a facility that previously hosted a Sony electronics manufacturing center located in the Miyagi Fukko Park	
	Development of the farm was subsidized by the Japanese Ministry of Economy, Trade and Industry (METI) and is part of the reconstruction effort as the region recovers from the impact of the Great East Japan Earthquake that occurred in 2011	
	Mirai worked with GE to develop LED lights tuned to the lettuce growing cycle. METI also encouraged the use of technology in the farm that uses customized sensors and control systems for all aspects of the agricultural environment. The partners say that the result is 50 % better plant production relative to FL farms along with a 40 % reduction in energy usage	
	The farm occupies 2300 m^2 of floor space and there are 17,500 LED lights in use. The lettuce is grown in cultivation racks to maximize the use of the high ceilings. The result is a 10,000-head daily harvest. Meanwhile, the success of the project has the partners looking to develop other LED-lit farms in Japan and to inquiries from interested parties in other regions of the globe	
4.	List of current product portfolio and a description of other areas of commercial horticulture that your company may be involved in	
	GE offers three LED product solutions, LED Batten and LED Inter-Lighting and LED Top Lighting. The LED Batten is designed to facilitate vertical farming/plant factories by providing the most suitable spectrum of light which allows plants to grow indoors efficiently under ideal conditions. The Inter-Lighting and LED Top Lighting platforms are designed for both vertical farming and greenhouses. These platforms also are available in three growth spectrums: reproductive, vegetative, and balanced. Reproductive has a higher red content to promote flowering and fruit generation. Vegetative has higher blue content to promote healthy and thick leafy plants. Balanced is an even mix of red and blue ratio to promote overall growth	
	We also offer Lucalox PSL high-pressure sodium lamp specially developed for greenhouses. These offer the benefits of stable lumen and micromole maintenance and full spectrum content that promotes photosynthesis. Photosynthesis active radiation (PAR, measured in μmols s^{-1}) is essential for plant growth. Lucalox™ is available in four wattages, 250 W, 400 W, 600 W, and 750 W	
5.	Comment of where your business is headed and its view of the future of urban agriculture	
	Vertical farming/plant factories provide a practical opportunity to address global challenges – from hunger to the depletion of natural resources. Seventy percent of freshwater is utilized in agriculture globally. With a rise in world population and depletion of natural resources such as freshwater, arable land, and phosphorous (for fertilizing crops), the need for alternative farming methods is dire. Vertical farms/plant factories use 98 % less water, 70 %	

(continued)

Table 24.12 (continued)

	less fertilizer, and absolutely no arable land. There is no need for pesticides and no water/fertilizer run-off due to closed water systems. Crops grow independent of climate, harvests are larger, and there is no threat of crop loss due to drought or other natural disasters. It is estimated that one 30-story vertical farm can supply more than 50,000 people a nutritionally sound, 2000 cal diet for an entire year
	Crop yield relative to minimizing the distance between the farm and consumers is another plus. Yield loss happens during harvest, transport, and shelf life. Minimizing the transport time and distance cuts yield loss and extends shelf life
	There are reasons behind the market potential beyond the plant productivity and energy savings discussed earlier. The farm-to-table and organic food trends are also drivers of indoor farming. Inside you can control the environment better for organics. And the farm-to-table premise essentially requires growers to operate near population bases
	With customized data analytics, energy, and lighting solutions, Current is well poised to help support the acceleration of this growing industry

For this publication, one of the LED lighting companies active in the global PFALs industry, Current powered by GE (Lead Product Manager, Sharee Thornton), was contacted in February 2016

Table 24.13 Illumitex

1.	Company name and location of company headquarters
	Illumitex, Inc.
	Austin, Texas, USA
2.	Vision for your company and/or view of the plant factory and vertical farming movement
	From vaccine-producing vertical farms to city-wide indoor farming operations, Illumitex is at the forefront of LED lighting control, precision, and accuracy. Illumitex provides a full solution approach to horticulture lighting with expert staff to help design, layout, and customize your grow operation for maximum yield and potential
	Outside providing full-service solutions for our customers, Illumitex is committed to the vertical farming movement through educational videos, webinars, and scientific papers
3.	List of key projects. Projects should be 1000 m^2 or greater and can exist anywhere in the world
	Project GreenVax: USA
	FarmedHere: USA
	Uriah Farms: USA
	Green Spirit Farms: USA
	Buckeye Farms: USA
	Sustenir: Singapore
4.	List of current product portfolio and a description of other areas of commercial horticulture that your company may be involved in
	Besides vertical farming solutions, Illumitex provides a wide array of horticulture lighting solutions for large indoor growing operations and greenhouse supplemental lighting. Several of the most prominent greenhouses in North America have selected Illumitex LEDs for their switch from legacy lighting to LED technology. In Africa, Syngenta is using our red LED Safari fixture on a large scale. Also, we have outfitted growth chambers at EGC and Conviron. Finally, we have several universities which are using our lights for research including Penn State, Cornell, the University of Nebraska-Lincoln, and the University of Arizona at the CEAC (Center for Environmental Agricultural Control) as of February 2016

(continued)

Table 24.13 (continued)

5.	Comment of where your business is headed and its view of the future of urban agriculture
	Continuing to improve fixture designs and make products more suited for vertical farming. Creating low wattage versions of our LED bars (for low shelving systems) and incorporating drivers and high-quality electronics into the bars for modularity. The ability to remote-control dynamic lighting is also being developed with smart-ready technology and remote sensing. Overall we continually strive to be worldwide leaders in horticultural lighting solutions

For this publication, one of the LED lighting companies active in the global PFALs industry, Illumitex (Director of Horticulture R&D and Marketing, Rebecca Knight, PhD), was contacted in February 2016

Fig. 24.7 Illumitex staff using Illumitex Eclipse bars in two different spectrums (F3 and X5) (Source: Illumitex)

Fig. 24.8 Power harvest in Mucci Farms (Source: Illumitex)

24.3 Economic Analysis

There has been an enormous amount of questions among people who are already involved or planning to be involved in the industry: if PFALs could be economically feasible. Economical aspect or cost structure of PFALs in general is discussed in another book, *Plant Factory: An Indoor Vertical Farming System for Efficient Quality Food Production* (Kozai et al. 2015).

Economical sustainability of PFALs largely depends on lighting costs, including initial capital and operating cost, among several other factors. Those costs depend on individual cases. Each product has different fixture cost and specifications, and each growers target different plants, cultivating with various designed farms and have various cultivation methods. This leads to various requirements on LED lighting that largely affect the amount of initial costs. As for operational cost, it should be emphasized that electricity rate differs in each country or area, which tremendously affects the operating cost.

In reality, almost all the growers face the challenges to compare economic feasibility of different lighting options. Table 24.14 shows a sample spreadsheet of comparing lighting options. In this table, readers are encouraged to consider their own cases for comparing lighting options. This is an example of the way which factors need to be taken into account while examining the economical sustainability.

Figure 24.9 shows some case studies of transition of built-up cash for lighting options. For this analysis, mainly (1) fixture cost, (2) life expectancy, (3) wattage, (4) crop weight, (5) quantity of lights, (6) electricity rate, (7) total operating hours

Table 24.14 Sample spreadsheet for comparing lighting options

			Product A	Product B	Product C	Product D	Product E
1.	Fixture cost		$	$	$	$	$
2.	Quantity of lights	Mounting height =					
3.	Target light levels (PPFD) (μmol m^{-2} s^{-1})						
4.	Total operating hours/day						
5.	DLI (mol m^2 d^{-1}) or photoperiod (h/day)						
6.	Operating hours/year						
7.	W/day						
8.	kWh/day						
9.	Electricity rate ($/kWh)		$	$	$	$	$
10.	Operating cost/day		$	$	$	$	$
11.	Operating cost/month		$	$	$	$	$
12.	Operating cost/year (e.g., at $0.10/kWh)		$	$	$	$	$
13.	Initial capital cost		$	$	$	$	$
14.	Life expectancy						

Source: Created by Chris Higgins
Notes:
1. Quantity of lights is always based on mounting position (or distance of light to plant).
2. Amount of supplemental light will vary by crop. 200 μmol m^{-2} s^{-1} is the average target for leafy greens and culinary herbs grown in the USA.
3. Light intensity and light quantity will be two of the primary factors in driving yield.
4. Operating hours may be dependent on variable or fixed energy pricing as well as basic plant requirements.
5. Initial capital cost is not a true representation of total cost of ownership.
6. DLI denotes daily light integral.
7. PPFD denotes photosynthetic photon flux density.

Assumptions
Size of farm/project.
All lights have or are made from a LED array that is suited for plant growth.
Target mols.day.m^2 is known by grower.
Cost does not include other equipment and installation (transformers, mounting accessories, etc.).
Grower knows cost kWh.
Growers understand that equipment wattage will impact cooling cost (i.e., operating cost).

Est Cost per Fixture
Product A: $
Product B: $
Product C: $
Product D: $
Product E: $

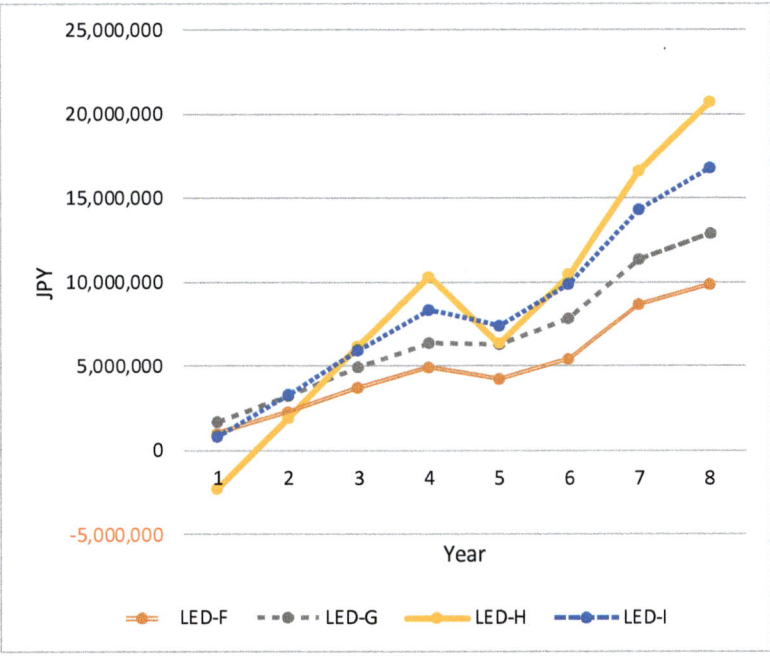

Fig. 24.9 Elapsed years and transition of built-up cash of lighting options (As compared to the case of FL, which is taken as 0) 1 USD = 105 JPY as of June 2016 (Source: Based on an Analysis by Takuji Akiyama (PlantX)). Note: The following main items were taken into account for economic analysis of lighting options. *1*. Fixture cost (JPY), *2*. life expectancy (year), *3*. wattage (W), *4*. crop weight (g/head), *5*. quantity of lights (quantity/shelf), *6*. electricity rate (JPY/kWh), *7*. total operating hours of light (hours/day), *8*. COP, *9*. production capacity (heads/shelf/day), *10*. sales price of plants (JPY/kg), *11*. quantity of cultivation racks (shelf)

of light, (8) COP, (9) production capacity, (10) sales price of plants, and (11) quantity of cultivation racks were examined in Japan by Takuji Akiyama (PlantX). In this study, four different LED lightings and FL were compared in order to figure out the economic feasibilities of each product. This figure shows that after 6–7 years of those lightings installations, LED-H starts to be most economical among other options. LED-I seems stably reasonable over a period of years.

24.4 Market Creation for PFALs

PFALs or vertical farming has been a remarkable phenomenon throughout the world particularly since around 2010. Despite many questions whether PFALs can be economically sustainable, people in the industry believe in the possibilities and needs of PFALs to address the challenges this globe has been and will be facing. The industry has been positively booming, and the number of PFALs has been increasing worldwide. However, it is also certain that behind the media attentions, there are also numerous PFALs that stopped operating while others started their farms. In order to truly develop the industry to be economically and technically sustainable, each player of the industry needs to be more networked in a sufficient way in order to share and utilize their experiences for the acceleration of technical and business development.

We can define the PFALs business into three major categories: (1) PFAL operating business, (2) system manufacturing and supplying business, and (3) PFAL-related consulting business. However, as more PFALs will be built with more varieties of PFAL design and business models, those three business categories will overlap one another and will rise in complexity. For instance, once the original farmers will accumulate knowledge on cultivating and operating PFALs, they may start consulting business or may initiate developing the unique systems with system companies based on their actual operating and cultivating experiences.

Such is the case with LED grow light business. While distributing their LED lightings, most companies will realize that, in order to be successful in lighting business, the capabilities of providing clients with advice on plant growth with lighting, such as growth recipes with their LED lightings, proposal on new kinds of plants suitable for their lightings, and ways to minimize the electrical costs, would be essential. Many lighting companies may start to have their own PFAL facilities to accumulate LED-related cultivation know-how. Some companies simultaneously may become PFAL farmers. Some may start to develop cultivation systems or PFAL-related systems equipped with their lighting products. As LED grow light firms expand their business domains, other system manufacturers or even farmers also expand into new business lines. This will eventually encourage fiercer competition on LED lighting business among not only increasing number of global lighting companies but also non-lighting companies including system manufacturers or farmers.

Many of the current growers have more or less three questions in common regarding LEDs: (1) what to grow with LEDs, (2) how to grow with LEDs, and (3) who to consult with when facing some challenges. Software business including consultancy or intelligent lighting business will be more important for LED-related business. There will be more or less fully custom-made LEDs, mass custom LEDs, and simple white light LEDs, and each LED will keep improving in various ways. It is assumed that LED lighting cost will be reduced and efficiency will also be improved as a consequence of the spread of consumer LEDs.

A notable significance is that PFALs are great mixtures of not only agriculture and engineering, but all the other related technologies. PFALs consist of numerous elemental technologies. This may mean that how to elect and connect with the right technical or business partners is greatly important in conjunction with considering real consumers and social needs in order to realize sustainable PFAL technology and business. Networked innovation: this might be the way to move the industry into the next-generation PFALs.

It should be noted that PFALs also need to apply for the solution of food issues particularly driven by global poverty issues. As challenges on urbanization or population expansion increase, who would indeed be affected and suffer from the food issues the most are poverty groups throughout the world. Therefore, it is extraordinarily important to consider sustainable PFAL business models along with developing extra technologies to apply for base of the pyramid sector, not just targeting top of the pyramid with pricy products. LED lighting might play much more crucial role to supply nutritious foods in the future. Whatever their business domains are, PFAL industry players that impact the industry, or global issues, with innovative mind and dynamic ways will be leading this industry in the near future.

Reference

Kozai TG, Niu M, Takagaki (eds) (2015) Plant factory, 1st edn: an indoor vertical farming system for efficient quality food production, ISBN:9780128017753, Elsevier, MA, USA, 351–386.

Chapter 25
Consumer Perception and Understanding of Vegetables Produced at Plant Factories with Artificial Lighting

Yuki Yano, Tetsuya Nakamura, and Atsushi Maruyama

Abstract Plant factories with artificial lighting (hereafter PFALs) garner considerable attention around the world as a way to solve problems related to agriculture, resources, and the environment. Although many companies in various regions have entered the PFAL business, they are faced with numerous technical and business challenges. Although many studies have been undertaken on technical issues concerning PFALs, only a few have been on the business and management aspects, including marketing. However, to succeed in business, understanding consumer perceptions and needs regarding PFALs and their products is essential. This chapter provides a summary of recent studies carried out in Japan and Hong Kong that attempt to explore consumers' awareness, perceptions, and understanding of leaf vegetables produced at PFALs. These studies also attempt to identify the factors that influence consumers' impressions of PFAL-produced vegetables. Moreover, the chapter contains a brief discussion of the potential demand for the products and the critical importance of marketing activities, such as market research and promotions, and education for the steady development of the PFAL industry.

Keywords Consumer perception • Plant factories • Marketing activities • Education • Consumer understanding

---

Y. Yano (✉) • A. Maruyama
Graduate School of Horticulture, Chiba University, Matsudo 648, Matsudo City, Chiba 271-8510, Japan
e-mail: y.yano@chiba-u.jp; yuki.yano.8911@gmail.com

T. Nakamura
Department of International Business Management, Kyoei University, 4158 Uchimaki, Kasukabe, Saitama 344-0051, Japan

25.1 Introduction

Plant factories with artificial lighting (hereafter PFALs), which have gained worldwide attention in recent years, are significantly different from traditional horticultural production systems in that the cultivation environment is highly controlled through a computer. By artificially controlling the indoor environment, a PFAL (or vertical farming) can grow high-quality leafy greens anywhere, even in office buildings in the center of a large city (Kozai 2013). New LED technology and special liquid nutrient solutions have made it possible to produce various types of leaf vegetables, such as low-potassium lettuce and vitamin-rich green vegetables, which have never been seen in the market. Moreover, because a multitier system equipped with lighting devices and automatic control system is employed, the productivity per land area and resource use efficiency of PFALs are remarkably higher than those of open-field culture and traditional house culture.

Because of these advantages of PFALs over traditional farming methods, many firms are attempting to enter this new market. For example, in Japan, the number of PFALs for commercial production has increased more than fivefold since approximately 2009, from 35 in December 2009 to 185 in March 2015 (Japan Greenhouse Horticulture Association 2015), and is expected to further increase. Japanese indoor plant factory systems have received recent attention and are already exported to countries in Asia and the Middle East. Although no data are available on the number of PFALs in foreign countries, the development and popularization of indoor vertical farming systems are promoted around the world.

Although PFALs are expected to contribute to make eating habits healthy and rich in the future, many technical and business challenges still need to be overcome. First, the high initial investment costs in constructing the facility make it difficult for firms or farmers to start a PFAL business. Moreover, the operating costs (primarily for electricity, even if LED is used) are extremely high. Not all existing PFALs are profitable, and a limited variety of crops can be produced, such as lettuce and spinach (Yamori et al. 2014). If further cost performance improvements can be realized through research and development (such as LED development), the management of PFALs will be more stable, and growing a wider variety of crops will be possible. According to Kozai (2014), further improvements in cost performance are possible within 5–10 years.

In addition to these technical issues, that only a few attempts have been made to date to understand consumers' perception and buying behavior toward vegetables produced at PFALs cannot be ignored (Kurihara et al. 2014; You et al. 2013a, 2013b). Namely, marketing efforts have been insufficient. Recently, PFALs in Japan using hydroponics have often been introduced by TV programs or other media; however, the market share of leaf vegetables produced at PFALs is still small, and consumers do not understand indoor hydroponic growing systems. Thus, for consumers to realize the real value of the products seems difficult. Additionally, some consumers seem to strongly resist the production of leafy greens without sunlight or soil and worry about their nutritional value or taste (Mitsubishi UFJ Research and Consulting Co. Ltd. 2013).

This chapter provides an overview of the research findings of recent studies on consumers' perception and understanding of PFAL-produced leaf vegetables. We first explain current Japanese consumers' images of PFALs and their products. Then, the results of an analysis of the impact of knowledge on consumers' impression are presented. Subsequently, a study that attempts to measure awareness, understanding, and anxiety regarding PFALs in Japan and to identify the factors that influence consumers' impressions in greater detail is summarized. Moreover, survey research on PFALs conducted in Hong Kong, where significant interest exists in food safety and available arable land is severely limited, is described. Finally, future prospects of the PFAL business are outlined.

25.2 Consumer Perception of PFAL-Produced Vegetables in Japan

This section explains the present impressions of Japanese people of PFALs and their products. To reveal consumers' innermost feelings about these products, including information not anticipated by a researcher, the free-word association method is useful (Isojima 2009). In recent years, this qualitative technique has become increasingly popular for exploring consumers' perceptions of the subject (Ares et al. 2008; Ares and Deliza 2010; Guerrero et al. 2010; Son et al. 2014). The technique calls for the respondent to be presented with a list of carefully chosen words or phrases (called stimulus words). The respondent is then required to provide the first thoughts or images that come to mind. The data can be obtained through the questionnaire or interview method, and then elicited images are analyzed using statistical methods.

25.2.1 Consumer Impressions of PFALs and Their Products

A questionnaire survey was conducted at Kyoei University, which is located in the city of Kasukabe, Saitama, approximately 20 miles north of Tokyo, Japan. Visitors to the campus festival held on November 3, 2014 were randomly asked to complete the survey (Yano et al. 2015). Eventually, 233 responses were collected, and 230 were used for the analysis, excluding three questionnaires with incomplete answers. Of the valid responses, 65 % were from females and 35 % were from males. This bias was viewed as not serious, considering that women are the primary buyers of food. Participants' ages (20s 19%, 30s 14%, 40s 28%, 50s 16%, and older than 60s 23 %), occupations (general private companies 25 %, public officials 14 %, housewife 33 %, student 11 %, and other 17 %), educational attainment (high school

34 %, junior/technical college 26 %, college/university 35 %, and others 5 %), and number of household members (one 6 %, two 16 %, three 25 %, four 28 %, five 17 %, or more than six 8 %) varied widely.

In the free-word association method, the survey participants were first asked to write down all of the words or phrases that came to mind when thinking of the stimulus phrases "vegetables grown using artificial light," and "vegetables grown hydroponically." These two stimuli were presented to understand consumers' images of vegetables grown at PFALs by decomposing them into two factors: "using artificial light rather than natural sunlight" and "using hydroponics rather than a soil culture." In addition to these stimuli, respondents were also presented with a stimulus phrase, "plant factory,[1]" to explore consumers' associations with that phrase.

A word association question represents an open-ended question that requires participants to supply their own answers. This type of question tends to elicit a number of blank responses. Therefore, respondents were asked to "write down at least three words or phrases" with which they associated each stimulus word. The questionnaire also included questions about knowledge of PFALs and their products and questions to investigate how consumers evaluate these products. Because these questions could have affected respondents' associations in some way, the word association questions were placed first to avoid such bias.

All of the responses were summed up for each stimulus word, and the total number of elicited words was 551 for "vegetables grown using artificial lighting," 546 for "vegetables grown hydroponically," and 528 for "plant factory." Despite asking the participant to write down at least three words or phrases, the average number of responses per respondent was only in the range of 2.27–2.40. This result proves that increasing the number of responses for these stimuli was very difficult.

Next, the elicited words or phrases that had the same or a similar meaning were merged into more inclusive (upper level) categories. For example, "clean" and "sanitary" were grouped into the "safe and reliable" category, "does not seem to be nutritious" and "nutrient deficiency" were included in the "anxious about nutritional value" category, and "many plants (can be grown)" and "mass production" were grouped in the "large-scale mass production" category. Of the categories created in this manner, those with a high appearance frequency are listed in Table 25.1. Note that the numbers in bold type indicate categories mentioned by more than 10 % of the participants for each stimulus word. These frequent categories can be interpreted as the images that many respondents have toward stimuli.

For all stimuli, the respondents are shown to have common positive images, such as "safe and reliable" and "stable supply." Therefore, the image of stable indoor production without any concerns about bugs or other external substances at PFALs has penetrated the market to some extent. However, the result also indicates that the frequencies of the other images differ across stimuli. For "vegetables grown using

---

[1] Here, we presented a phrase "plant factory," but in general, "plant factory" is considered the more general term, including plant factories with sunlight.

Table 25.1 Frequently mentioned images for all stimuli (Yano et al. 2015)

Frequent consumers' images	Vegetables grown using artificial light	Vegetables grown hydroponically	Plant factory
Safe and reliable	**96**	**81**	**61**
Stable supply	**76**	**41**	**50**
Anxious about nutritional value	**68**	**48**	6
Fresh and tasty	17	**55**	7
Large-scale mass production	9	5	**60**
Anxious about taste	**38**	17	7
Easy to grow	16	**25**	4
Artificial	9	6	**25**
Anywhere	13	11	15
Thoroughly controlled	8	2	**25**
Great possibility	8	3	18
Low cost and cheap	9	3	15
Harmful to the body	17	1	3
Average number of responses	2.40	2.34	2.27
Total sample size	551	546	522

The images elicited by more than 10 % of the survey participants are in bold type

artificial lighting," negative images such as "anxious about nutritional value," "anxious about taste," and "harmful to the body" were more frequently cited. Therefore, some consumers indicated concerns over vegetables grown without natural sunlight. In addition, "plant factory" was associated more frequently with images such as "low cost and cheap," "large-scale mass production," "artificial," and "thoroughly controlled (by computers and machines)," as related to the traditional concept of a so-called factory. The term "plant factory" may produce a negative impression of its product among consumers (e.g., cut off from nature). Furthermore, for "vegetables grown hydroponically," positive images such as "easy to grow" and "fresh and tasty" were more often mentioned than for other stimuli, but images concerning nutritional values were also cited. To be noted is that some images overlapped for "vegetables grown using artificial lighting" and "vegetables grown hydroponically," whereas differences in the frequencies of other images also existed. Such differences should be clearly recognized in marketing.

25.2.2 Impact of Knowledge on Consumer Impressions

Statistical analyses were conducted to investigate whether differences existed in frequently cited images by type of consumer. We found that no difference existed in images by individual characteristics, such as gender, age, education, and

Table 25.2 The impact of prior knowledge about PFALs on consumers' images

		Knowledge of PFALs		
	Highly frequent images	No	Yes	Significance
Vegetables grown using artificial light	Safe and reliable	25	72	**
	Stable supply	19	57	**
	Anxious about nutritional value	31	37	**
	Anxious about taste	23	15	***
	Total	98	181	
Vegetables grown hydroponically	Safe and reliable	15	68	**
	Fresh and tasty	22	35	*
	Anxious about nutritional value	17	31	n.s.
	Stable supply	6	36	**
	Total	60	170	
Plant factory	Safe and reliable	14	48	n.s.
	Large-scale mass production	23	38	*
	Stable supply	7	44	**
	Thoroughly controlled	10	16	n.s.
	Total	54	146	

p-value was computed by Chi-square test of independence
*$p < 0.1$, **$p < 0.05$, ***$p < 0.01$, n.s. indicates "not significant"

occupation, but that knowledge or prior information on PFALs had a statistically significant impact on consumer images.

Table 25.2 summarizes the relation between the images and the presence or absence of knowledge (prior information) of PFALs. "A consumer having knowledge" refers to a person who learned about PFALs from the media or who already purchases or has seen leaf vegetables produced at PFALs. A statistically higher proportion of the respondents with knowledge than those without it mentioned positive images, such as "safe and reliable" and "stable supply." At the same time, negative images, such as "anxious about nutritional value and/or taste," were cited more frequently by respondents without knowledge. Thus, the images of growing leafy greens using artificial lighting can be improved dramatically by acquiring knowledge of PFALs.

Additionally, with regard to "vegetables grown hydroponically," significantly more participants with knowledge than those without knowledge wrote down positive images, such as "safe and reliable" and "stable supply." However, no significant difference existed in images regarding nutritional value between two groups. Therefore, even if one knows PFALs, no improvement in images concerning nourishment was observed for hydroponically grown vegetables. For consumers to accurately understand the hydroponics systems (e.g., a nutrient solution management) from the information obtained through the TV or a product packaged at a store may be difficult. Additionally, for "plant factory," respondents with knowledge were more likely to mention the image of "stable supply" and were less likely to cite the image of "large-scale mass production."

To summarize, acquiring knowledge or prior information about PFALs has a positive impact on improving consumers' image of leaf vegetables grown at PFALs. Therefore, marketers should effectively provide information about PFAL and its products to consumers through mass media, product packaging, displays, tasting, and seminars. In particular, information regarding the nutritional value and taste of PFAL-produced vegetables must be conveyed accurately to consumers to eliminate any anxiety that they have about the products.

25.3 Consumer Understanding of PFAL-Produced Vegetables in Japan

In the previous section, we found that Japanese consumers often resisted growing leaf vegetables without natural sunlight and/or soil. In particular, they seemed to have negative images (anxiety) of the nutritional value or taste of vegetables produced at PFALs. We also revealed that prior knowledge of PFALs had a positive effect on improving consumers' images of the products. To date, however, the level of consumer understanding of PFAL and its products (and other factors) has yet to be considered.

This section summarizes the major findings from a study that attempted to measure the current level of consumer understanding, awareness, and anxiety about PFAL-produced vegetables. This section describes the statistical analyses conducted to investigate in greater detail the factors that can eliminate anxiety. To measure the level of understanding, true–false tests were useful and have been, in fact, widely used in the literature (Kunugita 2008; Suzuki 2014). The data collected from a survey are analyzed statistically.

25.3.1 Awareness and Recognition Process

We carried out a questionnaire survey at a farmers' market run by JA[2] Tozai Shirakawa in Fukushima, where leaf vegetables produced at a PFAL owned by the JA are sold. A total of 135 valid responses were obtained by asking visitors to complete the questionnaire survey at the JA farmers' market "Miryokumanten Monogatari" in Fukushima on June 19 and 20, 2015 (Friday and Saturday). The majority of the respondents (more than 80%) lived in Fukushima Prefecture, approximately 19% in Shirakawa City, 63% in other regions in Fukushima, and 18% were outside the prefecture. In terms of gender, most of the respondents (nearly 80%) were female. Additionally, of the respondents, 88% were over 40 years of age and 90% were high school or junior/technical college graduates.

---

[2] JA stands for Japan Agricultural Cooperatives.

The reason for this bias is that main customers of the farmers' market are local people who can visit there by car, even on weekdays.

Regarding awareness, approximately 60 % of the respondents were aware of PFAL and its products. In response to the question, "how did you come to know about vegetables produced at PFALs?", "at store" was the most frequent answer (nearly two-thirds). This reply was followed by "on TV." The most frequently chosen answer to the same question in another survey for high school students in Kasukabe City, Saitama (conducted on June 16, 2015), was "on TV" (89 %), followed by "on the Internet." These results show that recognition of PFAL-produced vegetables has increased among consumers since their more frequent coverage by the media in recent years. However, nearly 40 % of the respondents were still not aware of them even at locations at which they are sold.

25.3.2 Level of Understanding

To determine how well consumers understand PFAL and its products, we included 12 true–false questions in the questionnaire. The survey respondents were asked to choose one answer for each question from "true," "false," or "I don't know." Table 25.3 indicates the correct answer rates for the 12 questions on PFAL and its product.

Questions with high correct answer rates (higher than 75 %) were "Q1: Crops are unaffected by weather or season," "Q3: Can increase nutritive values such as vitamins by controlling the nutrient solution," "Q11: Workers have no dress code when entering the grow room," and "Q12: A very small portion of vegetables should be removed when using it." The results imply that the majority of the respondents understand that "PFALs ensure a stable supply of crops by carefully controlling the growing environment" and "leafy vegetables grown at PFALs are easier to use for cooking."

In contrast, questions concerning soilless culture systems or product cleanliness had low percentages of correct answers (lower than 50 %): "Q2: Rich soil is used in the facility," "Q4: Necessary to wash the products before cooking," "Q6: Workers do not touch vegetables with bare hands at all through production to shipment," and "Q8: PFALs should be built on land suitable for agriculture." This result indicates that more than half of the respondents do not have a clear grasp of "soilless and automated production systems" and "the degree of cleanliness of PFAL-produced vegetables." In particular, most respondents were unsure whether or not soil is used in PFALs, although almost all existing PFALs use water culture hydroponics systems.

For questions regarding the use of pesticides and the nutritional value of the products (Q5, Q7, and Q10), approximately two-thirds of the respondents provided the correct answers. Therefore, one-third of the respondents were mistaken about the safety and quality aspects of vegetables produced at PFALs or could not judge whether or not the statements were correct.

Table 25.3 True-or-false questions regarding PFALs and correct answer rates

Questions		Answer	Correct answer rate ($n = 135$)
Q1	Crops are unaffected by weather or season	○	90 %
Q2	Rich soil is used in the facility	×	19 %
Q3	Can increase nutritive values such as vitamins by controlling the nutrient solution	○	76 %
Q4	Necessary to wash the products (vegetables) before cooking	×	41 %
Q5	Being washed repeatedly prior to shipment, the products (vegetables) have little nutritional value	×	60 %
Q6	Workers do not touch vegetables with bare hands at all through production to shipment	○	34 %
Q7	The products have poor nutritional value due to lack of exposure to natural sunlight	×	66 %
Q8	PFALs should be built on land suitable for agriculture	×	50 %
Q9	Harvested vegetables are packaged inside the PFAL	○	70 %
Q10	Does not have to use pesticides at all	○	64 %
Q11	Workers have no dress code when entering the grow room	×	79 %
Q12	A very small portion of vegetable should be removed when using it	○	78 %

Overall, many survey participants appear to have the image of a traditional greenhouse for horticultural cultivation rather than a sophisticated hydroponic production system used in most PFALs. Yet, room for improvement exists in recognizing and understanding PFALs and their products.

25.3.3 Factors Affecting the Degree of Anxiety

We first present the results from measuring consumers' anxiety level with regard to three aspects of PFAL-produced leaf vegetables: sanitation, attached substances (safety), and nutritional value/taste. Then, we describe how the level of understanding and other factors influence the total anxiety level score, which is calculated by summing the anxiety level scores regarding three aspects. The anxiety level for each aspect was measured using a 5-point Likert scale, where "1" indicated no anxiety and "5" indicated high anxiety (i.e., (1) confident, (2) rather confident, (3) neutral, (4) rather anxious, (5) anxious).

The mean values of the anxiety levels for sanitation, attached substances, and nutritional value/taste of PFAL vegetables were 1.24, 1.44, and 1.79, respectively. Three mean values are less than 2, indicating that the degree of anxiety about vegetables produced at PFALs is, on the whole, low. However, we found that the mean of the anxiety level regarding nutritional value or taste was significantly higher than that regarding attached substances and sanitation (ANOVA and multiple comparisons: there was a statistically significant difference in the means ($F = 30.1$, $p < 0.001$)). Moreover, the mean degree of anxiety concerning attached substances is greater than that concerning sanitation. As in the previous section, these results also show the tendency that consumers relatively worry about the nutritive values or taste of PFAL-produced vegetables.

Table 25.4 summarizes the impact on consumers' anxiety toward PFAL-produced vegetables related to their understanding (the number of correct answers) and other factors, such as individual characteristics. Although whether or not a participant knows PFAL in advance has no impact on consumers' image of its products, an increased understanding of PFAL and its product has a positive effect (i.e., negative effect on the level of anxiety). Therefore, even if consumers know about PFALs, a lack of understanding will lead to a high level of anxiety. Interestingly, the individual characteristics of the respondents have no influence over the anxiety level. The point is that improvements in consumers' understanding through marketing activities, such as package design and tasting, and education will be very important to the continued development of the PFAL industry.

25.4 Case Study in Hong Kong

To this point, we have focused on Japanese consumers' perception and understanding of PFAL and its product. Although the number of PFALs (vertical farming facilities) around the world has increased, to the best of our knowledge, no research has previously attempted to explore consumers' image and understanding of PFALs in other countries.

This section focuses on a case study conducted in Hong Kong (Special Administrative Region, China) that attempts to know how Hong Kong consumers think of and have knowledge about PFALs. The reason we carried out a questionnaire survey concerning PFALs in Hong Kong was that people are interested in food safety, arable land is severely limited, and many tall buildings exist because of a lack of space. PFALs are expected to have numerous advantages in such a country. Incidentally, Mitsubishi Chemical Corporation has already sold a plant factory system employing LED lighting to the Vegetable Marketing Organization Ltd. (VMO) based in Hong Kong.

25.4.1 Awareness, Understanding, and Impression in Hong Kong

We implemented a questionnaire survey to find out what Hong Kong consumers think of and how well they understand PFAL and its products. Of the 324 responses collected, 282 were valid and used for statistical analysis. Males constitute 40.4 %

Table 25.4 Multiple regression result

Explanatory variables	Coefficients (t-value)	
Number of correct answers	−0.325 (−6.22)	***
Prior knowledge	0.013 (0.05)	
Gender (male = 0, female = 1)	−0.141 (−0.39)	
Age	−0.017 (−1.54)	
Number of children	−0.189 (−1.13)	
Education dummy (university/technical college = 1, other = 0)	0.410 (1.33)	
Housewife dummy (housewife = 1, other = 0)	0.257 (0.82)	
Constant	7.696 (9.94)	***
Adjusted R^2	0.22	

Asterisks *** represent 1 % significance level

and females represent 59.6 % of survey participants. Participants' ages were relatively widely distributed: 20s 18.4 %, 30s 29.8 %, 40s 24.8 %, 50s 16.3 %, and 60 or older 10.6 %. Approximately half of the participants were from the New Territories, 30 % from Kowloon, and 20 % from HK Island or other regions. In terms of education level, "university or junior college" was 40.1 %, "vocational school and high school" was 27 %, "junior high school" was 29.1 %, and others were 3.9 %. Interestingly, only 1.4 % of the participants live alone.

The questionnaire consists of several items to measure respondents' awareness, understanding, and anxiety level toward PFAL and its product. Regarding the level of understanding, we asked the survey respondents whether they already knew about the water culture system with artificial lighting that is used in most PFALs. We measured consumers' anxiety level regarding five aspects of PFAL-produced leaf vegetables: sanitation, attached substances (safety), nutritional value, taste, and effects on the body. As in the previous section, the level of anxiety for each aspect was measured using a 5-point Likert scale (i.e., (1) confident, (2). rather confident, (3) neutral, (4) rather anxious, (5) anxious; compared with that of vegetables grown outside).

The tabulation result shows that 66 % of the respondents (valid responses) already knew about PFALs, primarily through the television and newspapers and at stores. In response to the question, "Do you know that crops inside the PFAL are unaffected by weather, pests, and diseases because they are grown indoor?", 53 % of the respondents (approximately half) said "yes," whereas only 35 % said "yes" in response to the question: "Do you know that leaf vegetables are grown with liquid fertilizer in the (most) PFALs?" These results indicate that PFALs are being recognized among Hong Kong consumers, whereas a hydroponic culture system in facilities is still poorly understood. Similar to the previous section, we conclude that significant scope exists to improve the understanding of the plant cultivation system in a PFAL.

Moreover, the calculated average values of the anxiety levels for sanitation, attached substances, nutritional value, taste, and effects on the body of PFAL-produced vegetables were 2.23, 2.60, 2.71, 2.77, and 2.79, respectively. All mean values fall between 2 and 3, and the percentage of "neutral" was relatively high, particularly for three aspects (more than 70 %): nutritional value, taste, and effects on the body. These results suggest that consumers in Hong Kong do not yet have a detailed image of PFAL-produced vegetables. Interestingly, however, and similar to the case observed in Japan, we found that the mean values of anxiety levels for nutritional value, taste, and effects on the body were statistically significantly greater than those for sanitation and attached substances. Namely, Hong Kong consumers also tend to have relatively higher levels of anxiety over the ingredients or taste of PFAL-produced vegetables, compared with other aspects.

25.4.2 Relationship Between Knowledge and Confidence

We now attempt to investigate the factors that influence the level of consumer confidence in each aspect of PFAL-produced vegetables. To do so, we estimated five binary logistic regression equations for five different dependent variables, each of which takes the value of one if the respondent is confident in each aspect and zero if not (i.e., consumer confidence in the following five aspects: sanitation, safety regarding attached substances, nutritional value, taste and texture, or effects on the body). The independent variables included in the model are the same for all equations. Table 25.5 summarizes the results of the logit models.

Knowledge concerning indoor cultivation has a positive effect on respondents' confidence in all aspects. Therefore, the effective provision of information on the indoor cultivation through the media, for instance, can enhance consumers' general impression of PFAL-produced vegetables. Acquiring knowledge about the use of liquid fertilizer (soilless culture system) also has a positive impact on respondents' confidence in the safety of PFAL-produced vegetables, but has no impact on their confidence in nutritional value, taste and texture, and effects on the body. Offering information on products' nutritional components, taste, and texture is needed to further improve consumer impressions (in terms of nutritional value and taste). In addition, the results show the tendency for respondents with a habit of eating fresh vegetables to be more likely confident in the quality of PFAL-produced vegetables.

In summary, as with the previous survey conducted in Japan, an increased understanding of production systems in PFAL has a positive impact on consumer confidence in its product. Thus, enhancing consumers' understanding is effective in improving their impression of leaf vegetables grown at PFALs. At the same time, providing information on nutritional value, taste, or texture is required to further improve their impression. Marketing activities, such as package design, display, tastings, and seminars, also play an important role in the rapid diffusion and establishment of PFALs in Hong Kong.

Table 25.5 Results of the binary logistic regression model ($n = 282$)

Independent variables	Dependent variable (confident = 1, not confident = 0)				
	Safety (sanitation)	Safety (attached substances)	Nutritional value	Taste and texture	Effects on the body
Knowledge (hydroponics)	+	+			
Knowledge (indoor cultivation)	+	+	+	+	+
Gender (male = 0, female = 1)	+				
Regional dummy (Kowloon = 1)					
Regional dummy (New Territories)					
The young (20s or 30s = 1)					
Middle age (40s or 50s = 1)					
Number of household				−	
Number of children (under 10)					
Education (university/technical = 1)					
Frequency of cooking meals at home					
Frequency of eating fresh vegetables	+	+	+	+	
Interest in health maintenance					
Interest in natural environment					
Interest in scientific technology					
Interest in social problems					
Constant			−	−	−
Pseudo R^2	0.34	0.25	0.26	0.28	0.17

Only the signs (positive or negative) of estimated coefficients which are significant at 5 % level are given

25.5 Future Prospects of the Plant Factory Business

This section discusses the future prospects of the plant factory business. We first describe the significant potential demand for crops produced at PFALs around the world. Marketing activities such as market research and promotions and education are indispensable for actually discovering such latent demand and converting it into effective demand. We then discuss the critical importance of marketing activities and education for the steady development of the plant factory industry.

25.5.1 Potential Demand for PFAL-Produced Vegetables

The global population has reached 7.3 billion in 2015 and is projected to surpass 9.7 billion in 2050 (United Nations 2015). To meet the increased demand for food, global agricultural production must increase by 70 % during the next 35 years (FAO 2009). In emerging and developing countries, the demand for vegetables and fruits is expected to continue to grow as a result of rising incomes and urbanization. Meanwhile, land suitable for agriculture is unevenly distributed around the world, and barren lands are expanding because of a lack of water, desertification, salt damage, or soil contamination. Additionally, climate change may significantly affect food production and make the situation worse. Concerns over shortages in energy and natural resources are also increasing. We need to solve these global issues related to agriculture, food, the environment, and resources.

That a PFAL can contribute to simultaneously solving these issues is a strong possibility. Because crops can be produced with less water and agrochemicals, and outside air is shut out, a PFAL can stably supply safe and fresh food even in places in which traditional agriculture would have been impossible. Thus, a PFAL has an advantage in areas with short hours of sunlight and very low or high average temperatures, even in dry or contaminated areas. Hereafter, the demand for PFAL-produced crops will increase in those areas.

Additionally, a PFAL that can achieve the ultimate ideal form of local production and consumption also has a significant advantage in metropolitan areas. In particular, small PFALs can be set up in, for example, restaurants, business offices, shopping malls, convenience stores, and private residences (Kozai 2013), allowing nearby people to grow and eat fresh vegetables. Because urbanization is expected to advance rapidly throughout the world, the demand for unit type of small PFALs will continue to grow.

Moreover, as previously mentioned, new LED technology (e.g., a mix of red and blue LEDs) and special liquid nutrient solutions have recently enabled the production of functional vegetables, such as low-potassium lettuce that is good for kidney disease patients and vegetables rich in polyphenols and vitamin C. In Japan, for instance, the demand for such products from hospitals, nursing homes, and the elderly is increasing. In the future, a broader variety of healthy/nutritious vegetables and fruits will be produced in PFALs. Because the aging of the population is also expected to advance in both developed and developing countries, the demand for such functional foods will increase throughout the world.

To summarize, huge latent demand seems to exist for PFALs and their products in various areas. PFALs could dramatically change the way that food is grown and distributed throughout the world depending on whether cost performance is further improved through technological progress.

25.5.2 Necessity of Marketing Activities and Education

In Japan, PFAL-produced vegetables have been typically promoted as having high safety (e.g., "no need to wash" and "pesticide-free"). Consumers highly evaluate "low pesticide and bacteria" (Kurihara et al. 2014), which coincides with high consumer confidence in the safety aspects of PFAL-produced vegetables. However, Japanese consumers seem to have originally been confident in domestic agricultural products and, hence, have difficulty realizing such merits. Additionally, as previously seen, most consumers have difficulty obtaining knowledge about soilless production systems and, thereby, are incapable of understanding why PFAL-produced vegetables are so clean. Therefore, henceforth, differentiating the PFAL products on the basis of only "high safety" will presumably be difficult.

Additionally, as described in previous sections, we have found that some consumers have negative impressions of PFAL-produced vegetables. In particular, people in both Japan and Hong Kong are more likely to be anxious about the nutritional values, taste, or texture of these vegetables. To eradicate or reduce such anxiety, in the short term, the information should be provided efficiently through marketing, such as package design, display, tasting, and seminars. From a long-term point of view, education or seminars will be absolutely critical to enhancing consumers' overall understanding of PFAL and its product. The same may be said of other places at which PFALs will be installed in the future.

Furthermore, because the demand for functional foods is expected to increase further, exploring consumers' attitudes toward the products and capturing the needs of people, in, for example, hospitals, nursing homes, or elderly households, are inevitable. If we fail to understand how consumers evaluate functional vegetables (or fruits) produced at PFALs, the products cannot be easily differentiated from crops grown in open-field or traditional greenhouses, leading to the failure of the PFAL business. Also of significant importance is enhancing the understanding through education or other means of special crops (e.g., low-potassium lettuce) that can be produced at PFALs to enable consumers to realize their true value.

25.6 Conclusion

This chapter presents a summary of the findings of recent studies conducted in Japan and Hong Kong that attempts to explore consumers' perception and understanding of PFAL-produced leaf vegetables.

The main findings are as follows: (1) Many consumers have positive impressions of leaf vegetables produced at PFALs, such as "safe and reliable" and "stable supply"; however, others tend to occasionally feel anxious about nutritional components and/or taste. (2) The term "plant factory" produces impressions associated with the traditional concept of "factory," such as "low cost and cheap," "large-scale mass production," "artificial," and "thoroughly controlled." (3) Currently, Japanese consumers understand that crops inside PFALs are grown stably through careful control of the growing environment and that PFAL-produced vegetables are easier to use for cooking. However, these consumers do not seem to have much understanding of the soilless and automated production systems and the real degree of cleanliness of PFAL-produced vegetables. (4) In Hong Kong, consumers have begun to recognize PFALs, whereas the soilless hydroponic system that has been used in most existing PFALs is still poorly understood. (5) An increase in the level of understanding (acquiring knowledge) regarding PFALs has a positive impact on consumer confidence in their products.

From these findings, we see that plenty of room exists to improve the understanding of the plant cultivation system used in PFALs and the quality of their products. Enhancing consumers' understanding is imperative to improve their impressions of PFAL-produced products. Marketing activities, such as promotion and advertising, and education will play an important role in the future success of the PFAL business. Moreover, from a global perspective, finding an appropriate PFAL business model based on to consumers' needs in each region or country is important. Much still remains to be done to understand consumer perceptions, needs, and behavior toward PFALs and their products.

References

Ares G, Deliza R (2010) Studying the influence of package shape and colour on consumer expectations of milk desserts using word association and conjoint analysis. Food Qual Prefer 21:930–937
Ares G, Ana G, Adriana G (2008) Understanding consumers' perception of conventional and functional yogurts using word association and hard laddering. Food Qual Prefer 19:636–643
FAO (2009) How to feed the world: 2050. High expert forum, 12–13 October 2009. Food and Agriculture Organization of the United Nations, Rome
Guerrero L, Claret A, Verbeke W et al (2010) Perception of traditional food products in six European regions using free word association. Food Qual Prefer 21:225–233
Isojima A (2009) Consumer needs regarding purchasing agricultural products. Association of Agriculture and Forestry Statistics, Tokyo (in Japanese)
Japan Greenhouse Horticulture Association (2015) A report on introduction support of next generation horticulture. Japan Greenhouse Horticulture Association, Tokyo, pp 89–109
Kozai T (2013) Plant factory in Japan: current situation and perspectives. Chron Hortic 53(2):8–11
Kozai T (2014) Role of plant factory with artificial light (PFAL) in urban horticulture and next generation lifestyle. Presentation at the meeting of the international conference on vertical farming and urban agriculture, September 9–10, 2014, Nottingham
Kunugita N (2008) Investigation of the relationship between knowledge concerning radiation and the level of anxiety toward radiation in student nurses. J Univ Occup Environ Health 30 (4):421–429 (in Japanese)
Kurihara S, Ishida T, Suzuki M et al (2014) Consumer evaluation of plant factory produced vegetables. Focus Mod Food Ind 3(1):1–9
Mitsibishi UFJ Research and Consulting Co., Ltd (2013) Consumer survey on plant factory produced vegetables. (in Japanese)
Son JS, Do VB, Kim KO et al (2014) Understanding the effect of culture on food representations using word associations: the case of "rice" and "good rice". Food Qual Prefer 31:38–48

Suzuki T (2014) The problems of analyzing the factors behind people's anxiety based upon the use of true-false test in regards to their knowledge of radiation. Japan J Sci Commun 15:3–16 (in Japanese)

United Nations (2015) World population prospects: the 2015 revision. United Nations, New York

Yamori W, Zhang G, Takagaki M et al (2014) Feasibility study of rice growth in plant factories. J Rice Res 2(119):1–6

Yano Y, Nakamura T, Maruyama A (2015) Consumer perceptions toward vegetables grown in plant factories using artificial light – an application of the free word association method. Focus Mod Food Ind 4:11–18

You Z, Zhang X, Chen CH et al (2013a) Impact of relevant knowledge on purchase intention of plant-factory-produced plants: case study in both Singapore and Japan. Focus Mod Food Ind 2(2):63–69

You Z, Zhang X, Koyama S (2013b) Informational vs. emotional appeals of logo design in influencing purchase intentions for plant-factory-produced vegetables. Int J Adv Psychol (IJAP) 2(4):224–230

Part VII
Basics of LEDs and LED Lighting Systems for Plant Cultivation

Chapter 26
Radiometric, Photometric and Photonmetric Quantities and Their Units

Kazuhiro Fujiwara

Abstract The article provides a simple explanation to facilitate systematic understanding of quantities, especially relating to "light intensity", and the SI units necessary to describe the specifications of light output of artificial light sources and of light environments for plant cultivation. Important quantities similarly situated in radiometry, photometry and photonmetry, respectively, are explained, contradistinguished from one another.

Keywords Artificial lighting • Illuminance • Irradiance • Light environment • Radiometry • Photometry • Photonmetry • Photon flux density • Plant cultivation • SI unit • Terminology

26.1 Introduction

For plant cultivation, several fundamental quantities, especially those related to "light intensity", and their units for artificial light sources, including LEDs, must be understood appropriately and effectively. This article simply explains fundamental and important quantities relating to "light intensity" and their SI units necessary to specify light output of artificial light sources and light environments for plant cultivation. For that purpose, those quantities similarly situated in radiometry, photometry and photonmetry are explained, respectively, and are mutually contradistinguished.

Terminology definitions are mostly quoted from International Electrotechnical Commission (IEC) 60050-845-01 (1987). Some terms associated with the fundamental quantities are also explained using illustrations and graphs. Part of this chapter includes partially reconstituted and modified contents of the author's previous works (Fujiwara 2013, 2015).

K. Fujiwara (✉)
Graduate School of Agricultural and Life Sciences, The University of Tokyo, 1-1-1 Yayoi, Bunkyo-ku, Tokyo 113-8657, Japan
e-mail: afuji@mail.ecc.u-tokyo.ac.jp

26.2 Importance of Photonmetric Quantities for Plant Cultivation

Quantities related to "light intensity" are classified into three measurement categories: radiometry, photometry and photonmetry. Radiometry, photometry and photonmetry can be restated, respectively, as energy-based, luminosity-based and photon-based measurements. Radiometry and photometry are well known. Their explanations are found with relative ease in books and articles (e.g. Meyer-Arendt 1968). In contrast, "photonmetry" is rarely found in books and articles, but quantities in "photonmetry" are those which must be used to describe "light intensity" for plant responses to light because many plant responses to light are observed as a result of photochemical reactions.

26.3 Fundamental Quantities in Radiometry, Photometry and Photonmetry and Their SI Units

Fundamental quantities in radiometry, photometry and photonmetry and their SI units are shown in Table 26.1. Four fundamental quantities, with their SI units, in radiometry and photonmetry are placed, respectively, in both side columns of the following four fundamental quantities in photonmetry: luminous intensity, luminous flux, quantity of light and illuminance so that the relations among the quantities and SI units can be readily understood. Relational expressions among the quantities in the same column are provided in the rightmost column to indicate quantitative relations. Quantities in the same row are mutually equivalent in a metric sense.

Definitions of these terms are presented clearly in IEC 60050-845-01 (1987). The following definition in the first sentence for each term is fundamentally a direct quote from IEC 60050-845-01 (1987). Quantity symbols are deleted from the original definitions, and SI units are added. Supplementary explanations are subsequently added for most terms to elucidate their definitions.

26.3.1 Radiant Intensity $[W\ sr^{-1}]$

Quotient of the radiant flux [W] leaving the source and propagated in the element of solid angle (see Fig. 26.2 for an explanation of "solid angle") containing the given direction, by the element of solid angle [sr]. In other words, it is the radiant flux [W] leaving the source per unit solid angle [sr] in the direction considered.

Table 26.1 Fundamental quantities in radiometry, photometry and photonmetry and their SI units

Radiometry (Energy basis)	Radiant intensity [W sr^{-1}]	Radiant flux [(W sr^{-1}) sr] = [W]	Radiant energy [W s] = [J]	Irradiance [W m^{-2}]
Photometry (Luminosity basis)	Luminous intensity [cd]	Luminous flux [cd sr] = [lm]	Quantity of light [lm s]	Illuminance [lm m^{-2}] = [lx]
Photonmetry (Photon basis)	Photon intensity [mol s^{-1} sr^{-1}]	Photon flux [(mol s^{-1} sr^{-1}) sr] = [mol s^{-1}]	Photon number [(mol s^{-1}) s] = [mol]	Photon flux density (Photon irradiance) [(mol s^{-1}) m^{-2}] = [mol m^{-2} s^{-1}]
Relations	A	A·sr = B	A·sr·s = B·s	A·sr·m^{-2} = B·m^{-2}

Modified from Fujiwara (2013)

26.3.2 Radiant Flux (Radiant Power) [W] (= [J s^{-1}])

Power [W] emitted, transmitted or received in the form of radiation. The amount of radiant energy [J] emitted, transmitted or received per unit time [s].

26.3.3 Radiant Energy [J]

Time integral of the radiant flux [W] over a given duration [s].

26.3.4 Irradiance [W m^{-2}]

Quotient of the radiant flux [W] incident on an element of the surface containing the point, by the area [m^2] of that element.

26.3.5 Luminous Intensity [cd]

Quotient of the luminous flux [lm] leaving the source and propagated in the element of solid angle containing the given direction, by the element of solid angle [sr]. In other words, it is the luminous flux [lm] leaving the source per unit solid angle [sr] in the direction considered.

26.3.6 Luminous Flux [lm]

Quantity derived from radiant flux [W] by evaluating the radiation according to its action upon the CIE standard photometric observer. The luminous flux is calculated by integrating from 0 nm to infinity (actually 360–830 nm for photopic vision) the product of spectral radiant flux [W nm^{-1}], the CIE standard spectral luminous efficiency function for photopic vision (usually denoted as $V(\lambda)$, with value ranges of 0–1 according to wavelength λ), and the maximum spectral luminous efficacy (of radiation) for photopic vision [lm W^{-1}] (usually denoted as K_m; approximately 683 lm W^{-1} at 555 nm (540 THz)). Explicit and detailed descriptions of the relation between radiometric and photometric quantities are presented by Ohno (1997).

26.3.7 Quantity of Light [lm s]

Time integral of the luminous flux [lm] over a given duration [s].

26.3.8 Illuminance [lx]

Quotient of the luminous flux [lm] incident on an element of the surface containing the point, by the area [m^2] of that element.

26.3.9 Photon Intensity [mol s^{-1} sr^{-1}]

Quotient of the photon flux [mol s^{-1}] leaving the source and propagated in the element of solid angle containing the given direction, by the element of solid angle [sr]. In other words, it is the photon flux [mol s^{-1}] leaving the source per unit solid angle [sr] in the direction considered.

26.3.10 Photon Flux [mol s^{-1}]

Quotient of the number of photons [mol] emitted, transmitted or received in an element of time [s], by that element.

26.3.11 Photon Number (Number of Photons) [mol]

Time integral of the photon flux [mol s^{-1}] over a given duration [s].

26.3.12 Photon Flux Density (Photon Irradiance) [mol m^{-2} s^{-1}]

Quotient of the photon flux [mol s^{-1}] incident on an element of the surface containing the point, by the area [m^2] of that element. In IEC 60050-845-01 (1987), the term "photon flux density" is not listed, but the term "photon irradiance" is listed. In fields related to plant cultivation and related sciences such as horticultural science and environmental plant physiology, "photon flux density" has been widely used instead of "photon irradiance" (Fig. 26.1).

26.4 Spectral Distribution of Radiometric, Photometric and Photonmetric Quantities

Spectral distribution is a generic term used to describe a quantity as a function of wavelength or quantity per unit wavelength, e.g. 1 nm, arranged in wavelength sequence. In this chapter, the spectral distribution is used exclusively in the latter meaning. The adjective "spectral" is useful for the 12 terms described above and the other terms describing quantities pertaining to radiation when the property or wavelength dependency of the quantity is described, such as spectral irradiance.

The relations of spectral irradiance, illuminance and photon flux density for wavelengths of 300–800 nm are presented in Fig. 26.2 in the case of spectral irradiance, taking a constant value of 0.1 W m^{-2} nm^{-1}. Values outside and inside of parentheses presented in the upper right of each graph are the irradiance, illuminance and photon flux density, respectively, for wavelengths of 300–800 nm and 400–700 nm. Actually they are the integrals of spectral irradiance, illuminance and photon flux density in the respective wavelength ranges. As a useful example, we also prepared Fig. 26.3 showing the reference solar spectral irradiance (IEC 60904-3-2 2008) and its corresponding spectral illuminance and photon flux density for wavelengths of 300–800 nm. The reference solar spectral irradiance was prepared by IEC as a reference for terrestrial photovoltaic solar devices.

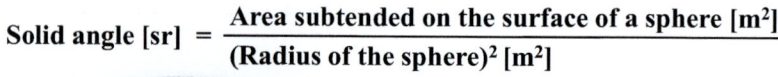
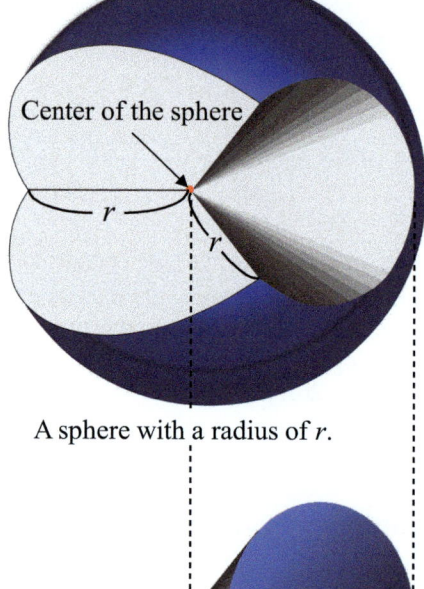

Fig. 26.1 Solid angle and its mathematical explanation

26.5 Quantitative Relations of Radiometric, Photometric and Photonmetric Quantities

The spectral illuminance curves in Figs. 26.2 and 26.3 were drawn using the CIE standard spectral luminous efficiency function for photopic vision ($V(\lambda)$) by CIE Standard (2005), in which the values of the function at 1-nm increments are tabulated. The value of spectral illuminance (for photopic vision) at a wavelength of λ [nm], $L_{is,\lambda}$ is calculable as

$$L_{is,\lambda} = K_m \cdot R_{is,\lambda} \cdot V(\lambda) \tag{26.1}$$

Fig. 26.2 Spectral irradiance taking a constant value of 0.1 W m^{-2} nm^{-1} and its corresponding spectral illuminance and photon flux density for wavelengths of 300–800 nm. Values outside and inside of parentheses shown in the upper right of each graph, respectively, represent the irradiance, illuminance and photon flux density for respective wavelengths of 300–800 nm and 400–700 nm

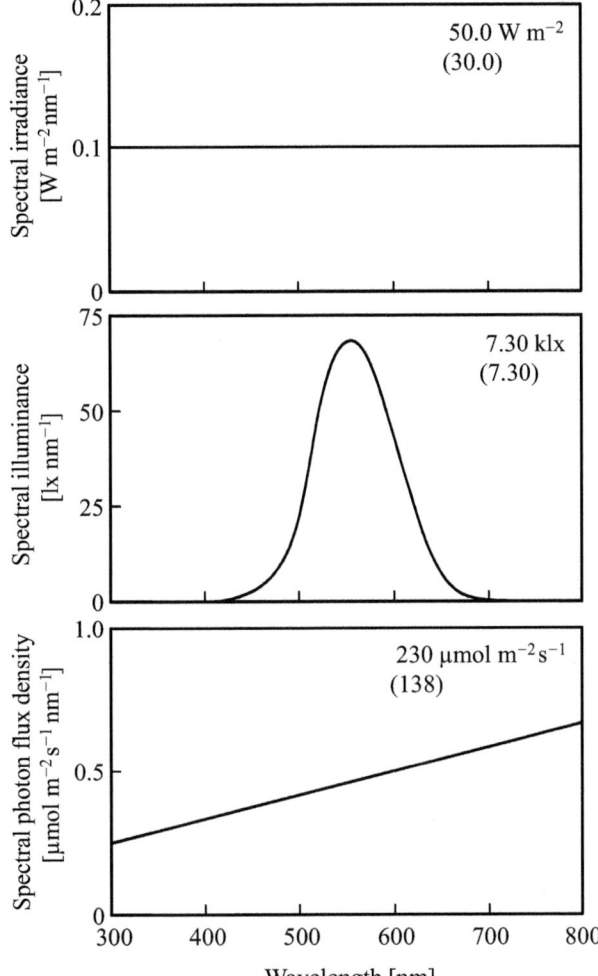

where $R_{is,\lambda}$ stands for the spectral irradiance at wavelength λ [W m^{-2} nm^{-1}] and where K_m is a constant relating the radiometric quantities and photometric quantities, called the "maximum spectral luminous efficacy of radiation for photopic vision". The value of K_m is given by the 1979 definition of candela, which defines the spectral luminous efficacy of light at 540×10^{12} Hz (at 555-nm wavelength) to be 683 lm W^{-1} (Ohno 1997).

The curves of spectral photon flux density in Figs. 26.2 and 26.3 were drawn using a theoretically simple equation described below. Energy, E [J], that n moles of photons with a wavelength of λ [nm] possess is given as the following equation:

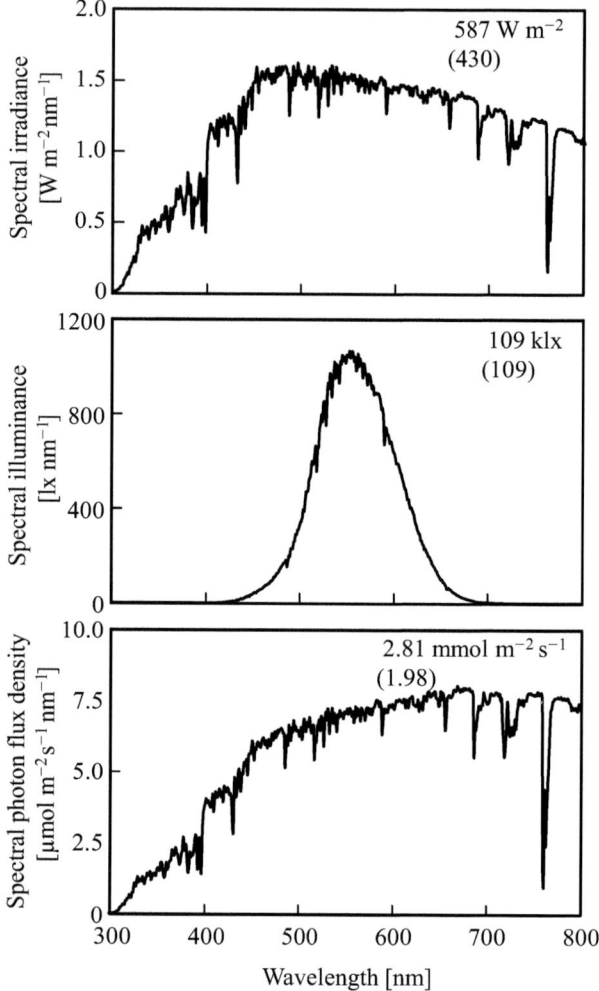

Fig. 26.3 Reference solar spectral irradiance (IEC 60904-3-2 2008) and its corresponding spectral illuminance and photon flux density for wavelengths of 300–800 nm. Values outside and inside of parentheses shown in the upper right of each graph, respectively, represent the irradiance, illuminance and photon flux density for respective wavelengths of 300–800 nm and 400–700 nm

$$E = n \cdot N_A \cdot h \cdot \nu = \frac{n \cdot N_A \cdot h \cdot c}{\lambda \times 10^{-9}} \tag{26.2}$$

Therein, h stands for Planck's constant (6.626×10^{-34} J s), c is the speed of light (2.998×10^8 m s^{-1}) and N_A is Avogadro's number (6.022×10^{23} mol^{-1}).

Provided that spectral irradiance $R_{is,\lambda c}$ [W m^{-2} nm^{-1}], which is the irradiance [W m^{-2}] per wavelength range [nm] with its central wavelength at λ_c [nm], is given, then the spectral photon flux density at λ_c is calculable by substituting $R_{is,\lambda c}$ into E and $P_{is,\lambda c}$ into n in Eq. 26.2.

$$R_{is,\lambda c} = \frac{P_{is,\lambda c} \cdot N_A \cdot h \cdot c}{\lambda_c \times 10^{-9}} \quad (26.3)$$

$$\therefore P_{is,\lambda c} = R_{is,\lambda c} \frac{\lambda_c \times 10^{-9}}{N_A \cdot h \cdot c} \quad (26.4)$$

For example, the spectral photon flux density for λ_c of 600 nm, $P_{is,600}$ in Fig. 26.2, is calculable by substituting $R_{is,600}$ (=0.1 in Fig. 26.2) into $R_{is,\lambda c}$ in Eq. 26.4.

$$\begin{aligned}
P_{is,600} &= R_{is,600} \frac{\lambda_c \times 10^{-9}}{N_A \cdot h \cdot c} \\
&= 0.1 \times \frac{600 \times 10^{-9}}{(6.022 \times 10^{23}) \times (6.626 \times 10^{-34}) \times (2.998 \times 10^8)} \\
&= 0.5016 \times 10^{-6} \text{mol m}^{-2}\text{s}^{-1}\text{nm}^{-1}
\end{aligned}$$

It is to be noted that the spectral photon flux density at λ_c of 600 nm, taking an approximate value of 0.5 µmol m^{-2} s^{-1} nm^{-1}, means that the photon flux density within a 1-nm wavelength range of 599.5–600.5 nm is 0.5 µmol m^{-2} s^{-1}.

Quantitative relations among the three metrics are held for the other categories (i.e. for the other columns in Table 26.1), e.g. spectral radiant flux, luminous flux and photon flux.

26.6 Photosynthetically Active Radiation

Radiation in the wavelength range of 400–700 nm is regarded as "photosynthetically active radiation". The photon flux density in the wavelength range is therefore designated as "photosynthetic photon flux density" (see Chap. 28 for a detailed explanation). The photosynthetic photon flux density is a frequently measured quantity when plants are cultivated under artificial lighting. Figure 26.4 presents a typical spectral response of a quantum sensor and ideal quantum response for measurement of the photosynthetic photon flux density.

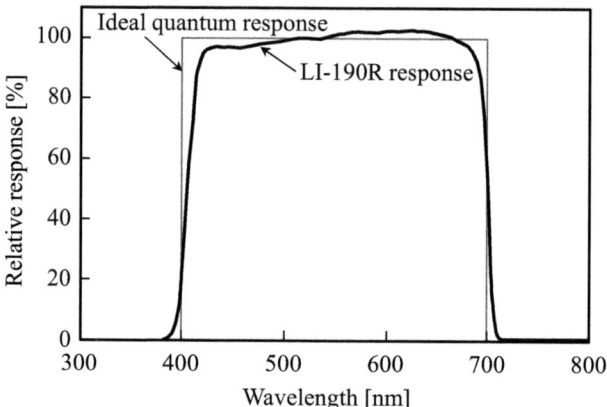

Fig. 26.4 Typical spectral response of LI-190R quantum sensors and ideal quantum response for the measurement of photosynthetic photon flux density (photosynthetic photon irradiance). The figure was redrawn from an LI-190R catalogue

References

CIE Standard (2005) Photonmetry—the CIE system of physical photonmetry, CIE S 010/E:2004. CIE Central Bureau, Vienna

Fujiwara K (2013) Fundamentals of light on plant cultivation and LED light irradiation technology. Refrigeration 88:163–168 (in Japanese)

Fujiwara K (2015) Light sources. In: Kozai et al (eds) Plant factory: an indoor vertical farming system for efficient quality food production. Academic, London, pp 118–128

IEC (International Electrotechnical Commission) 60050-845-01 (1987) International Electrotechnical Vocabulary, Chapter 845: Lighting

IEC (International Electrotechnical Commission) 60904-3-02 (2008) Photovoltaic devices – Part 3: measurement principles for terrestrial photovoltaic (PV) solar devices with reference spectral irradiance data

Meyer-Arendt J (1968) Radiometry and photometry: units and conversion factors. Appl Opt 7(10):2081–2084

Ohno Y (1997) NIST Measurement services: photometric calibrations. NIST Special Publication 250–37, Gaithersburg, Maryland, 66 pp

Chapter 27
Basics of LEDs for Plant Cultivation

Kazuhiro Fujiwara

Abstract LEDs have received remarkable attention recently as a light source for plant factories and greenhouses. This chapter presents a simple explanation of LED basics with special reference to plant cultivation. The following items are selected and highlighted in this chapter for that purpose: (1) definitions for LED product terms; (2) light-emitting principle of an LED; (3) LED package configuration types; (4) basic terms for expressing optical, electrical, and radiational characteristics of an LED; (5) optical, electrical, and radiational characteristics in LED operation; (6) lighting methods; (7) radiant flux control methods; (8) special requirements for LED lamps to cultivate plants; (9) advantages and disadvantages of the use of LED in plant cultivation; and (10) luminous efficacy and energy-photon conversion efficacy for plant cultivation.

Keywords Artificial lighting • LED characteristics • LED lighting • LED types • Light-emitting diode • Terminology • Wavelength

27.1 Introduction

An LED is a semiconductor diode that emits light when an electric current is applied in the forward direction. Recently, LED lamps have often been introduced and installed in plant factories as a main light source and in greenhouses for supplemental lighting. This trend is primarily attributable to the steadily improved efficacy and rapidly reduced price of LED lamps in recent years. For several cases of LED application in plant factories and greenhouses, refer to Parts IV and VI.

The basics of LEDs necessary to use LEDs as a light source for plant cultivation are the following: (1) definitions of LED product terms; (2) light-emitting principle of an LED; (3) LED package configuration types; (4) basic terms for expressing optical, electrical, and radiational characteristics of an LED; (5) optical, electrical, and radiational characteristics in LED operation; (6) lighting methods; (7) radiant

K. Fujiwara (✉)
Graduate School of Agricultural and Life Sciences, The University of Tokyo, 1-1-1 Yayoi, Bunkyo-ku, Tokyo 113-8657, Japan
e-mail: afuji@mail.ecc.u-tokyo.ac.jp

flux control methods; and (8) special requirements for LED lamps to cultivate plants. This chapter first briefly describes the items described above and then explains some (9) advantages and disadvantages of the use of LED in plant cultivation and (10) the luminous efficacy and energy-photon conversion efficacy for plant cultivation.

The terminology used in this chapter is mainly referred from and based on recently published books by Mottier (2009), Kitsinelis (2011), Khan (2013), Japanese Industrial Standards (JIS) handbook no. 20-2 (2013), and International Electrotechnical Commission (IEC) 62504 (2014). A few segments of the chapter include partially reconstituted and modified contents of the author's previous works (Fujiwara 2014, 2015).

27.2 Definitions of LED Product Terms

At the beginning of this chapter, LED product terms must be presented clearly to avoid confusion or misunderstanding. The primary LED product terms are LED, LED package, LED module, LED lamp, LED light source, LED luminaire, and LED lighting system. Most definitions for these terms are direct quotes from IEC 62504 (2014).

27.2.1 LED

A solid device embodying a p-n junction, emitting incoherent optical radiation when excited by an electric current (IEC 62504 2014).

In the definition above, an LED chip and LED die are used to mean the same thing. Generally speaking, most people use the term "LED" to refer to an LED package, LED module, LED lamp, and LED light source, without regard to their classification.

27.2.2 LED Package

A single electrical component encapsulating principally one or more LED dies, possibly including optical elements and thermal, mechanical, and electrical interfaces (IEC 62504 2014).

27.2.3 LED Module

An LED light source having no cap, incorporating one or more LED packages on a printed circuit board, and possibly including one or more of the following: electrical, optical, mechanical, and thermal components, interfaces, and *control gear* (IEC 62504 2014).

27.2.4 LED Control Gear

A unit inserted between the electrical supply and one or more LED modules, which serves to supply the LED module(s) with its (their) rated current, and may consist of one or more separate components and may include means for dimming, correcting the power factor and suppressing radio interference, and further control functions (IEC 62504 2014).

Most single LEDs are low-current devices (mostly less than 1 A). Generally, LED modules consist of a combination of series and parallel connection strings. Commercial electric power supplies rarely match the voltage or current requirements of most of LED modules. Control gear for regulating the voltage or current for the LED module(s) is needed.

27.2.5 LED Lamp

An LED light source provided with (a) cap(s) incorporating one or more LED modules and possibly including one or more of the following: electrical, mechanical, and thermal components, interfaces, and control gear (IEC 62504 2014).

27.2.6 LED Light Source

An electrical light source based on LED technology (IEC 62504 2014).

The term "electrical light source" refers to a device emitting light by transforming electric energy into light. An electrical light source comprises one or more electrical lighting devices/lamps, possibly together with components designed to distribute the light, to position and protect the lamps, and to connect the lamps to the electric power supply.

27.2.7 LED Luminaire

A luminaire designed to incorporate one or more LED light sources (IEC 62504 2014).

The term "luminaire" refers to the entire electrical light fitting, including all the components needed to mount, operate, and protect the lamp. The luminaires protect the lamp, distribute its light, and prevent it from causing glare. However, a luminaire is sometimes defined as a term referring to a complete lighting unit, including one or more lamps, reflective surfaces, protective housings, electrical connections, and circuitry.

27.2.8 LED Lighting System

A system composed of one or more LED light sources with their luminaires connected to an instrument to control the quantity or quality of light such as a timer, dimmer, or computer-programmed control system. No literature-based definition is accessible for this term.

27.3 Light-Emitting Principle of an LED

As described above, an LED is a semiconductor diode formed by contacting p-type and n-type materials. Application of forward voltage within an appropriate range to the diode to move holes (having a positive electric charge) in the p-type material side and electrons (having a negative electric charge) in the n-type material side toward the other side allows the holes and electrons to recombine in the junction. The recombination can produce a photon, which has energy equivalent to (or in some cases less than) the amount that an electron released by transitioning when recombined (Fig. 27.1). The wavelength of photons produced by the recombination is dependent on the energy released by transitioning.

27.4 LED Package Configuration Types

Lamp (also called indicator or round) type, surface-mount device (SMD)type, and high-power type are the general configurations for LED packages. A so-called flux type has two anodes and two cathodes shaped like four legs of a table. Photographs of samples (Fig. 27.2) of the four LED types and simplified cross-sectional anatomies (Fig. 27.3) of the lamp, SMD, and high-power-type LEDs are shown.

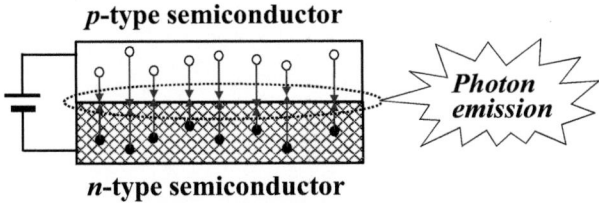

Fig. 27.1 LED photon/light emission from the junction of a *p*-type and *n*-type semiconductors when electricity is supplied. *Open* and *solid circles*, respectively, denote holes (positive charged) and electrons (negatively charged)

27.5 Basic Terms for Expressing Optical, Electrical, and Radiational Characteristics of an LED

The following are basic terms (Fujiwara 2015) that often appear in an LED specification sheet provided by a manufacturer. These are terms that people who use LEDs must remember.

27.5.1 Forward Current [A]

Electric current flowing through an LED from the anode to the cathode. An LED only emits photons/light in forward bias. In an LED specification sheet, a rated maximum forward current and rated maximum "pulse" forward current are generally described. The former is the maximum value allowed for continuous operation. The latter is that allowed when producing pulsed light. Incidentally, the term "rated" refers to a manufacturer or responsible vendor-declared value for a given type of electrical device or product as a specification or performance when operated under specified conditions; in some cases it is used in referring to "maximum" or "limit."

27.5.2 Half Width (at Half Maximum) [nm]

The width of the spectral radiant flux distribution curve (spectrum) at 50 % of the maximum spectral radiant flux. It is a measure of the light monochromaticity (Fig. 27.4). Sometimes HWHM is used as an acronym for this term.

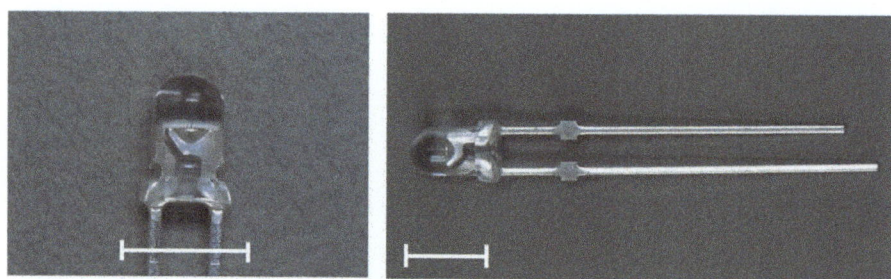

Lamp type

SMD type

High power type

Flux type

Fig. 27.2 Lamp (NSPW310DS; Nichia Corp.), surface-mount device (SMD) (NESW115T; Nichia Corp.), high-power (NS3W183AT; Nichia Corp.), and flux (NSPWR60CS-K1; Nichia Corp.)-type LED packages. *White bars* represent a scale of approximately 5 mm

Fig. 27.3 Schematic cross-sectional diagram of a lamp, surface-mount device (SMD), and high-power-type LED packages

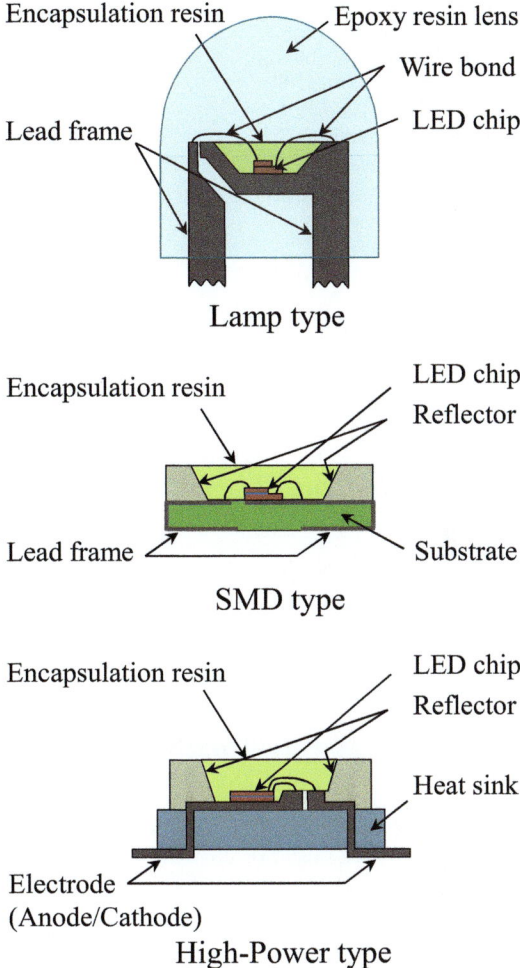

27.5.3 *Luminous Intensity [cd]*

The luminous flux [lm] per unit solid angle [sr] in a specific direction (generally optical axis direction as for LED packages, in which the value shows the maximum one). Radiant intensity [W sr^{-1}] is used for LEDs with a peak wavelength outside the visible wavelength range. Furthermore, see Chap. 26 for the definition of *luminous intensity* by IEC and for an explanation of solid angle.

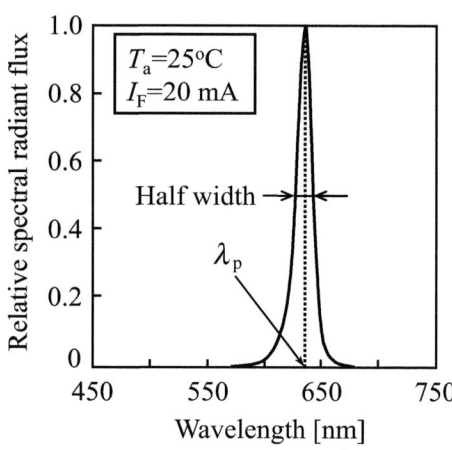

Fig. 27.4 Peak wavelength (λ_p) and half width using a sample spectral distribution of a *red* LED (NSPR310R; Nichia Corp.) (The figure was redrawn from the specification sheet)

27.5.4 Radiant Flux [W] (= [J s^{-1}])

The amount of radiant energy [J] emitted per unit time [s]. "Radiant power" [W] is occasionally used instead of "radiant flux". Furthermore, see Chap. 26 for the definition by IEC.

27.5.5 Peak Wavelength [nm]

The wavelength at which the spectral radiant flux [W nm^{-1}] in the spectral radiant flux distribution curve of an LED is the maximum (Fig. 27.4).

27.5.6 Viewing Half Angle [°]

The angle from the LED optical axis, at which LED radiant intensity [W sr^{-1}] is the maximum, to which the radiant intensity is reduced to half of the maximum value at the LED optical axis.

27.6 Optical, Electrical, and Radiational Characteristics in LED Operation

LEDs have common operating characteristics to which those who use LEDs for both plant cultivation and light environmental effects research should devote attention. The most important ones are itemized as shown below (Fujiwara 2015):

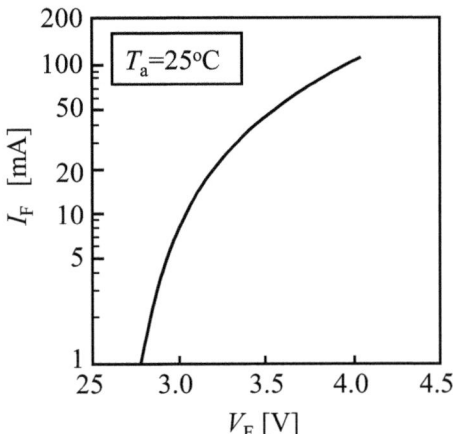

Fig. 27.5 Relation between forward voltage (V_F) and forward current (I_F) of a *blue* LED (NSPB310B; Nichia Corp.) in operation (The figure was redrawn from the specification sheet)

1. Electric current through an LED increases exponentially with increasing applied voltage to the LED up to around its standard forward voltage (Fig. 27.5).
2. Relative luminous intensity of an LED package is roughly proportional to the forward current when the ambient temperature is constant (Fig. 27.6).
3. Even if forward current is constant, the relative luminous intensity of an LED package decreases with the ambient temperature increase (Fig. 27.7).
4. The rated maximum forward current value allowed (allowable forward current) drops sharply when the ambient temperature increases over a certain temperature (for most LEDs around 40 °C), and maximum ambient temperature exists for the use of LEDs (for most LEDs around 80 °C) (Fig. 27.8).

27.7 Lighting Methods

Two lighting methods are used for plant cultivation in respect of "truly" or "apparently" continuous light during the photoperiod. The common method is "truly" continuous lighting by which light is emitted continuously in a strict sense during the photoperiod. The other method is intermittent lighting in which light is emitted intermittently in a certain cycle. Intermittent lighting is sometimes designated as pulsed lighting, especially when the cycle is less than or equal to 1 s. An LED can produce pulsed light with a cycle shorter than 10 μs.

27.8 Radiant Flux Control Methods

A simple and general method for radiant flux control or dimming of an LED is to control the forward current through the LED. The method is called constant-current operation. A constant-voltage operation is a formally similar method, but it can be

Fig. 27.6 Effect of forward current on relative luminous intensity of a *blue* LED (NSPB310B; Nichia Corp.) in operation (The figure was redrawn from the specification sheet)

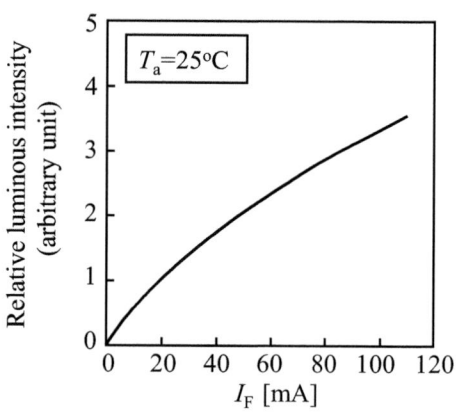

Fig. 27.7 Effect of ambient air temperature (T_a) on the relative luminous intensity of a *blue* LED (NSPB310B; Nichia Corp.) in operation (The figure was redrawn from the specification sheet)

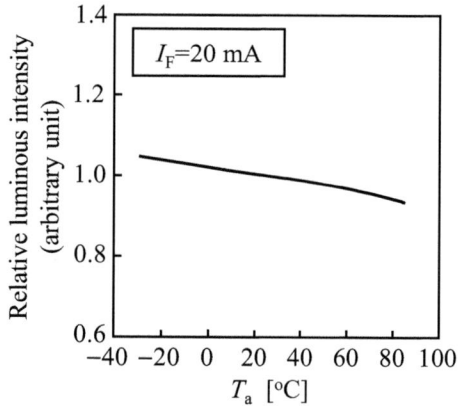

Fig. 27.8 Allowable forward current (I_F) against ambient air temperature (T_a) of a *blue* LED (NSPB310B; Nichia Corp.) in operation. (The figure was redrawn from the specification sheet)

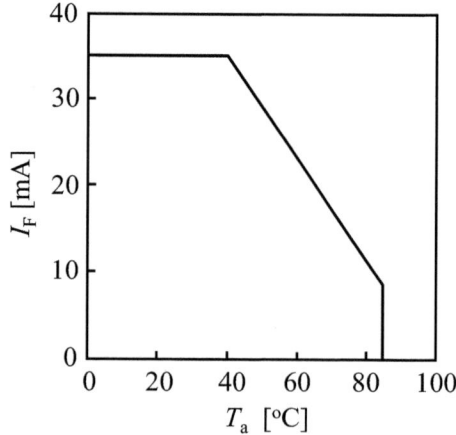

taken only under the conditions where low and constant temperatures of the LED chip and ambient air are guaranteed or the radiant flux from an LED can change according to those temperatures.

The other radiant flux control method is the so-called pulse width modulation (PWM): an LED is operated to repeat light on and off in a very short time period, respectively, by application of and breaking a constant forward current though the LED. The LED then produces a pulsed light. The pulse width refers to the time period of the current applied and light on. Incidentally, the ratio of light-on time period to one cycle time period is called the duty ratio (Fig. 27.9); and it is sometimes expressed in percentage. Radiant flux from the LED can be increased by increasing the duty ratio. At a 50 % duty ratio, the LED will theoretically emit half the radiant flux of the LED at a 100 % duty ratio. Pulsed light with a frequency (reciprocal of light-on time period plus light-off time period) higher than roughly 50–60 Hz can be perceived as a continuous light by the human eye.

Pulsed light does not promote the plant photosynthetic rate as continuous light does when the average PPFD is the same during the photoperiod, which means the photosynthetic photon numbers received per day at the plant canopy are equal, as explained theoretically by Rabinowitch (1956) and as demonstrated experimentally by Tennessen et al. (1995) for tomato and by Jishi et al. (2012, 2015) for cos lettuce. Sometimes, the results obtained by Emerson and Arnold (1932) have been misinterpreted as suggesting that pulsed light can increase the net photosynthetic rate, but they merely demonstrate that the light use efficiency was increased compared to continuous light by intermittently chopping off the continuous light using a rotating segmented wheel. It should be no surprise that the light use efficiency in pulsed light was increased compared to continuous light in their experiment because the average PPFD during the photoperiod was lower in pulsed light than in continuous light in the experiments conducted by Emerson and Arnold (1932).

27.9 Special Requirements for LED Lamps to Cultivate Plants

Some special requirements exist for LED lamps to cultivate plants that differ from those for general lighting. The most important requirement is the relative spectral distribution of light from LED lamps. With respect to photosynthesis, it is desirable that most photons emitted from LED lamps are photosynthetic photons, i.e., photons with wavelength of 400–700 nm.

Moisture and drip-proof LED lamps are necessary because plant factories and greenhouses have high humidity. The possibility exists for LED lamps installed in such places to be splashed or sprayed with water or a nutrient solution. When used in greenhouses with a water sprinkler or sprayer or in an open field, LED lamps with a high-ranked drip proof must be used.

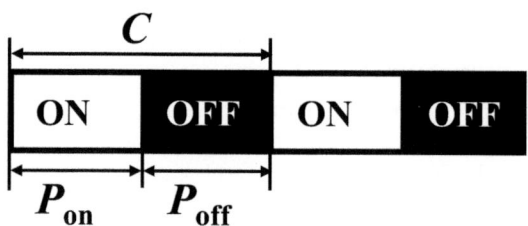

Fig. 27.9 Pulse cycle, frequency, and duty ratio in pulsed light

Pulse cycle (C): the time period required to complete one cycle from the beginning of one pulse to the beginning of the next.
$$C = P_{on} + P_{off}$$

Pulse frequency (F): the reciprocal of pulse cycle.
$$F = 1/C$$

Duty ratio (DR): the ratio of the time period of signal ON to that of signal ON + OFF, often expressed in percentage.
$$DR = P_{on} / (P_{on} + P_{off}) = P_{on} / C$$

In some cases, DR is expressed in percentage:
$$DR = 100 \times P_{on} / C \, [\%]$$

For greenhouses in which any type of sulfur fumigation (used for fungus disease control) is carried out, special attention must be devoted to LED sulfurization. Sulfurization corrodes the silver used in the LED package (e.g., electrode/lead frame, reflector) by sulfur derived from sulfide gases (Suzuki 2015). LED lamps with sulfurization corrosion resistance must be used for greenhouses in which sulfur fumigation is carried out.

27.10 Advantages and Disadvantages of the Use of LED Lamps in Plant Cultivation

27.10.1 Advantages of the Use of LED Lamps

As described in many reports of the literature (e.g., Massa et al. 2008; Fujiwara 2015), LEDs/LED lamps have various advantages over traditional forms of lighting for plant cultivation. Their advantages are that they are compact/small and robust, with a long lifetime, cool emitting temperature, rapid turn on and turn off, giving stable output immediately after turn on with controllable light output, and include no hazardous substances. In addition to those generally known advantages, LED lamps have unknown advantages related to their use for plant cultivation.

Electrical lamps used for plant cultivation have mainly been fluorescent lamps (FLs) and high-pressure sodium lamps (HPSLs); FLs include a general type, high-frequency type, and cool cathode type. Presuming a normal plant cultivation for which the light source position is fixed, easily controllable parameters in terms of lighting are the start and end times of the photoperiod of the day. The photon flux density can be controlled at the plant level only when dimmable lamps are used; dimmable FLs of a few kinds are available in the market, but they are expensive compared to the other FLs. Moreover, the relative spectral power distribution of light cannot be controlled even with a dimmable lamp.

In contrast to the use of FLs and HPSLs, the use of LED lamps enables control not only of photon flux density at the plant level but also (relative) spectral power distribution of light because of the variety of producible light when using LEDs of different types (Fig. 27.10), even according to the time of day. Actually, Jishi et al. (2016) reported that the shoot fresh weights of cos lettuce plants grown under irradiation patterns with red LED light irradiation starting time delayed 4 and 7 h from the blue LED light irradiation starting time were significantly greater than those grown with blue LED light irradiation and red LED light irradiation starting simultaneously. Some advantages of the use of LED lamps are therefore the greater possibility of controlling the light environment compared to the conventional light sources such as FL and HPSL. Recently, LED light source systems with 5 (Fujiwara et al. 2011; Yano and Fujiwara 2012), 6 (Fujiwara and Yano 2013), and 32 (Fujiwara and Sawada 2006; Fujiwara and Yano 2011; Fujiwara et al. 2013) wavelength LEDs have been developed, although they were developed for research purposes. In light of these development efforts, most effective or appropriate light environments for plant cultivation are expected to be found through studies conducted using LED lamps as a light source. In turn, LED lamps are anticipated for use eventually as a light source to realize such effective and appropriate light environments in plant cultivation.

In recent years, LED packages with wavelength conversion phosphor(s) of one or a few types have been developed to produce a wide variety of white light (with various spectral radiant flux distributions), as shown in Fig. 27.11. Typical white LEDs with a yellow phosphor that converts some blue light from the LED chip into yellow light have been used widely for general lighting applications since around 2009. Recently, LED packages with a red phosphor that converts blue light into red light are used to liven up the color of the cut of meat in the showcase. If an LED package with a phosphor facilitating production of light with a conversion peak wavelength of around 660 nm is produced inexpensively, it can be useful for plant factories and greenhouses because the light covers not only the red waveband but also the far-red waveband to some extent.

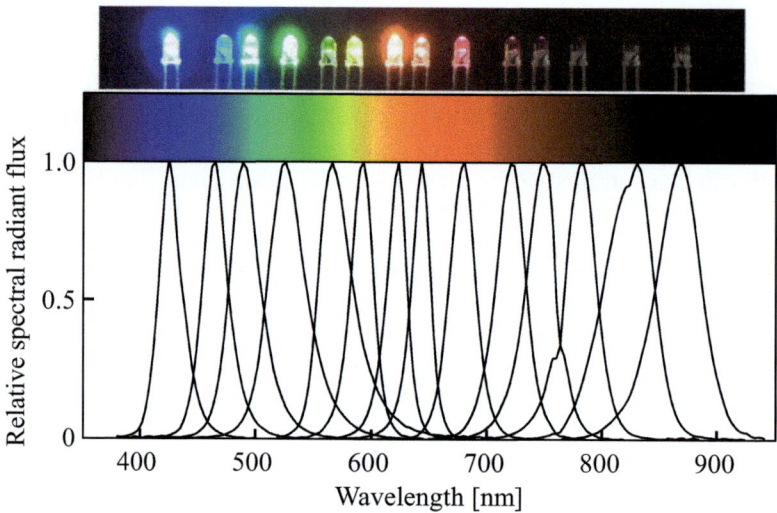

Fig. 27.10 Relative spectral radiant fluxes of light from various peak-wavelength lamp-type LEDs and their corresponding photographs

27.10.2 Disadvantages of the Use of LED Lamps

A readily apparent and most important disadvantage related to the use of LED lamps in plant cultivation would be higher initial cost of the lighting system "at this moment." Because the cost performance of LED lamps has been improved significantly every year, this disadvantage may be resolved in the next several years.

Another apparent disadvantage when using LED lamps as a light source for plant cultivation is that the related equipment is expensive compared to equipment that uses conventional FLs and HPSLs. However, that does not present a severe impediment to the introduction of LED lighting systems if it is possible to reduce running costs (including electricity, light sources, and luminaire maintenance and replacement) of the LED lighting system much lower than those of FL and HPSL lighting system.

Okada (2012) presented an estimation of the yearly electricity cost for lighting in a plant factory. It can be reduced by exchanging 3000 40-W straight-tube FLs each with a luminaire consuming 4 W for the same number of 22.5 W bar-type LED light sources with blue and red LEDs, which can provide the same level of PPFD on the shelf as the FLs. For the estimation, the condition of 1 kWh = 20 yen ($\approx$ US$0.19) and a photoperiod of 16 h d^{-1} were assumed. According to his estimation, the yearly electricity cost for lighting can be reduced by about 10 million yen ($\approx$ US$95,000) for a plant factory. He pointed out that the use of LED light sources instead of FLs with a luminaire even reduces electricity costs for cooling.

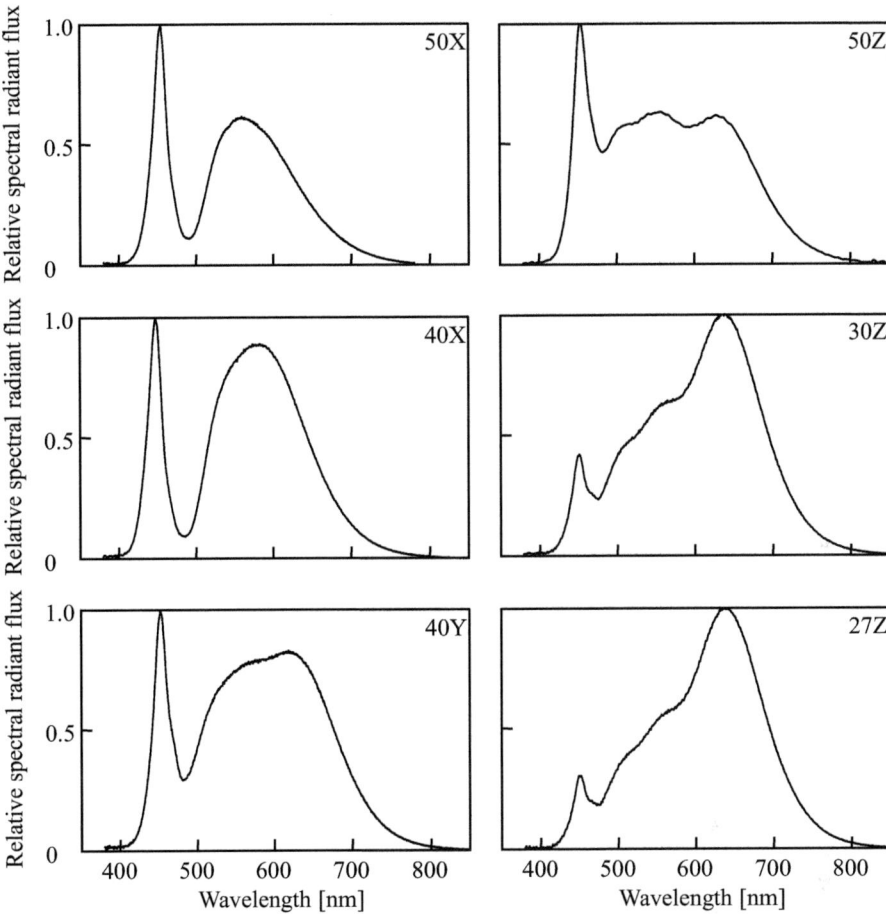

Fig. 27.11 Relative spectral radiant fluxes of phosphor-based *white* LEDs (GTEW1656JTE series; Stanley Electric Co. Ltd.) with phosphor(s) of one or a few types. *Numbers* with a capital letter shown in the *upper right* of each graph represent the *color* temperature (27: 2700 K; 30: 3000 K, 40: 4000 K; 50: 5000 K) and *color* rendering index (CRI) range of produced light (X: 70 ≤ CRI <85; Y: 85≤ CRI <95; Z: 95≤ CRI). Here, the CRI is a measure of how well the light source reproduces the color of any object in comparison to a reference source (Kitsinelis 2011) (The figures were redrawn from the specification sheet)

A less-noted disadvantage is that there is no general-lighting LED lamp that is widely available for plant cultivation. In contrast, general-lighting FLs and HPSLs have been available for plant cultivation using drip-proof and moisture-proof luminaires.

27.11 Luminous Efficacy and Energy-Photon Conversion Efficacy for Plant Cultivation

27.11.1 Luminous Efficacy

In general, the luminous flux [lm] obtained per unit electric power [W] input to a lamp (or light source) is defined as the "luminous efficacy (of a source)" [lm W^{-1}]; IEC(1987) defined it as the quotient of the luminous flux emitted by the power consumed by the source (IEC 60050-845-01-55). The definition is widely applied to conventional incandescent lamps, FLs, HPSLs, etc. In contrast, the luminous efficacy for LED packages/modules/lamps is defined as the rated luminous flux, which is measured in an atmosphere of 25 °C, divided by the rated electric power input in many instances. The term "total efficacy" [lm W^{-1}] is used for an LED lamp with its luminaire, which is defined as luminous flux [lm] obtained per unit electric power [W] input to the LED lamp with its luminaire.

Incidentally, the theoretical maximum luminous efficacy of an ideal lamp (with no energy conversion loss) is attained as 683 lm W^{-1} with a truly monochromatic light at a wavelength of 555 nm (light color appears to be yellow green) based on the definition of the luminous intensity of the SI base units. A feasible maximum luminous efficacy of a yellow-phosphor white LED package is predicted to be around 260 lm W^{-1} (Okubo and Nozawa 2012).

27.11.2 Energy-Photon Conversion Efficacy for Plant Cultivation

Luminous efficacy is a term that has significance when illuminating an object or its surroundings so that the human can perceive them visually. For plant cultivation, "energy-photon conversion efficacy" [mol J^{-1}] should be used: it is defined as the number of photons [mol] emitted per unit of electric energy [J] input to a lamp or as photon flux [mol s^{-1}] obtained per unit of electric power [W] input to a lamp because plant light-responses are a photochemical reaction that is started by photoreceptors receiving photons each with a corresponding level of energy (or wavelength) for the reaction. When emphasizing photosynthesis, photons that should be counted are the photosynthetic photons with wavelength of 400–700 nm; the energy-photon conversion efficacy should be termed as "photosynthetic photon number efficacy" (see also Chap. 28).

References

Emerson R, Arnold W (1932) A separation of the reactions in photosynthesis by means of intermittent light. J Gen Physiol 15:391–420

Fujiwara K (2014) Application of LEDs for plant cultivation and LED light irradiation techniques. In: Takatsuji M, Kozai T (eds) Important considerations and countermeasures for the plant factory management. Jyohokiko, Tokyo, pp 127–136 (in Japanese)

Fujiwara K (2015) Light sources. In: Kozai T et al (eds) Plant factory: an indoor vertical farming system for efficient quality food production. Academic, London, pp 118–128

Fujiwara K, Sawada T (2006) Design and development of an LED-artificial sunlight source system prototype capable of controlling relative spectral power distribution. J Light Vis Environ 30(3):170–176

Fujiwara K, Yano A (2011) Controllable spectrum artificial sunlight source system using LEDs with 32 different peak wavelengths of 385–910 nm. Bioelectromagnetics 32(3):243–252

Fujiwara K, Yano A (2013) Prototype development of a plant-response experimental light-source system with LEDs of six peak wavelengths. Acta Hortic 970:341–346

Fujiwara K, Yano A, Eijima K (2011) Design and development of a plant-response experimental light-source system with LEDs of five peak wavelengths. J Light Vis Environ 35(2):117–122

Fujiwara K, Eijima K, Yano A (2013) Second-generation LED-artificial sunlight source system available for light effects research in biological and agricultural sciences. In: Proceedings of the 7th LuxPacifica2013, The imperial queen's park hotel, Bangkok, 6–8 Mar 2013. p 140–145

IEC (International Electrotechnical Commission) 60050-845-01 (1987) International Electrotechnical Vocabulary, Chapter 845: Lighting

IEC (International Electrotechnical Commission) 62504 (Edition 1.0 2014−06) (2014) General lighting—Light emitting diode (LED) products and related equipment—Terms and definitions

Japanese Standard Association (2013) JIS (Japanese Industrial Standards) Handbook No. 20-2: Electrical equipment III. Japanese Standard Association, Tokyo (in Japanese)

Jishi T, Fujiwara K, Nishino K et al (2012) Pulsed light at lower duty ratios with lower frequencies is less advantageous for CO_2 uptake in cos lettuce compared to continuous light. J Light Vis Environ 36(3):88–93

Jishi T, Matsuda R, Fujiwara K (2015) A kinetic model for estimating net photosynthetic rates of cos lettuce leaves under pulsed light. Photosynth Res 124:107–116

Jishi T, Kimura K, Matsuda R, Fujiwara K (2016) Effects of temporally shifted irradiation of blue and red LED light on cos lettuce growth and morphology. Sci Hortic 198:227–232

Khan MN (2013) Understanding LED illumination. CRC Press, Boca Raton

Kitsinelis S (2011) Light sources: technologies and applications. CRC Press, Boca Raton

Massa GD et al (2008) Plant productivity in response to LED lighting. HortSci 43(7):1951–1956

Mottier P (ed) (2009) LEDs for lighting applications. ISTE and Wiley, London

Okada T (2012) LED lighting technology. In: Goto E (ed) Agriphotonics II. CMC Publishing, Tokyo, pp 191–200 (in Japanease)

Okubo S, Nozawa T (2012) The goal over 200 lm/W. p 41–48, In: Nikkei Electronics (ed) LED2012-2013, Nikkei Bussines Publications Tokyo (in Japanease)

Rabinowitch EI (1956) Photosynthesis and related processes, vol II. Interscience, New York, part 2

Suzuki Y (2015) Analysis of LED sulfurization and sulfur outgassing using gas chromatograph FPD method. Oki Tech Rev 82(1):76–79 (in Japanese)

Tennessen DJR, Bula J, Sharkey TD (1995) Efficiency of photosynthesis in continuous and pulsed light emitting diode irradiation. Photosynth Res 44:261–269

Yano A, Fujiwara K (2012) Plant lighting system with five wavelength-band light-emitting diodes providing photon flux density and mixing ratio control. Plant Methods 8:46

Chapter 28
Measurement of Photonmetric and Radiometric Characteristics of LEDs for Plant Cultivation

Eiji Goto

Abstract The spectral parameters and intensity of light received by a plant are widely used to describe experimental conditions, evaluate light environments, and compare light sources. They include the spectral distribution, the photosynthetic photon flux density (PPFD), and the ratio of the photon fluxes of specific wavelength ranges. As for LED lighting systems, spectral distribution, angular distribution of luminous intensity, photosynthetically active radiant energy efficiency, and photosynthetic photon number efficacy are necessary measurements.

Keywords Angular distribution of luminous intensity • Photosynthetic photon number efficacy • Photosynthetically active radiant energy efficiency • PPFD • Spectral distribution

28.1 Plant Light Environment

Several spectral parameters of light received by a plant are widely used to describe experimental conditions, evaluate light environments, and compare light sources. Each parameter is significant and, in most cases, considering single parameters in isolation is insufficient for a sound interpretation of plant responses to varying light environments.

28.1.1 Spectral Distribution Curve

Photomorphogenic responses include growth effects (such as seed germination, phototropism, and organ elongation) and differentiation (e.g., flower bud and leaf

---

E. Goto (✉)
Graduate School of Horticulture, Chiba University, Matsudo 648, Matsudo, Chiba 271-8510, Japan
e-mail: goto@faculty.chiba-u.jp

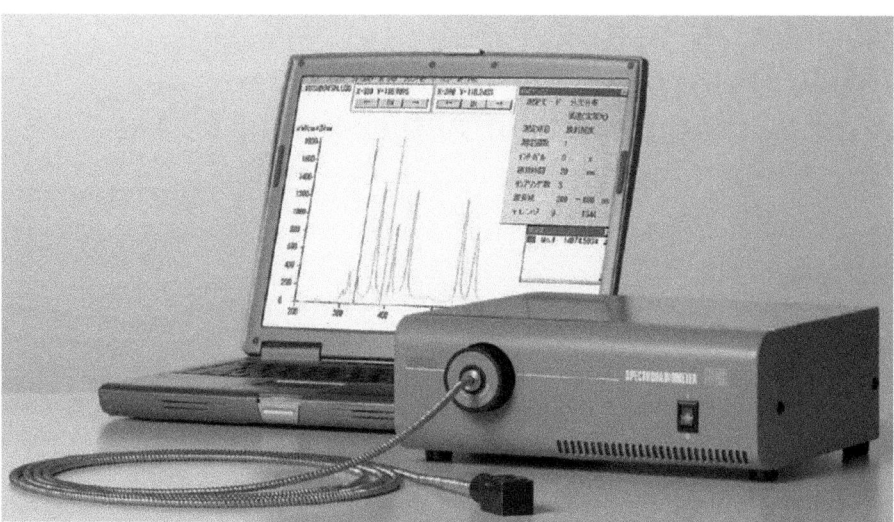

Fig. 28.1 A spectroradiometer system

formation and the regulation of photosynthetic pigments). Spectral distribution is measured by a spectroradiometer (Fig. 28.1). It is necessary to measure not only the photosynthetically active radiation (PAR) wavelength range (400–700 nm) but also the photomorphogenic response wavelength range (300–800 nm). In Fig. 28.2, the y axis may represent spectral irradiance (W m^{-2} nm^{-1}) or spectral photon flux density (μmol m^{-2} s^{-1} nm^{-1}). The value of spectral irradiance measured can be converted to spectral photon flux density (see Chap. 26).

28.1.2 Photosynthetic Photon Flux Density (PPFD)

PPFD (μmol m^{-2} s^{-1}) is the most frequently employed estimate of PAR in μmol of photons m^{-2} s^{-1}. PPFD weighs each photon between 400 and 700 nm equally. It is calculated by integrating photon flux density at each wavelength (μmol m^{-2} s^{-1} nm^{-1}) from the spectral photon flux density distribution curve (Fig. 28.3).

Quantum sensors (e.g., LI-COR LI-190 series, Fig. 28.4) are widely used to measure PPFD. It must be noted, though, that the values measured by a PPFD sensor contain errors compared with ideal values measured by a spectroradiometer.

Since photosynthetic efficiency is not homogenous across this wavelength range (McCree 1972; Inada 1976), yield photon flux (YPF) was proposed as an alternative measure of PAR (Fig. 28.5). YPF weighs different wavelengths according to the average leaf photosynthetic efficiency curve originally developed by McCree (1972). The YPF can be calculated from photon flux data and wavelength-specific

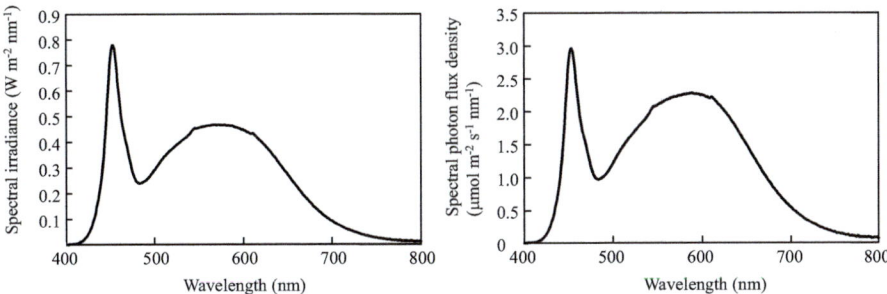

Fig. 28.2 Spectral irradiance curve of a white LED (5000 K type) measured by a spectroradiometer and calculated spectral photon flux density curve

Fig. 28.3 Spectral photon flux density curve of sunlight and calculated photosynthetic photon flux density (PPFD)

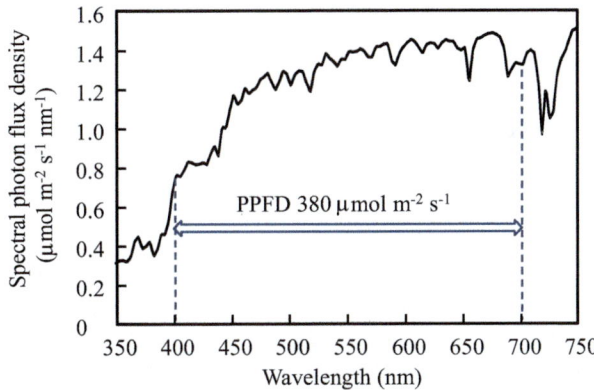

Fig. 28.4 A quantum sensor and light meter

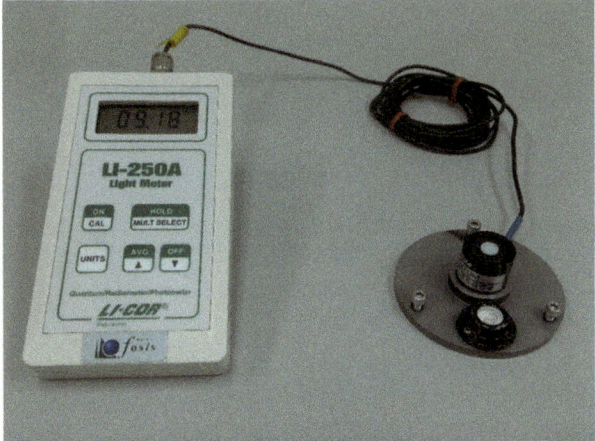

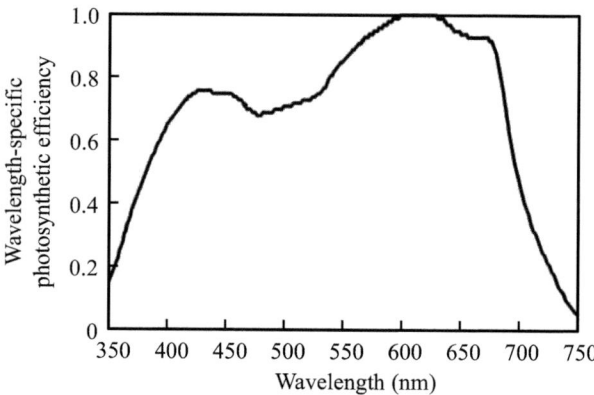

Fig. 28.5 Wavelength-specific photosynthetic efficiency used for calculation of yield photon flux (YPF)

photosynthetic efficiency, of which digital files are available in the literature (Sager et al. 1988).

28.1.3 Ratio of the Photon Fluxes of a Specific Wavelength Range

The ratios of the photon fluxes in the blue and red (B:R) and in the red and far-red (R:FR) portions of the spectrum influence photomorphogenic responses such as stem elongation and leaf structure. The R:FR ratio is directly related to the spectral properties of the phytochrome system.

28.1.4 Creation of a Summary Table of Characteristics of Light Quality Environment

Table 28.1 is an example of a table summarizing the characteristics of light quality environment. In many cases, the PPFD, 100-nm range in photomorphogenic response wavelength range, and B:R and R:FR ratios are listed. The table is useful to evaluate the light quality conditions and compare the light conditions created under different LED lights.

28.2 Characteristics of LED Lighting System

A manufacturing company that sells an LED luminaire or an LED lighting system is required to provide the following data for lighting design for growing plants.

Table 28.1 Spectral characteristics of light sources calculated from the spectral radiant flux distribution of each light source

Characteristic	White fluorescent lamp (5000 K)	White LED (5000 K)	White LED (3000 K)
Wavelength range (nm)			
400–700 (PPFD)	100.0	100.0	100.0
300–400 (UV)	1.0	0.1	0.0
400–500 (blue)	26.5	22.9	19.8
500–600 (green)	43.2	43.8	36.6
600–700 (red)	30.3	33.2	43.6
700–800 (far red)	3.6	4.6	3.5
B:R	0.88	0.69	0.45
R:FR	8.45	7.20	12.55
YPF[a]	87.7	86.8	88.6

[a]Yield photon flux. The value of YPF was calculated according to Sager et al. (1988)

28.2.1 Spectral Distribution (Spectral Radiant/Luminous Flux Distribution)

Spectral distribution (300–800 nm) is measured by a spectroradiometer or a total luminous flux measurement system (described later).

28.2.2 Angular Distribution of Luminous Intensity (Luminous Intensity Distribution)

This is a value measured by a goniophotometric measurement system. The measurement system measures not only angular distribution of luminous intensity but also spectral radiant flux and color rendering index (Ra). The angular distribution of luminous intensity is used to design or evaluate the goniophotometric characteristics of a lighting system.

A user can estimate light intensity distribution of plant canopy in a greenhouse or plant cultivation area in a plant factory using a data file of angular distribution of luminous intensity (Fig. 28.6) such as the Illuminating Engineering Society (IES) format data file.

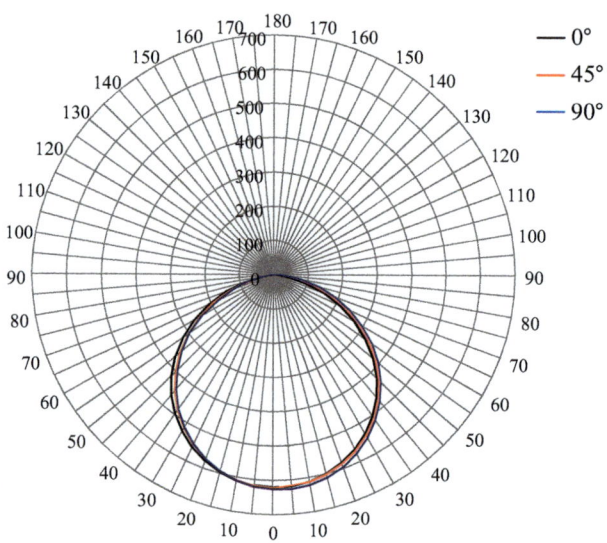

Fig. 28.6 An example of angular distribution of luminous intensity measured by a goniophotometric measurement system

28.2.3 Photosynthetically Active Radiant Energy Efficiency ($J\ J^{-1}$)

This is a calculated value obtained by dividing the radiant flux (W = J s^{-1}) of PAR by power consumption (W = J s^{-1}). The radiant flux is measured by a total luminous flux measurement system. The power consumption is measured by a power meter. The power consumption should include electric power consumed by a lighting unit, a power supply, and an instrument to control the quantity or quality of light, such as a timer, a dimmer, or a computer-programmed control system. The total luminous flux measurement system also measures luminous flux (lm), color temperature (K), and color rendering index (Ra).

The value obtained is useful for comparing photosynthetic light emitting efficiency of lighting units on the basis of energy.

28.2.4 Photosynthetic Photon Number Efficacy (μmol J^{-1})

This is a calculated value obtained by dividing the photon flux (mol s^{-1}) of PAR by power consumption (W = J s^{-1}). The photon flux is calculated by integrating photon flux at each wavelength (μmol nm^{-1}). The photon flux is calculated from radiant flux at each wavelength (W nm^{-1}), measured by a total luminous flux measurement system. The power consumption is measured by a power meter.

The power consumption should include electric power consumed by a lighting unit, a power supply, and an instrument to control the quantity or quality of light, such as a timer, a dimmer, or a computer-programmed control system.

The value obtained is useful for comparing photosynthetic light emitting efficacy of lighting units on the basis of photons.

28.2.5 Creation of a Summary Table of Characteristics of LED Lighting System

Table 28.2 is an example of a table summarizing the characteristics of an LED lighting system. In addition to optical and energy characteristics, the table includes characteristics of the system and other items measured by a total luminous flux measurement system. The table is useful to evaluate a commercial LED lighting system and compare lighting systems when a user designs a new light environment using LEDs.

Table 28.2 An example of a summary table of the characteristics of an LED lighting system

Item	Unit	Value
Measurement condition of temperature	°C	25
Power		
Voltage	V	200
Current	A	0.16
Effective power consumption	W	32.0
Light characteristics		
Photosynthetic photon flux	$\mu mol\ s^{-1}$	48.0
Photosynthetically active radiant flux	$W\ (=J\ s^{-1})$	8.0
Luminous flux	lm	450
Color temperature	K	3000
Color rendering index	Ra	87.0
Efficiency/efficacy		
Photosynthetically active radiant energy efficiency	$J\ J^{-1}$	0.25
Photosynthetic photon number efficacy	$\mu mol\ J^{-1}$	1.50

References

Inada K (1976) Action spectra for photosynthesis in higher plants. Plant Cell Physiol 17:355–365
McCree KJ (1972) The action spectrum, absorbance and quantum yield of photosynthesis in crop plants. Agric Meteorol 9:191–216
Sager JC, Smith WO, Edwards JL, Cyr KL (1988) Photosynthetic efficiency and phytochrome photoequilibria determination using spectral data. Trans ASAE 31:1882–1889

Chapter 29
Configuration, Function, and Operation of LED Lighting Systems

Akira Yano

Abstract Light-emitting diodes (LEDs) have been exploited from laboratory plant researches to commercial cultivations in greenhouses or plant factories with artificial lighting (PFAL). Benefits of LEDs for plant cultivation are their ability to emit thermal-free monochromatic light, solidity, minute size, minimal electricity consumption, longevity, and controllability of light emission. Varieties of LED lamps are available in the market, specialized respectively for PFAL, greenhouse, and laboratory applications. Those commercial products are well designed and useful, but users also can design and fabricate crop-specific LED lighting systems according to their purposes. To maximize the merits of LEDs for plant cultivations, the knowledge about fundamental principles of LED engineering would help growers. In this chapter, the light emission from a semiconductor p–n junction, the regulation of emission, heat treatments, and the regulation of light distribution on a culture bed are described.

Keywords Circuit • Control • Cultivation • Heat • Light-emitting diode • Plant

29.1 Introduction

Light-emitting diodes (LEDs) have been exploited for a wide range of applications, from basic plant studies (Fig. 29.1) to commercial cultivations in greenhouses or plant factories with artificial lighting (PFAL). A salient benefit of LEDs is their ability to emit monochromatic light. This benefit enables growers to design their own lighting recipes for cultivation. Broad spectrum lighting can be synthesized if some LED types with different peak wavelength emissions are combined (Schubert and Kim 2005; Fujiwara and Sawada 2006; Fujiwara and Yano 2011). LEDs that do not emit infrared radiation enable plant irradiation without heating the plants. This makes proximity LED light irradiation possible for plant cultivation (Massa et al. 2006, 2008; Stutte 2009; Poulet et al. 2014; Zhang et al. 2015). The

A. Yano (✉)
Faculty of Life and Environmental Science, Shimane University, 1060 Nishikawatsu, Matsue, Shimane 690-8504, Japan
e-mail: yano@life.shimane-u.ac.jp

Fig. 29.1 Lettuce irradiated with light from LEDs

thermal-free feature of LED lighting becomes particularly beneficial in multi-tier cultivation strategies in PFALs (Watanabe 2011; Goto 2012; Zhang et al. 2015). If far-red radiation is also necessary for managing plant morphogenesis (Higuchi et al. 2012; Kubota et al. 2012), co-irradiation of far-red and other visible LEDs can be a solution. Solidity of LEDs (Bourget 2008) allows them to be used in environments with human workers, although other lighting setups can be vulnerable to contact with workers. In addition, LEDs are movable (Yang et al. 2012; Li et al. 2014) because of its mechanical reliability. Another particular merit of LEDs is their minute size compared to the whole plant size. Azuma et al. (2012) irradiated only a bunch of grapes using blue and red LEDs during the night and demonstrated enhanced accumulation of anthocyanins on the fruit skin. It must be emphasized for this case that the electricity consumption for irradiation was minimal because light was emitted only to the necessary parts of the plant. Moreover, the direction of irradiation is not limited to downward incidence (Massa et al. 2008; Trouwborst et al. 2010). Upward or lateral irradiation can be applied using LEDs positioned near the plants (Massa et al. 2006; Zhang et al. 2015). Those irradiations can supplement light doses to lower leaves of high-density crops (Massa et al. 2008).

Through-hole and surface-mount-type LED packages are used. An LED chip is mounted on a light reflector and embedded in transparent epoxy or silicon resin, which functions as a lens to focus light emissions (Schubert 2006). Encapsulation prevents contact and invasion of moisture and dust (Lafont et al. 2012). A general lighting apparatus often comprises of a combination of LED modules that incorporates some LED chips in each module (Lafont et al. 2012). For instance, a linear type lamp has several LED chips in a straight line formation (Landis et al. 2013). A bulb-type LED lamp (Yasuda et al. 2010; Landis et al. 2013) arranges chips in a circular or lattice pattern on a circuit board in the bulb. Varieties of LED lamps are available in the market, specialized respectively for PFALs, greenhouses, chambers, and tissue culture vessels. Those commercial products are well designed, considering, for example, electricity consumption, hanging mechanisms, light emission directions, humidity resistance, suitability of wavelengths, and photon densities of light for plant cultivation. Readers can access information related to the products immediately using the Internet. Actually, related information is increasing and is updated rapidly and continually (Stutte 2015). Therefore, the author does not introduce any commercial products in this chapter. Instead, the author emphasizes that a user can purchase a single LED package from a market: anyone can freely design crop lighting systems (see, e.g., Fujiwara and Yano 2011; Poulet et al. 2014). This benefit is incomparably greater than those of other conventional lamps. A grower might want to irradiate only a part of an orchard (Azuma et al. 2012) or a tissue culture seedling in a culture vessel (Fang et al. 2011). Another grower might want to irradiate high-frequency pulse light (Tennessen et al. 1995; Jishi et al. 2015) or approximated sunlight (Fujiwara and Sawada 2006; Fujiwara et al. 2007; Fujiwara and Yano 2011) to plants. Functions of commercial LED lighting systems for plant cultivations might not be sufficient to satisfy users' demands. Greenhouse plant producers might want to check the efficacy of LED systems on their greenhouse crop production before investing in a large-scale lighting system (Mitchell 2015). For situations of this kind, small-scale preliminary experiments are desired. LEDs may be effective in such experiments. To maximize the merits of LEDs for plant cultivations, fundamental principles of LED engineering would help growers, in addition to knowledges on crop biology, nutrient chemistry, environmental physics, and economics related to products. In the next paragraph, fundamental principles of LED lighting engineering are described.

29.2 Semiconductor p–n Junction and Light Emission

A semiconductor is an intermediate material between conductors and insulators. A pure semiconductor is an intrinsic semiconductor. Si is a typical semiconductor with a tetravalent element. Four valence electrons form covalent bonds with neighbor Si elements in a crystal structure. An electron is lacking in the crystal structure if a Si atom is replaced with a trivalent element. This absence of an electron caused by the constraints of the crystal structure is called a hole. Semiconductors of this type are classified as p-type. In contrast, an electron is redundant if a Si atom is replaced with a pentavalent element. This type of semiconductor is n-type. Electrons in an n-type semiconductor move to the p-type side, and holes in a p-type semiconductor move to the n-type side when forward voltage is supplied to a p–n junction semiconductor. The electrons meet holes in the p-type semiconductor and they are recombined. The holes meet electrons in the n-type semiconductors as well. Electromagnetic waves can be emitted during the recombination because the

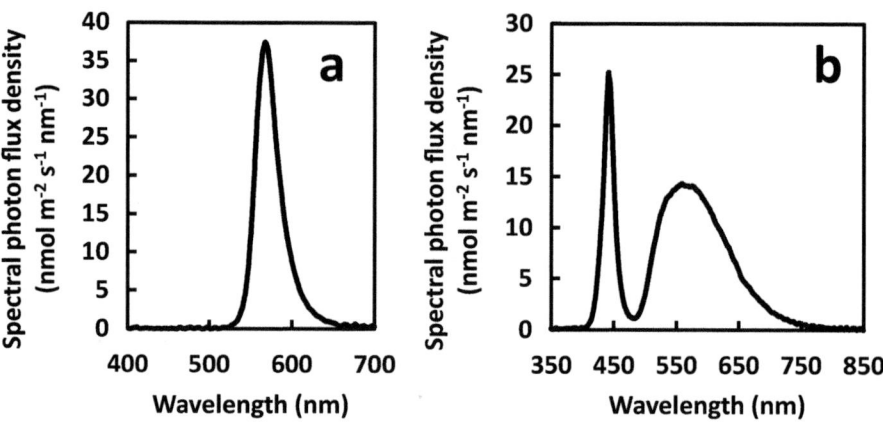

Fig. 29.2 Spectral photon flux densities measured at a surface irradiated from a GaP LED with a peak emission at 565 nm (**a**) and of a white LED (**b**). Standard half-value width of the GaP spectrum is 30 nm at $I_F = 20$ mA

accelerated electric carriers fall in a stable and lower energy level. If the energy gap corresponds to visible light energy, then the emission can be visible light (see more detail, e.g., Nakamura 2015). Although Si is suitable for explaining the basics of the p–n junction semiconductor, it might not be used for visible light-emitting diodes because the energy emission during the recombination is lower than that of visible light. InGaN (blue), GaP (green), and AlGaAs (red), for example, are used as visible light-emitting diodes.

The spectrum of light emitted from a single-type LED has a monochromatic peak with around ±30 nm half-value width of the wavelength (Fig. 29.2a). White light can be emitted if a blue LED is covered with a transparent but yellow phosphor-inclusive material (Schlotter et al. 1997; Pimputkar et al. 2009; Narukawa et al. 2010). The sharp peak at a shorter wavelength in Fig. 29.2b is emitted from the blue LED itself. The broad peak in the longer wavelength band is emitted from the phosphors.

29.3 Regulation of Emission

When a forward voltage V_F is supplied to an LED, forward current I_F is

$$I_F = I_s \exp\frac{e(V_F - V_D)}{kT}$$

where I_s stands for the saturation current when a reverse voltage is supplied to the LED, e signifies the elementary electric charge, V_D denotes the threshold voltage, k is Boltzmann's constant, and T represents temperature (Schubert 2006). Figure 29.3 shows the relation between I_F and V_F of LEDs of three types. Irradiation

Fig. 29.3 Relation between LED forward current I_F and forward voltage V_F

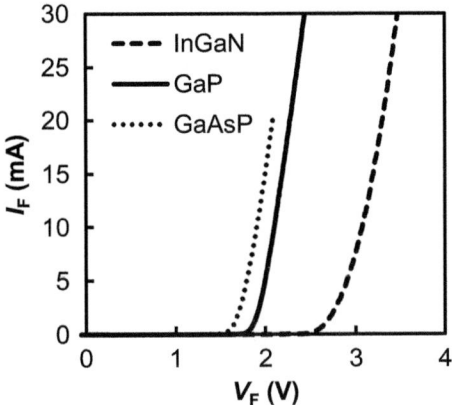

increases linearly or saturates with increasing I_F (Koide et al. 1991; Nakamura et al. 1993, 1995).

29.3.1 Basic LED Drive Circuits

LEDs can be operated using a conventional dry cell. Assume, for instance, that one is driving three equivalent LEDs of I_F with 20 mA and V_F with 2.0 V using a 9 V dry cell (Fig. 29.4a). Six volts (=2.0 + 2.0 + 2.0) are necessary to drive the LEDs when they are connected in series. Therefore, one can supply 6.0 V to the LEDs using a resistor that is connected in series with the LEDs and carries 3.0 V on it. According to the Ohm's law,

$$V_R = I_F \times R = 0.02 \times R$$

where R stands for the resistance and V_R denotes the voltage drop at the resistor. Also, the following equation holds:

$$V_R = 9.0 - 6.0 = 3.0 \text{ V}$$

Therefore,

$$R = \frac{V_R}{I_F} = \frac{3.0}{0.02} = 150 \, \Omega$$

which fulfills the requirement of voltage and current balances of the circuit (Fig. 29.4a). The resistor must be appropriate for the power consumption, as

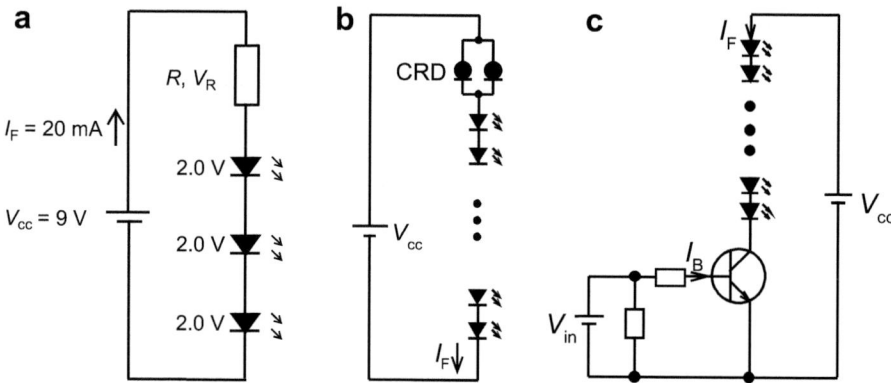

Fig. 29.4 Concepts of some conventional LED drive circuits based on (**a**) resistor, (**b**) CRD, and (**c**) transistor regulations

$$P = I_F^2 R = 0.02^2 \times 150 = 0.06\,\text{W}$$

In addition, current-regulating diodes (CRDs) are useful for providing constant current to LEDs (Lin et al. 2013; He et al. 2015) (Fig. 29.4b). A transistor is also useful to provide controlling current to LEDs (Fig. 29.4c). Regulating the base current I_B, then one can control I_F. In the transistor circuit, V_{in} can be supplied from a computer through a voltage output interface. Thereby, the irradiation is programmable (Shimada and Taniguchi 2011; Yano and Fujiwara 2012). The peak wavelength of emitted light shifts with I_F value because of band filling and the quantum-confined Stark effect, so some color shift can be expected when using analog dimming (Chang et al. 2012).

A DC power supply such as a series or switching regulator equipped with a current and voltage control is valuable as an LED power source. Many other conventional LED drive circuits are available in the market. Related information is readily available from the Internet. Power line frequency AC electricity might drive a string of LEDs directly or after simple rectification to DC electricity (Held 2009; Tan and Narendran 2014). For this case, the influence of a ripple of emitting light on plant responses should be studied preliminarily.

29.3.2 LED Lighting Mode for Plant Cultivations

A lighting mode of LED light irradiation can be designed according to a cultivation strategy. LED appliances designed for use in human life are not always suitable for plant cultivation. For PFALs, light and dark cycles do not necessarily follow that of the outside diurnal pattern. In addition, LED illumination can be designed to regulate plant photomorphogenesis (Higuchi et al. 2012; Liao et al. 2014). Irradiation that increases gradually with time and then decreases, resembling the pattern of sunlight on the earth, is also producible using LEDs because of the ability to control photon emissions (Fujiwara and Sawada 2006; Fujiwara et al. 2013). Spectral irradiance of the sunlight at dawn or dusk (Fujiwara and Sawada 2006; Fujiwara et al. 2013) or cloudy sky (Fujiwara and Yano 2011) is also approximately producible using a combination of varieties of LEDs.

Plant growth can be regulated precisely by controlling irradiance from LEDs using the feedback parameters of plant conditions. Fujiwara et al. (2005) succeeded in maintaining the tomato seedling weight during storage by minimizing the net photosynthetic rate of plants by controlling photosynthetic photon flux density (PPFD) supplied from the LEDs in response to the CO_2 concentration in the plant container. Li et al. (2012) developed a control system of LED output for plant irradiation based on information related to chlorophyll fluorescence from the plants. As these examples demonstrate, dynamic LED control is possible and useful for plant lighting systems.

Particularly for human life application, light dimming using pulse width modulation (PWM) is often applied (Held 2009; Chiu et al. 2010; Choi et al. 2015). In principle, the ratio of the on- and off-duration of lighting is controlled. Humans feel that it is becoming dark as the ratio of the off-duration increases, even though the photon flux density is constant during the on-duration. Usually, the on–off frequency of PWM is chosen as faster than that of human perception ability so that humans do not notice the flicker. Plants are exposed to the series of light pulses if a grower uses the PWM LED lighting system for plant cultivation. A new and interesting topic is the investigation of how plants respond to PWM lighting (Shimada and Taniguchi 2011; Dong et al. 2015; Jishi et al. 2015). To achieve a certain level of PPFD using PWM, one might increase the ratio of the on-duration or increase the photon flux. The latter might present some limitations because the plant photosynthetic rate saturates in greater PPFD values. Care should also be taken in relation to the measurement of the PPFD value of PWM light. A fast response PPFD sensor and a high time-resolution monitor that is suitable for the PPFD sensor output are necessary (Fig. 29.5).

29.4 Heat Dissipation

The photosynthetically active radiation (PAR) range LEDs are superior in terms of energy efficiency for plant cultivation and nonthermal plant irradiation compared to incandescent lamps which accompany infrared radiation. However, heat generation in the LED chip is an ineluctable problem because the input electricity cannot be converted into light with 100 % efficiency (Humphreys 2008). As I_F increases, LED junction temperature increases (Xi and Schubert 2004). The chip comprises semiconductors that are unstable against heat. The chip volume is too small to radiate heat outside smoothly. Removing heat from the chip embedded in resin is not easy. As the chip temperature increases, radiant flux from the LED decreases (Cho et al. 2010). Continuous removal of heat from the chip is necessary for stable

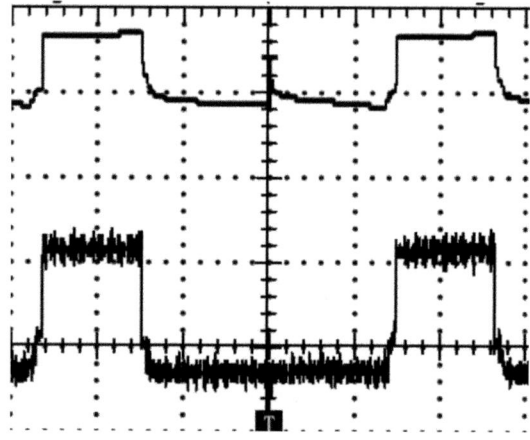

Fig. 29.5 Oscilloscope observation of PWM LED drive current (*top*) and the output voltage (*bottom*) of the PPFD sensor which was irradiated by the LED PWM light. The on-time pulse duration was 3 ms

LED illumination (Ye et al. 2013). The photon flux of an LED reaches the maximum immediately after switching on and decreases with elapsed time until a stable value is reached (Fig. 29.6).

Figure 29.7 presents a sketch of heat diffusion of through-hole and surface-mount type LEDs (for details, see Lafont et al. 2012). The electrode conductors are important heat convection routes for the through-hole LED. LEDs should be operated in a range of the rated p–n junction temperature. The junction temperature of the chip affects optical characteristics such as the color and dominant wavelength (Chang et al. 2012; Meneghini et al. 2012). An LED chip often accompanies a heat radiator, which is generally heavier and greater than the LED chip. Fans are also useful to cool the LEDs (Fujiwara and Yano 2011; Yano and Fujiwara 2012; Wollaeger and Runkle 2013; Poulet et al. 2014) (Fig. 29.6), although the electricity consumption of the lighting system is thereby increased. A deliberate heat design of LEDs is fundamentally important for long-term and large-scale cultivation in PFALs (Bourget 2008).

According to the definition proposed by the Alliance for Solid-State Illumination System and Technology, failure of an LED is regarded as taking place when the luminosity decays to 70 % (L70) or 50 % (L50) of its initial value (Chang et al. 2012; Lafont et al. 2012). When operated at favorable temperatures, individual LEDs generally have a predicted lifetime of up to 50,000 h, which corresponds to about 16.7 years when used an average of 8 h per day or 3000 h per year (Nelson and Bugbee 2014). As I_F increases, the radiant flux of an LED increases, but the lifetime decreases. Among the causes of failures are a temperature rise induced by loss in hole–electron recombination according to an imperfect crystal structure, reverse voltage bias, electrostatic discharge, as well as damages to junctions and separate components according to the different thermal expansion rates of the respective materials (Chang et al. 2012; Lafont et al. 2012).

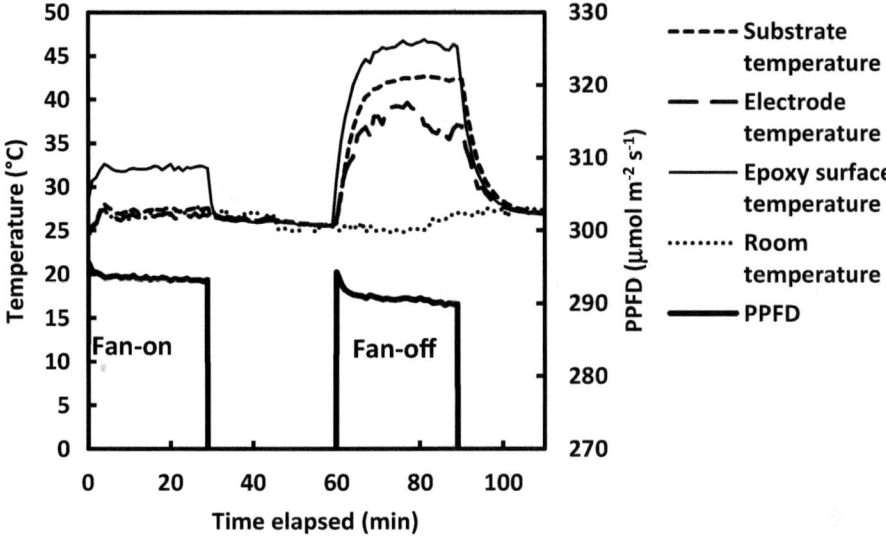

Fig. 29.6 Effects of fan cooling on LED temperature and PPFD irradiated at a surface from the LED module

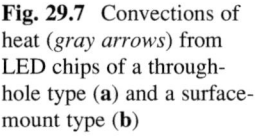

Fig. 29.7 Convections of heat (*gray arrows*) from LED chips of a through-hole type (**a**) and a surface-mount type (**b**)

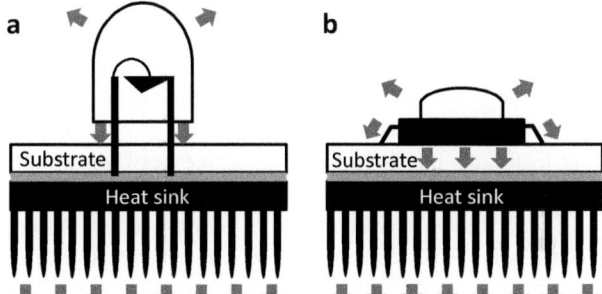

29.5 Distribution of Photon Flux Densities on an Irradiated Surface

A uniform PPFD distribution on a culture bed is desirable for a plant lighting design, although it is difficult to achieve. Irradiating a culture bed from some height by a lamp or lamps which are mutually separated at a constant interval (e.g., Fig. 29.8a) produces a higher PPFD at the center which decreases toward the edges of the irradiated culture bed (Fig. 29.8b, c). This deviation is affected by the distance between the lamp and the culture bed (Bornwaßer and Tantau 2012). For this reason, the PPFD distribution should be designed carefully in PFALs because the PPFD deviation produces plant growth deviation. A grower can reduce the PPFD at the central part of the culture bed, providing less electric current to the

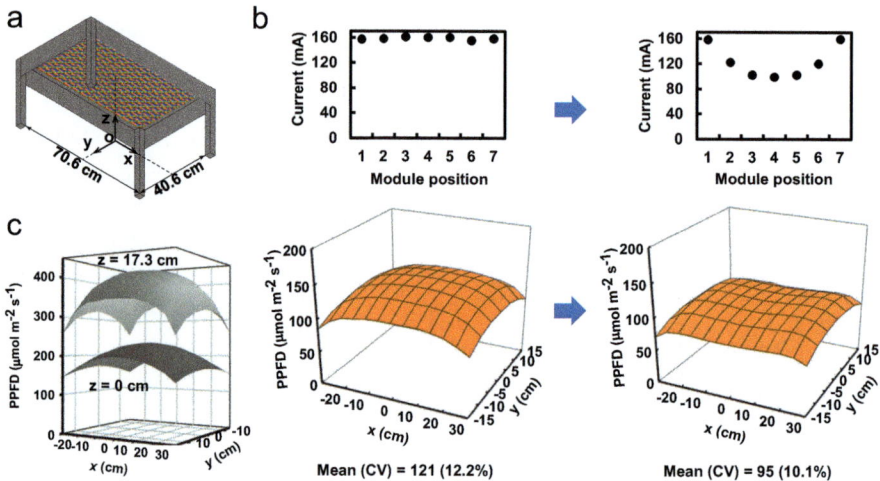

Fig. 29.8 Distribution of photosynthetic photon flux density (PPFD) (**b**) below an LED panel (**a**) before (*left*) and after current regulation (*right*) and the distribution of PPFD at the two heights from the floor (**c**). Details of the LED lighting system (**a**) are described in Fujiwara and Yano (2013)

central LEDs on the lighting panel. Figure 29.8b presents a related example in which less current was provided to the central LEDs so that the PPFD distribution becomes uniform. Of course, a user can provide a nonuniform PPFD distribution deliberately by an LED current control to produce inhomogeneous plant responses (Yano and Fujiwara 2012). Poulet et al. (2014) developed the target LED lighting that irradiates plants but not the space between the plants. This method promises better electrical energy use efficiency in PFALs.

The distance between the LED lamps and a culture bed of a typical multi-tier system of PFAL can be 30–40 cm. Plants approach the lamps as they grow. Accordingly, PPFD at the upper leaves of plants increases as time elapses. In contrast, seedlings that are delayed during their initial growth stage are shaded by other fast-growing plants. Growth of the delayed plants can be delayed further according to the shading effect. The degree of the delay can be influenced by the ratio of red and far-red light photon flux densities (Shibuya et al. 2013). Similar phenomena are visible with sunlight or in greenhouse cultivations, but the tendency is less than that in PFALs because a vertical PPFD gradient is negligible under the sunlight. The vertical PPFD gradient (Fig. 29.8c) in PFALs is unavoidable because of the limited distance separating the lamps and plants.

References

Azuma A, Ito A, Moriguchi T et al (2012) Light emitting diode irradiation at night accelerates anthocyanin accumulation in grape skin. Acta Hortic 956:341–347
Bornwaßer T, Tantau H-J (2012) Evaluation of LED lighting systems in vitro cultures. Acta Hortic 956:555–562
Bourget CM (2008) An introduction to light-emitting diodes. HortScience 43:1944–1946
Chang M-H, Das D, Varde PV et al (2012) Light emitting diodes reliability review. Microelectron Reliab 52:762–782
Chiu H-J, Lo Y-K, Chen J-T et al (2010) A high-efficiency dimmable led driver for low-power lighting applications. IEEE Trans Ind Electron 57:735–743
Cho J, Yoon E, Park Y et al (2010) Characteristics of blue and ultraviolet light-emitting diodes with current density and temperature. Electro Mater Lett 6:51–53
Choi J-h, Cho E-b, Ghassemlooy Z et al (2015) Visible light communications employing PPM and PWM formats for simultaneous data transmission and dimming. Opt Quant Electron 47:561–574
Dong C, Shao L, Liu G et al (2015) Photosynthetic characteristics, antioxidant capacity and biomass yield of wheat exposed to intermittent light irradiation with millisecond-scale periods. J Plant Physiol 184:28–36
Fang W, Chen CC, Lee YI et al (2011) Development of LED lids for tissue culture lighting. Acta Hortic 907:397–402
Fujiwara K, Sawada T (2006) Design and development of an LED-artificial sunlight source system prototype capable of controlling relative spectral power distribution. J Light Vis Environ 30:170–176
Fujiwara K, Yano A (2011) Controllable spectrum artificial sunlight source system using LEDs with 32 different peak wavelengths of 385–910 nm. Bioelectromagnetics 32:243–252
Fujiwara K, Yano A (2013) Prototype development of a plant-response experimental light-source system with LEDs of six peak wavelengths. Acta Hortic 970:341–346
Fujiwara K, Sawada T, Kimura Y et al (2005) Application of an automatic control system of photosynthetic photon flux density for LED-low light irradiation storage of green plants. HortTechnology 15:736–737
Fujiwara K, Sawada T, Goda S et al (2007) An LED-artificial sunlight source system available for light effects research in flower science. Acta Hortic 755:373–380
Fujiwara K, Eijima K, Yano A (2013) Second-generation LED-artificial sunlight source system available for light effects research in biological and agricultural sciences. In: Proceedings of the 7th LuxPacifica2013, The imperial queen's park hotel, Bangkok, 6–8 Mar 2013, pp 140–145
Goto E (2012) Plant production in a closed plant factory with artificial lighting. Acta Hortic 956:37–50
He Y, Qiao M, Dai G et al (2015) A vertical current regulator diode with trench cathode based on double epitaxial layers for LED lighting. In Proceedings of the 27th international symposium on power semiconductor devices & IC's, pp 157–160
Held G (2009) Introduction to light emitting diode technology and applications. Taylor & Francis, Florida
Higuchi Y, Sumitomo K, Oda A et al (2012) Day light quality affects the night-break response in the short-day plant chrysanthemum, suggesting differential phytochrome-mediated regulation of flowering. J Plant Physiol 169:1789–1796
Humphreys CJ (2008) Solid-state lighting. MRS Bull 33:459–470
Jishi T, Matsuda R, Fujiwara K (2015) A kinetic model for estimating net photosynthetic rates of cos lettuce leaves under pulsed light. Photosynth Res 124:107–116
Koide N, Kato H, Sassa M et al (1991) Doping of GaN with Si and properties of blue m/i/n/n^+ GaN LED with Si-doped n^+-layer by MOVPE. J Cryst Growth 115:639–642

Kubota C, Chia P, Yang Z et al (2012) Applications of far-red light emitting diodes in plant production under controlled environments. Acta Hortic 952:59–66

Lafont U, van Zeijl H, van der Zwaag S (2012) Increasing the reliability of solid state lighting systems via self-healing approaches: a review. Microelectron Reliab 52:71–89

Landis TD, Pinto JR, Dumroese RK (2013) Light emitting diodes (LED) – applications in forest and native plant nurseries. For Nursery Notes 33:5–13

Li Z, Ji J, Zou Q et al (2012) In situ monitoring system for chlorophyll fluorescence parameters of tomato at greenhouse in northern China. Acta Hortic 956:583–590

Li K, Yang Q-C, Tong Y-X et al (2014) Using movable light-emitting diodes for electricity savings in a plant factory growing lettuce. HortTechnology 24:546–553

Liao Y, Suzuki K, Yu W et al (2014) Night-break effect of LED light with different wavelengths on shoot elongation of *Chrysanthemum morifolium* Ramat 'Jimba' and 'Iwa no hakusen'. Environ Control Biol 52:51–55

Lin W-K, Uang CM, Wang P-C et al (2013) LED strobe lighting for machine vision inspection. In IEEE International Symposium on Next-Generation Electronics (ISNE), pp 345–346

Massa GD, Emmerich JC, Morrow RC et al (2006) Plant-growth lighting for space life support: a review. Grav Space Biol 19:19–30

Massa GD, Kim H-H, Wheeler RM et al (2008) Plant productivity in response to LED lighting. HortScience 43:1951–1956

Meneghini M, Lago MD, Trivellin N et al (2012) Chip and package-related degradation of high power white LEDs. Microelectron Reliab 52:804–812

Mitchell CA (2015) Academic research perspective of LEDs for the horticulture industry. HortScience 50:1293–1296

Nakamura S (2015) Nobel lecture: background story of the invention of efficient blue InGaN light emitting diodes. Rev Mod Phys 87:1139–1151

Nakamura S, Senoh M, Mukai T (1993) High-power InGaN/GaN double-heterostructure violet light emitting diodes. Appl Phys Lett 62:2390–2392

Nakamura S, Senoh M, Iwasa N et al (1995) High-brightness InGaN blue, green and yellow light-emitting diodes with quantum well structures. Jpn J Appl Phys 34:L797–L799

Narukawa Y, Ichikawa M, Sanga D et al (2010) White light emitting diodes with super-high luminous efficacy. J Phys D-Appl Phys 43:354002

Nelson JA, Bugbee B (2014) Economic analysis of greenhouse lighting: light emitting diodes vs. high intensity discharge fixtures. PLoS One 9:e99010

Pimputkar S, Speck JS, DenBaars SP et al (2009) Prospects for LED lighting. Nat Photonics 3:180–182

Poulet L, Massa GD, Morrow RC et al (2014) Significant reduction in energy for plant-growth lighting in space using targeted LED lighting and spectral manipulation. Life Sci Space Res 2:43–53

Schlotter P, Schmidt R, Schneider J (1997) Luminescence conversion of blue light emitting diodes. Appl Phys A 64:417–418

Schubert EF (2006) Light-emitting diodes, 2nd edn. Cambridge University Press, Cambridge

Schubert EF, Kim JK (2005) Solid-state light sources getting smart. Science 308:1274–1278

Shibuya T, Takahashi S, Endo R et al (2013) Height-convergence pattern in dense plant stands is affected by red-to-far-red ratio of background illumination. Sci Hortic 160:65–69

Shimada A, Taniguchi Y (2011) Red and blue pulse timing control for pulse width modulation light dimming of light emitting diodes for plant cultivation. J Photochem Photobiol B-Biol 104:399–404

Stutte GW (2009) Light-emitting diodes for manipulating the phytochrome apparatus. HortScience 44:231–234

Stutte GW (2015) Commercial transition to LEDs: a pathway to high-value products. HortScience 50:1297–1300

Tan J, Narendran N (2014) An approach to reduce AC LED flicker. J Light Vis Environ 38:6–11

Tennessen DJ, Bula RJ, Sharkey TD (1995) Efficiency of photosynthesis in continuous and pulsed light emitting diode irradiation. Photosynth Res 44:261–269

Trouwborst G, Oosterkamp J, Hogewoning SW et al (2010) The responses of light interception, photosynthesis and fruit yield of cucumber to LED-lighting within the canopy. Physiol Plant 138:289–300

Watanabe H (2011) Light-controlled plant cultivation system in Japan – development of a vegetable factory using LEDs as a light source for plants. Acta Hortic 907:37–44

Wollaeger HM, Runkle ES (2013) Growth responses of ornamental annual seedlings under different wavelengths of red light provided by light-emitting diodes. HortScience 48:1478–1483

Xi Y, Schubert EF (2004) Junction–temperature measurement in GaN ultraviolet light-emitting diodes using diode forward voltage method. Appl Phys Lett 85:2163–2165

Yang Z-C, Kubota C, Chia P-L et al (2012) Effect of end-of-day far-red light from a movable LED fixture on squash rootstock hypocotyl elongation. Sci Hortic 136:81–86

Yano A, Fujiwara K (2012) Plant lighting system with five wavelength-band light-emitting diodes providing photon flux density and mixing ratio control. Plant Methods 8:46

Yasuda T, Bessho M, Naoki S et al (2010) Light sources and lighting circuits. J Light Vis Environ 34:176–194

Ye H, Mihailovic M, Wong CKY et al (2013) Two-phase cooling of light emitting diode for higher light output and increased efficiency. Appl Therm Eng 52:353–359

Zhang G, Shen S, Takagaki M et al (2015) Supplemental upward lighting from underneath to obtain higher marketable lettuce (*Lactuca sativa*) leaf fresh weight by retarding senescence of outer leaves. Front Plant Sci 6:1110

Chapter 30
Energy Balance and Energy Conversion Process of LEDs and LED Lighting Systems

Akira Yano

Abstract Light energy irradiated on plants cultivated in a plant factory with artificial lighting (PFAL) is supplied from electrical energy through lamps that convert electricity into light. Plants exploit light as the energy source for driving photosynthesis. Plants also use light as a signal to trigger photomorphogenesis. An efficient use of both light and electrical energy is necessary for PFALs. This chapter presents a discussion of efficiencies of conversions step-by-step, from electrical input energy to the final chemical energy stored primarily as carbohydrates in crops in a PFAL.

Keywords Crop • Cultivation • Efficacy • Efficiency • Electricity • Light-emitting diode • PAR • Plant

30.1 Luminous Efficacy and the Conversion Efficiency from Electrical Input into Photosynthetically Active Radiation

The luminous efficacy (lm W^{-1}) of a light source is the luminous flux in lumens (light perceived by the human eye) per unit of electrical input power (Schubert and Kim 2005). Luminous efficacies greater than 130 lm W^{-1} and 100 lm W^{-1} have been achieved, respectively, for an LED chip and package (Narukawa et al. 2007; Humphreys 2008; Pimputkar et al. 2009). Luminous efficacies of 249 lm W^{-1} (Narukawa et al. 2010) and then 303 lm W^{-1} (see Nakamura 2015) by white LEDs have been reported. Luminous efficacies of 425 lm W^{-1} and 320 lm W^{-1} could potentially be achieved with dichromatic and trichromatic sources, respectively, if solid-state sources with perfect characteristics could be fabricated (Schubert and Kim 2005).

The conversion efficiencies η_{el_PAR} (Fig. 30.1) from electrical input into photosynthetically active radiation (PAR) are 0.4, 0.26, and 0.38, respectively, for an

---

A. Yano (✉)
Faculty of Life and Environmental Science, Shimane University, 1060 Nishikawatsu, Matsue, Shimane 690-8504, Japan
e-mail: yano@life.shimane-u.ac.jp

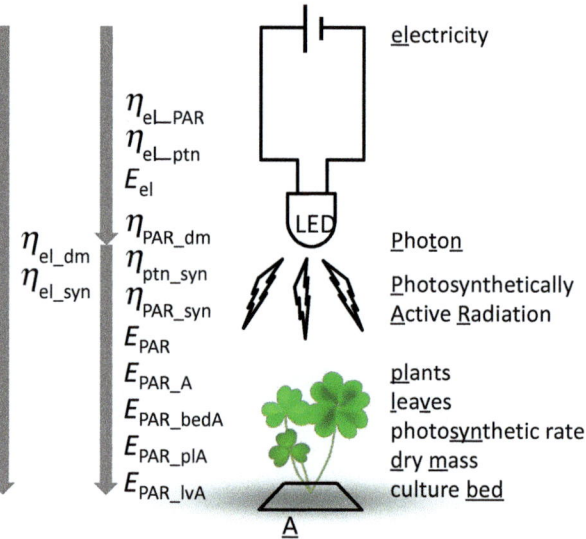

Fig. 30.1 Illustration showing the definition of efficiency η and energy E. Subscripts after η and E are denoted according to the reality shown at right (*underlined*)

LED lamp, a fluorescent lamp, and a high-pressure sodium lamp (Gislerød et al. 2012; Gómez et al. 2013; Kozai 2013a).

The number of photosynthetically active photons with wavelengths of 400–700 nm per unit of input electrical energy (mol J^{-1}) is an important criterion. It is therefore a widely used indicator for evaluating the efficiency of an artificial lighting for plant cultivation (see Nelson and Bugbee 2014). However, this author uses the same unit in the numerator and the denominator for efficiency calculation in this article to estimate the energy transfer from the input electricity to the final chemical energy stored in crops. The following equation

$$E = h\nu = \frac{hc}{\lambda} \tag{30.1}$$

interchanges the relation of these two ways of energy efficiency representation, where E (J) denotes energy, h (J s) is Planck's constant, and ν (s^{-1}) represents the frequency of the light. c (m s^{-1}) and λ (m) are, respectively, the speed and the wavelength of the light. However, the interchange occurs by a complicated process because an emission spectrum from an LED is actually spread in a wavelength range, thereby ν or λ in Eq. (30.1) does not represent a single specific value. The energy of each photon with a specific wavelength has to be summed up throughout the whole wavelength range of the spectrum if one calculates the E value from the emission spectrum of the LED.

30.2 Light and Electricity Use Efficiencies

30.2.1 PAR Energy Use Efficiency and Electrical Energy Use Efficiency Based on Plant Dry Mass

PAR energy use efficiency based on plant dry mass η_{PAR_dm} is

$$\eta_{PAR\_dm} = \frac{k \times \Delta D_n}{E_{PAR}} \quad (30.2)$$

where $k \times \Delta D_n$ (J m^{-2}) stands for the increment of chemical energy accumulated in plants and E_{PAR} (J m^{-2}) denotes the PAR energy irradiated from a lamp per cultivation area during a culture period (Kozai 2013a, b). The numerator is derived by multiplying the increment of plant dry mass per cultivation area during culture period ΔD_n (kg m^{-2}) by chemical energy stored in unit plant dry mass k (J kg^{-1}). The value of k is approximately 20 MJ kg^{-1}, although it deviates according to the concentrations of carbohydrates, proteins, and lipids in plants (Kozai 2013a, b). The dry mass of plants increases during a photoperiod by photosynthesis and decreases during a dark period by dark respiration. Whole plants or salable parts of plants can be chosen as the origin of the dry mass, according to the purpose of the estimation of η_{PAR_dm}. E_{PAR} = PAR (W m^{-2}) × photoperiod per day (s d^{-1}) × number of cultivation days (d) if the PAR is constant during photoperiods.

Chang and Chang (2014) harvested a dry mass of 2 g per upper part of a lettuce plant that had been cultivated in a chamber with an LED lighting system for 28 days. The PAR irradiated from the LEDs was 75 W m^{-2} comprised of red (623–673 nm), cyan (466–532 nm), and blue (427–478 nm) spectral peaks. The photoperiods and dark periods were, respectively, 18 h and 6 h per day during 28 days' cultivation. ΔD_n was $0.002/(142 \times 10^{-4})$ kg m^{-2} because the leaf area was 142×10^{-4} m^2. Assigning these values in the Eq. (30.2), $\eta_{PAR\_dm} = 0.021$.

Electrical energy use efficiency based on plant dry mass η_{el_dm} is the increment of chemical energy accumulated in plants per electrical input energy to the lamp (Kozai 2013a).

$$\eta_{el\_dm} = \eta_{el\_PAR} \times \eta_{PAR\_dm} \quad (30.3)$$

Martineau et al. (2012) cultivated Boston lettuce in a greenhouse with supplemental LED lighting with a spectral peak in red wavelength range. They estimated the dry mass increment of 16 g m^{-2} according to the supplemental lighting. Using this value, $k \times \Delta D_n = 20$ MJ kg^{-1} × 0.016 kg m^{-2} = 0.32 MJ m^{-2}. LED lamps consumed 82 W m^{-2} for 10.5 h supplemental photoperiod during 28 days of cultivation. This value corresponds to 87 MJ m^{-2}. Therefore, $\eta_{el\_dm} = 0.32/87 = 0.004$. Poulet et al. (2014) cultivated leaf lettuce in a walk-in chamber with a well-controlled hydroponic system for which lighting was supplied from LEDs. During the exponential growth phase of the lettuce plants, the ratio of the dry mass production to the electricity consumption of the lighting was 2.78 g kWh^{-1}. Using this value, $\eta_{el\_dm} = k \times 7.72 \times 10^{-10} = 0.015$.

30.2.2 Photosynthetic Photon Use Efficiency and Electrical Energy Use Efficiency Based on Net Photosynthetic Rate

Photosynthetic photon use efficiency η_{ptn_syn} based on net photosynthetic rate C_f (μmol m^{-2} s^{-1}) is described as

$$\eta_{ptn\_syn} = \frac{C_f}{PPFD} \quad (30.4)$$

where PPFD (μmol m^{-2} s^{-1}) represents photosynthetic photon flux density. η_{ptn_syn} can be derived continuously during the cultivation because both C_f and PPFD can be measured continuously. CO_2 released from plants during the dark respiration of the dark period is not considered in the Eq. (30.4). η_{ptn_syn} is affected by physiological and nutritional conditions of plants. Physical environment of crops also affects η_{ptn_syn}. Yorio et al. (2001) reported the η_{ptn_syn} values of 0.01, 0.04, and 0.02, respectively, for the single leaves of radish, spinach, and lettuce irradiated with red LEDs. C_f increases linearly and then saturates with PPFD increment (e.g., Taiz and Zeiger 1998). η_{ptn_syn} reaches a maximum value when PPFD is slightly higher than 0. It subsequently decreases with the PPFD increment because the slope of the PPFD–C_f curve corresponds to η_{ptn_syn}.

The PAR energy use efficiency η_{PAR_syn} based on C_f is described as Eq. (30.5) (Kozai 2011).

$$\eta_{PAR\_syn} = \frac{b \times C_f}{PAR} \quad (30.5)$$

In that equation, b (475 kJ mol^{-1}) denotes the conversion coefficient from photosynthetically fixed CO_2 into chemical energy (Kozai 2011, 2013a). Accordingly, the electrical energy use efficiency based on net photosynthetic rate (Kozai 2011) η_{el_syn} is defined as shown below:

$$\eta_{el\_syn} = \eta_{el\_PAR} \times \eta_{PAR\_syn} \quad (30.6)$$

30.3 Factors Affecting Electrical Energy Use Efficiency Based on Plant Dry Mass

Improvement of η_{el_dm} is necessary to reduce the cost of electricity in PFALs. Particular cares might be devoted on photonic η_{el_ptn} or radiation η_{el_PAR} efficiencies of lamps. However, improving only these efficiencies does not result in substantial cost reductions. A fraction of the input electrical energy supplied to the lamps in a PFAL is converted into light energy. Some of the light energy then arrives on the plant canopy in the PFAL. A part of that incident light energy is absorbed by leaves and is then partly converted into chemical energy via photosynthesis. Finally, a part of the harvested plant becomes the final salable product after harvestable plants have been selected from all plants (Kozai 2011). In PFALs, around 35 % of the light energy emitted from the lamps is incident on leaves. The remainder of the light energy (i.e., 65 %) is converted into heat after photons hit the culture bed, floor, and walls (Kozai 2011). This fact suggests that η_{PAR_dm} can be doubled if the 35 % efficiency can be increased to 70 % by improving the PFAL lighting design.

Possible strategies to improve η_{PAR_dm} are discussed hereafter. The following Eqs. (30.7), (30.8), (30.9), (30.10), (30.11), (30.12), (30.13), (30.14), and (30.15) are based on the cultivation area but not the floor area because multi-tier cultivations are often used in PFALs (Watanabe 2011; Goto 2012; Zhang et al. 2015). The following equations are related with (1) the electricity-PAR energy conversion efficiency, (2) the ratio of PAR exposure on plants, (3) the ratio of PAR exposure on leaves, (4) the ratio of PAR absorption by leaves, (5) the ratio of chemical energy fixation, (6) the ratio of weight losses by photorespiration and dark respiration, and (7) the ratio of crop salability. Improvement of coefficients in all the processes results in the improvement of η_{el_dm} (Kozai 2011).

30.3.1 Electrical Energy Consumed by Lamps During Cultivation per Cultivation Area and PAR Energy per Cultivation Area Emitted from the Lamps

Electrical energy consumed by lamps during cultivation per cultivation area E_{el} (J m^{-2}) is

$$E_{el} = n \times P \times T \tag{30.7}$$

where n (m^{-2}) stands for the number of lamps per cultivation area A (m^2), P (W) denotes the power consumption of a single lamp, and T (s) represents the total photoperiods during the cultivation. E_{el} can be reduced if the cultivating plants require low PAR or the crops need short cultivation duration until harvest. For this reason, suitable crops for the PFAL cultivation are expected to be leafy vegetables, seedlings, herbs, and medicinal herbal plants that require a moderate level of PAR (200–300 W m^{-2}) and shorter cultivation periods within 2 months (Kozai 2011, 2013b).

PAR energy E_{PAR_A} (J m^{-2}) per cultivation area A emitted from the lamps is described as

$$E_{PAR\_A} = \eta_{el\_PAR} \times E_{el} \qquad (30.8)$$

The η_{el_PAR} value of 0.40 has been reported for LED chips (Kozai 2013a). The total improvement of the system efficiencies and the reduction of the system cost are necessary together with the improvement of η_{el_PAR} because the cost of LED chips accounts for only about 15 % of the cost of the total lighting system, including such as LED lamps, a drive circuit, and a power source (Kozai 2011, 2013b). The condition of the lamp degrades as the duration of use accumulates. The failure of an LED is regarded as taking place when the luminosity decays to 70 % of its initial value (Chang et al. 2012; Lafont et al. 2012). The reduction of 30 % PAR is expected to result in nearly 30 % reduction of the photosynthetic rate of plants because photosynthetic rates of typical PFAL plants are related proportionally with PAR up to about 300 W m^{-2}.

30.3.2 PAR Energy Received in Cultivation Area with or Without Plants

PAR energy E_{PAR_bedA} (J m^{-2}) received in cultivation area A without plants is described as

$$E_{PAR\_bedA} = U \times E_{PAR\_A} \qquad (30.9)$$

where U denotes the coefficient of light utilization that is affected by reflectance of materials surrounding the lamps, the distance between the lamps and the culture bed, the size of the culture bed, and the directivity of the light emission from the lamps (Kozai 2011; Nelson and Bugbee 2014; Tong et al. 2014). In other words, U is the ratio of the impinging luminous flux on the irradiated surface and the total luminous flux emitted from the lamps or luminaires (Ryckaert et al. 2010; Dubois and Blomsterberg 2011). Yano and Fujiwara (2012) and Fujiwara and Yano (2013) used light reflector panels surrounding the LED panel to increase the U value at the culture bed (Fig. 30.2). Chin and Chong (2012) used small reflectors attached to each high-power LED to collimate rays. Although these reflectors might increase the U value, those prevent air circulations around LED packages and possibly plants. Thereby, thermal care should be considered. Li et al. (2014) demonstrated that the adjustment of the distance between the lamps and the canopy improves light use efficiency in a PFAL. Poulet et al. (2014) suggested that the light reflection on the culture bed may increase PAR exposure on plants.

Fig. 30.2 Use of glossy light reflector surrounding the LED panel to increase light utilization

PAR energy E_{PAR_plA} (J m^{-2}) received by plants in the cultivation area A can be described as

$$E_{PAR\_plA} = q \times E_{PAR\_bedA} \quad (30.10)$$

Coefficient q indicates the proportion of plant cover on A. q is almost 0 immediately after sowing the seed. The value of q increases with the increment of the leaf area index (LAI) (Larcher 1995) and reaches 1 at the harvest. The LAI value is around 3 when $q = 1$ (Kozai 2011). It is ideal to maintain LAI = 3 during cultivation by manipulating the plant density. However, the cost must be optimized considering the necessary manual labor or automated apparatuses necessary for LAI manipulation. Although E_{PAR_plA} increases as the distance between the lamp and the below plant decreases, the distributions of PAR and temperature across the cultivated plants become increasingly variable. Poulet et al. (2014) demonstrated that plant-target illumination reduces the waste of the lighting electricity consumption. Thereby, the q value can be maintained at a higher value from the sowing to the harvest. This is a particular merit of compact, heatless light emission, and controllability of LEDs.

PAR energy E_{PAR_lvA} (J m^{-2}) absorbed by plant leaves in the cultivation area A can be described as

$$E_{PAR\_lvA} = \alpha \times E_{PAR\_plA} \qquad (30.11)$$

where α represents the light absorption factor of leaves. The value of α is around 0.85 that is a specific optical characteristic of leaves (Kozai 2011).

30.3.3 Light Energy Fixation as Chemical Energy of Plant Dry Mass

The increment of plant dry mass during the photoperiod (i.e., excluding dark periods) of the cultivation ΔD_g (kg m^{-2}) is

$$\Delta D_g = f \times E_{PAR\_lvA} \qquad (30.12)$$

where f (kg J^{-1}) denotes the conversion coefficient of E_{PAR_lvA} into the chemical energy stored as the plant dry mass. The value of f is largely affected with the photosynthetic performance of plants and the physical environment of the cultivation. For this reason, controlling temperature, CO_2 concentration, PAR, light spectrum, air circulation, vapor pressure deficit, and nutrients is expected to be beneficial to optimize the f value at minimum cost (Kozai 2011; Tong et al. 2014). Kozai and Niu (2016) assumed 0.09 as a conventional f value and 0.10 for a possible maximum value.

The increment of plant dry mass at the end of the cultivation ΔD_n (kg m^{-2}) is

$$\Delta D_n = (1 - \beta) \times \Delta D_g \qquad (30.13)$$

where β stands for the ratio of the reduction of the plant dry mass by dark respiration during the dark period. $\beta = 0.1$ is possible (Kozai and Niu 2016) but variable with temperature (Kozai 2011).

The usable or salable plant dry mass ΔD_v (kg m^{-2}) is

$$\Delta D_V = HI \times \Delta D_n \qquad (30.14)$$

where HI represents the harvest index (Taiz and Zeiger 1998; Kozai 2011). HI can be increased to nearly 1.0 (Kozai and Niu 2016) by the cultivation management and environment control, including nutrient application. Finally, the following equation is derived:

$$\Delta D_V = HI \times (1 - \beta) \times f \times \alpha \times q \times U \times \eta_{el\_PAR} \times E_{el} \qquad (30.15)$$

Concerns might be focused on η_{el_ptn} or η_{el_PAR} efficiency of LEDs for PFAL lighting. However, all variables and coefficients in Eqs. (30.7), (30.8), (30.9), (30.10), (30.11), (30.12), (30.13), (30.14), and (30.15) contribute to electrical energy consumption (Fig. 30.3). The factors q, α, f, β, and HI vary with plant

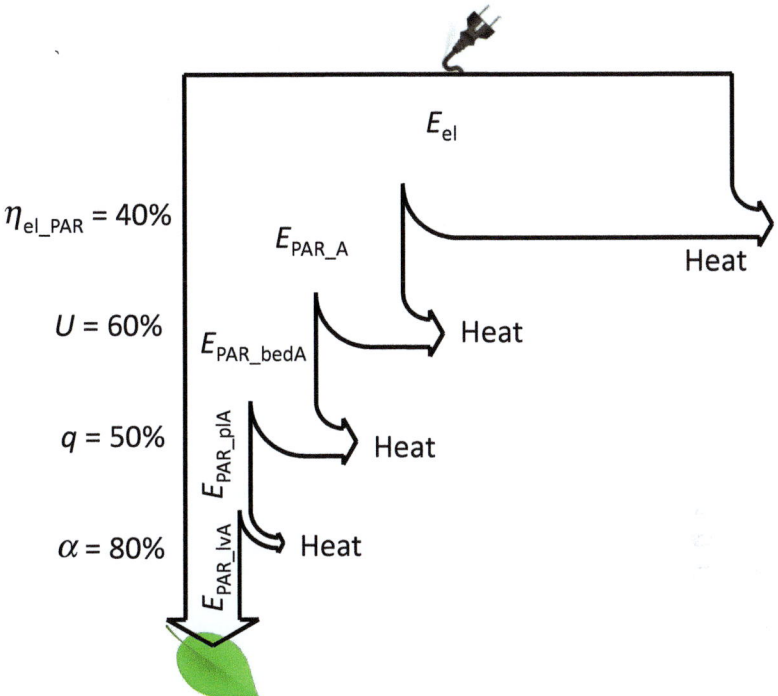

Fig. 30.3 Energy transfer from the lamps to leaves assuming possible coefficient values. E_{el}, E_{PAR_A}, E_{PAR_bedA}, E_{PAR_plA}, and E_{PAR_lvA} (J m^{-2}) represent electrical energy consumed by lamps during cultivation, PAR energy emitted from the lamps, PAR energy received on the culture bed without plants, PAR energy received by plants, and PAR energy absorbed by plant leaves per cultivation area A, respectively. η_{el_PAR}, U, q, and α represent conversion efficiency from electrical input into PAR, coefficient of light utilization that is affected by reflectance of materials surrounding the lamps, the distance between the lamps and the culture bed, the size of the culture bed, and the directivity of the light emission from the lamps, proportion of plant cover on cultivation area, and light absorption factor of leaves, respectively

growth stages. Particularly, improvement is anticipated for the coefficients in equations U, q, and f according to technological advancements and cultivation management.

References

Chang C-L, Chang K-P (2014) The growth response of leaf lettuce at different stages to multiple wavelength-band light-emitting diode lighting. Sci Hortic 179:78–84

Chang M-H, Das D, Varde PV et al (2012) Light emitting diodes reliability review. Microelectron Reliab 52:762–782

Chin L-Y, Chong K-K (2012) Study of high power light emitting diode (LED) lighting system in accelerating the growth rate of *Lactuca sativa* for indoor cultivation. Int J Phys Sci 7:1773–1781

Dubois M-C, Blomsterberg Å (2011) Energy saving potential and strategies for electric lighting in future North European, low energy office buildings: a literature review. Energy Build 43:2572–2582

Fujiwara K, Yano A (2013) Prototype development of a plant-response experimental light-source system with LEDs of six peak wavelengths. Acta Hortic 970:341–346

Gislerød HR, Mortensen LM, Torre S et al (2012) Light and energy saving in modern greenhouse production. Acta Hortic 956:85–98

Gómez C, Morrow RC, Bourget M et al (2013) Comparison of intracanopy light-emitting diode towers and overhead high-pressure sodium lamps for supplemental lighting of greenhouse-grown tomatoes. HortTechnology 23:93–98

Goto E (2012) Plant production in a closed plant factory with artificial lighting. Acta Hortic 956:37–50

Humphreys CJ (2008) Solid-state lighting. MRS Bull 33:459–470

Kozai T (2011) Improving light energy utilization efficiency for a sustainable plant factory with artificial light. In: Proceedings of Green Lighting Shanghai Forum, pp 375–383

Kozai T (2013a) Resource use efficiency of closed plant production system with artificial light: concept, estimation and application to plant factory. Proc Jpn Acad Ser B 89:447–461

Kozai T (2013b) Sustainable plant factory: closed plant production systems with artificial light for high resource use efficiencies and quality produce. Acta Hortic 1004:27–40

Kozai T, Niu G (2016) Plant factory as a resource-efficient closed plant production system. In: Kozai T, Niu G, Takagaki M (eds) Plant factory an indoor vertical farming system for efficient quality food production. Academic, London, pp 69–90

Lafont U, van Zeijl H, van der Zwaag S (2012) Increasing the reliability of solid state lighting systems via self-healing approaches: a review. Microelectron Reliab 52:71–89

Larcher W (1995) Physiological plant ecology, 3rd edn. Springer, Heidelberg

Li K, Yang Q-C, Tong Y-X et al (2014) Using movable light-emitting diodes for electricity savings in a plant factory growing lettuce. HortTechnology 24:546–553

Martineau V, Lefsrud M, Naznin MT et al (2012) Comparison of light-emitting diode and high-pressure sodium light treatments for hydroponics growth of Boston lettuce. HortScience 47:477–482

Nakamura S (2015) Nobel lecture: background story of the invention of efficient blue InGaN light emitting diodes. Rev Mod Phys 87:1139–1151

Narukawa Y, Narita J, Sakamoto T et al (2007) Recent progress of high efficiency white LEDs. Phys Stat Sol 204:2087–2093

Narukawa Y, Ichikawa M, Sanga D et al (2010) White light emitting diodes with super-high luminous efficacy. J Phys D Appl Phys 43:354002

Nelson JA, Bugbee B (2014) Economic analysis of greenhouse lighting: light emitting diodes vs. high intensity discharge fixtures. PLoS One 9:e99010

Pimputkar S, Speck JS, DenBaars SP et al (2009) Prospects for LED lighting. Nat Photon 3:180–182

Poulet L, Massa GD, Morrow RC et al (2014) Significant reduction in energy for plant-growth lighting in space using targeted LED lighting and spectral manipulation. Life Sci Space Res 2:43–53

Ryckaert WR, Lootens C, Geldof J et al (2010) Criteria for energy efficient lighting in buildings. Energy Build 42:341–347

Schubert EF, Kim JK (2005) Solid-state light sources getting smart. Science 308:1274–1278

Taiz L, Zeiger E (1998) Plant physiology, 2nd edn. Sinauer, Sunderland

Tong Y, Yang Q, Shimamura S (2014) Analysis of electric-energy utilization efficiency in a plant factory with artificial light for lettuce production. Acta Hortic 1037:277–284

Watanabe H (2011) Light-controlled plant cultivation system in Japan – development of a vegetable factory using LEDs as a light source for plants. Acta Hortic 907:37–44

Yano A, Fujiwara K (2012) Plant lighting system with five wavelength-band light-emitting diodes providing photon flux density and mixing ratio control. Plant Methods 8:46

Yorio NC, Goins GD, Kagie HR et al (2001) Improving spinach, radish, and lettuce growth under red light-emitting diodes (LEDs) with blue light supplementation. HortScience 36:380–383

Zhang G, Shen S, Takagaki M et al (2015) Supplemental upward lighting from underneath to obtain higher marketable lettuce (*Lactuca sativa*) leaf fresh weight by retarding senescence of outer leaves. Front Plant Sci 6:1110

Chapter 31
Health Effects of Occupational Exposure to LED Light: A Special Reference to Plant Cultivation Works in Plant Factories

Motoharu Takao

Abstract Light-emitting diode (LED) is one of the greatest inventions that changed plant production practices in the plant factories. Two main advantages of LED lighting are high-energy efficiency and arbitrary controllability of spectral design. In spite of the rapid advancements in LED lighting technology, workability and occupational health are not considered during plant factory design. In this chapter, I depict the impaired workability and the possible health problems associated with LED lighting. Plant utilizes blue light (BL) for photosynthesis, while BL may cause serious chronic diseases such as diabetes mellitus and breast cancer in humans by abnormal regulation of the human circadian rhythm. High-intensity BL also damages the lens and retina of the human eye, resulting in chronic visual impairment. The germicidal UV-C light causes severe chronic eye diseases in the laborers. Glare from the luminaires reduces visual performance and visibility, resulting in decrease of the laborers' productivity. The possible improvements that can be made to regulate the lighting environment will be discussed with special reference to the plant factories and the greenhouses with luminaires. Regulation of the lighting in the working environment will improve feasibility and comfort as well as raise productivity and health of the laborers.

Keywords Blue light • Ultraviolet light • Circadian rhythm • Retinal damage • Glare • Occupational health

31.1 Introduction

Light-emitting diode (LED) lighting revolutionized plant factories. The advantages of LED lighting can be summarized into two main facts: high-energy efficiency and arbitrary controllability of spectral design. Owing to these advantages, the plant

M. Takao (✉)
Department of Human and Information Science, Tokai University School of Information Science and Technology, Hiratsuka, Kanagawa 259-1292, Japan
e-mail: takao@keyaki.cc.u-tokai.ac.jp

factories can reduce the cost of plant production. The number of plant factories has been increasing year by year.

In spite of successful development of plant factories, the working environment of laborers has not been ergonomically well investigated in plant factories (Okahara et al. 2014). Especially, no study has been conducted for the investigation of the effect of artificial lighting on feasibility, labor productivity, and occupational health in both plant factories and greenhouses equipped with artificial lighting. In fact, so far, no occupational regulations have been issued by public authorities and industrial organizations.

In this chapter, I initially introduce psychophysical aspects of color perception required for plant cultivation works. Subsequently, I review the deteriorating effects of LED lighting on color perception of laborers in plant factories and greenhouses and the possible harmful effects on occupational health. Lastly, I discuss the desirable lighting environment for the laborers and the occupational health and safety regulations of lighting in plant factories and greenhouses equipped with artificial lighting.

31.2 Color Perception and Plant Cultivation Work

31.2.1 Neural Basis of Color Perception

Color perception is a crucial brain function for plant cultivation work. The farmers distinguish between the ripe and immature tomatoes by their superficial red color and identify the health of the grape trees by the color of their leaves. If color perception is hindered, plant cultivation work becomes difficult to perform. In this section, I will introduce the neural basis of color perception in *human* retina, as well as the psychophysical laws of color perception.

In the *human* retina, three types of cone photoreceptors capture the photons of light transmitted through the cornea, lens, and vitreous body (Figs. 31.1 and 31.2). The specialized cells convert the information provided by the photons into electrical energy by intracellular biophysical and electrochemical processes using photopigments and ion channels (Rodieck 1998). The amount of photons corresponds to the intensity of light. Wavelength is detected as a color by the neural circuit of the retina. Each type of cone photoreceptor has a different photopigment that has its own spectral photosensitivity. In *human* retina, the cone photoreceptors cover the wavelength range between 400 and 700 nm (Fig. 31.3). Each type of cone photoreceptor covers different wavelength ranges and has a different peak spectral sensitivity. Long-wavelength sensitive L-cones, middle-wavelength sensitive M-cones, and short-wavelength sensitive S-cones are maximally sensitive to maximum wavelengths of around 560 nm (red), 530 nm (green), and 420 nm (blue), respectively (Bowmaker and Dartnell 1980; Bowmaker et al. 1980).

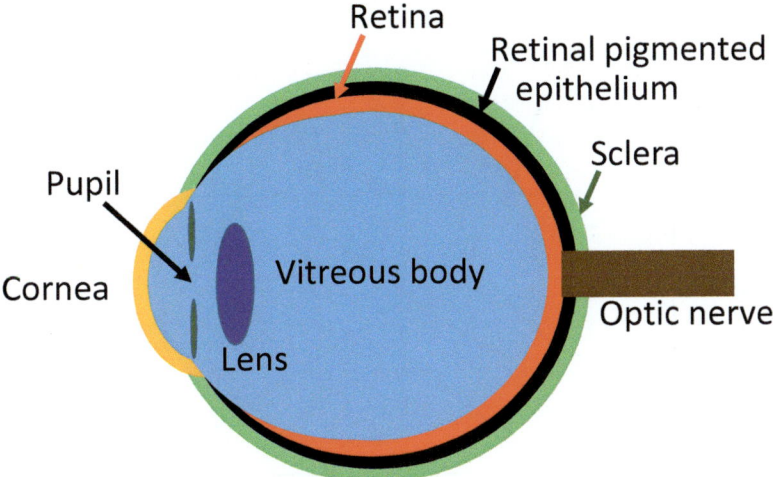

Fig. 31.1 Schematic diagram of the *human* eye

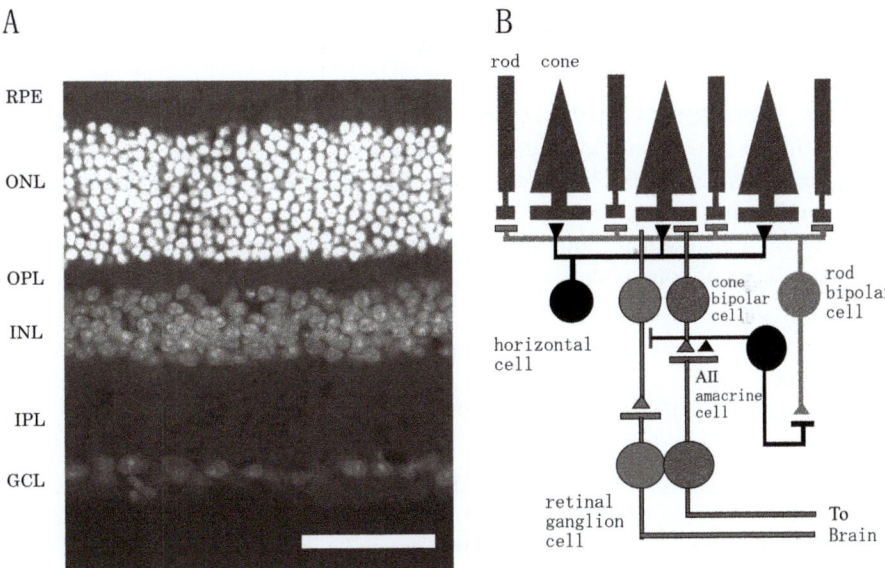

Fig. 31.2 The transected *mouse* retina (**a**) and the neural circuit of the retina (**b**). The basic structure of the *mouse* retina is the same as that of the *human*. The layers of the retina from the outermost surface to innermost surface: retinal pigmented epithelium (RPE), outer nuclear layer (ONL) that contains cone and rod photoreceptors; outer plexiform layer (OPL) in which the photoreceptors make synapses; inner nuclear layer (INL) that contains horizontal, cone, and rod bipolar, amacrine cells; outer plexiform layer (IPL) in which the retinal ganglion cells make synapses; and retinal ganglion cell layer (GCL) that mainly contains retinal ganglion cells. The scale bar, 50 μm. The cells were stained with DAPI, a fluorescent dye. The DAPI preferentially stains the cell nuclei (The image was taken by the author)

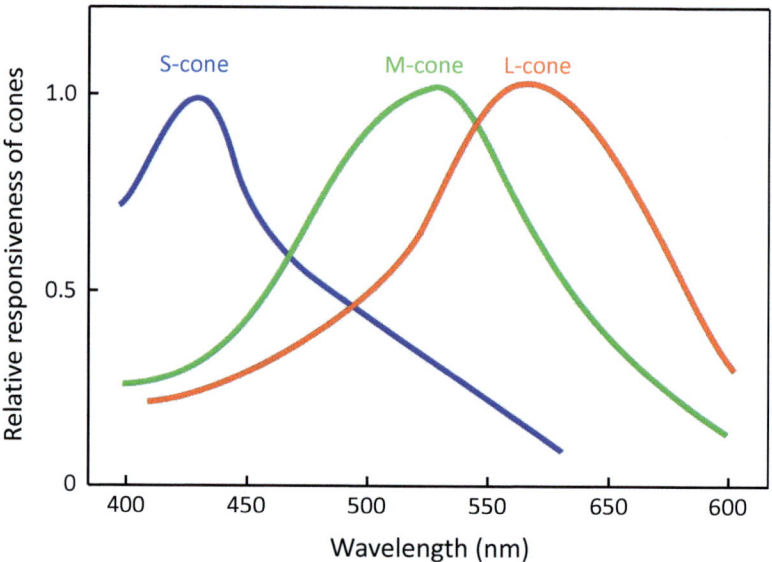

Fig. 31.3 The action spectra of cone photoreceptors. This graph shows how the three types of cone photoreceptors respond to different wavelengths of light

Cone photoreceptors make synaptic connections with horizontal and bipolar cells (Fig. 31.2b). Then, the spectral signals are integrated in amacrine and retinal ganglion cells (RGCs). The RGCs send color information to the brain regions through the optic nerve. In these neural processes, color perception is caused by a mixture of light intensity and spectrum, which are analyzed originally by three different types of cone photoreceptors.

31.2.2 Psychophysical Laws of Color Perception

A television color display generates adjacent pixels of red, green, and blue light (BL) and enables us to perceive color by stimulating all three types of our cone photoreceptors (Fig. 31.4). Each color is separated with a band path filter and prism and digitized by a charge-coupled device (CCD) or complementary metal oxide semiconductor (CMOS) sensor in a video camera. Through variation of the degrees of red, green, and BL intensities, we can create almost all the visible colors on a color display. These colors of light are called as the primary colors. The creation of color by mixing these primary colors of light is additive color synthesis. The combinations of red and green lights, red and BL, and green and BL make yellow, magenta, and cyan lights, respectively. These colors of light are sometimes referred to as secondary colors of light. Finally, the mixture of three primary colors of light in equal intensities produces white light.

Fig. 31.4 Additive color mixing

However, if blue and yellow lights or green and red lights are mixed, the result is not "bluish yellow" or "greenish red." For example, if we gradually add more of yellow light to BL, the mixed color is "paler (more unsaturated) blue" and finally becomes "white." These pairs of colors are called complementary colors. In the late eighteenth century, Edward Hering (1834–1918), a German physiologist, systematized these color phenomena as opponent process theory. This color opponency can be experienced as complementarity of afterimages. For a demonstration of this phenomenon, follow the instructions in the figure legend of Fig. 31.5. You will see the very faint afterimages on a white paper. These afterimages usually last for seconds to minutes. You will see the complementary colors as shown in Fig. 31.5c.

31.2.3 The Standard Chromaticity Diagram

The mixture of these three primary colors of light with varying degrees of intensity can be calculated by some equations and illustrated in the standard chromaticity diagram shown in Fig. 31.6 (Wandell 1995). The wavelengths for the primary colors of light are 460 nm (blue), 530 nm (green), and 620 nm (red). On the diagram, the proportions of red and green primaries are shown on horizontal and vertical axes, respectively. To match any given color, the proportion of blue primary can be calculated by subtracting the other two primaries from 1.0.

The facts depicted on this diagram do not reflect physical or optical laws but psychological processes of the *human* visual system. For other animals like dogs, fish, butterflies, and bees, independent diagrams that match the perceived color are required, because the spectral sensitivity and the types of cone photoreceptor are different among species. This fact is important for insect-pollinated flower growing.

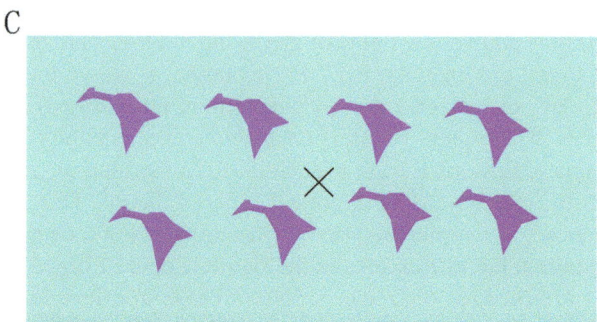

Fig. 31.5 Complementarity of afterimage. Stare at the image in the panel (**a**) for at least 20s. When you finish, you will move to the white area on the panel (**b**) and see the afterimage as shown in the panel (**c**). The "X" on each panel is an eye fixation point. If you do not see anything, you can try this again

The digits on the periphery of the diagram show the wavelengths. Any wavelength matches the perceived color in the visible spectral range. It is an interesting fact that mixture of the primary colors can also yield the same color as any visible wavelength of light. Therefore, we must keep in mind that the perceived color is not necessarily equivalent to the wavelength of light in plant factories. The term "metamers" is used to describe the stimuli that are physically different but perceptually equivalent (Sekuler and Randolph 2002).

Fig. 31.6 The standard chromaticity diagram (CIE 1931 color space)

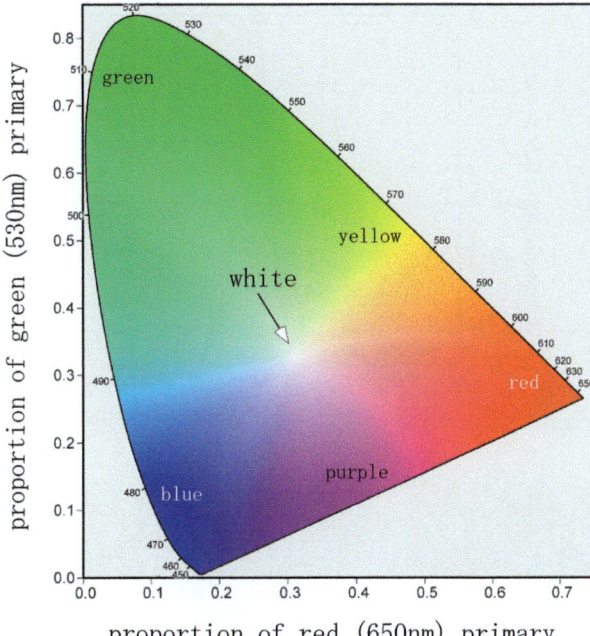

31.2.4 Color Constancy

Color constancy is also an important visual function for plant cultivation works. For example, the ripe tomato looks red in greenhouses. However, its color remains red even under candescent (tungsten) lighting in the dining room. The spectral composition of ordinary candescent light differs from that of sunlight. The candescent light contains less short-wavelength light. Therefore, the candescent light and the sunlight must yield different spectral compositions of the light reflected from the surface of a ripe tomato. The spectral composition of reflected light is called as reflectance spectrum.

Even though the light reaching to the retina varies, the perceived color of the ripe tomato remains red. The term, color constancy, means the propensity for the object's color to be constant even though the distribution of spectral reflectance changes. Even the spectral composition of sunlight varies with the time of day. The sunlight in late afternoon contains more long-wavelength light than the sunlight at noon. If color constancy is lost, screening of the ripeness of tomatoes in greenhouses will be very difficult because the spectral distribution of sunlight changes the reflectance spectrum of the surface of tomatoes depending on the time of day. Owing to color constancy, plant cultivation work can be performed in spite of light changes. However, this is possible under an adequate lighting with spectral distribution of which is broad enough for color discrimination.

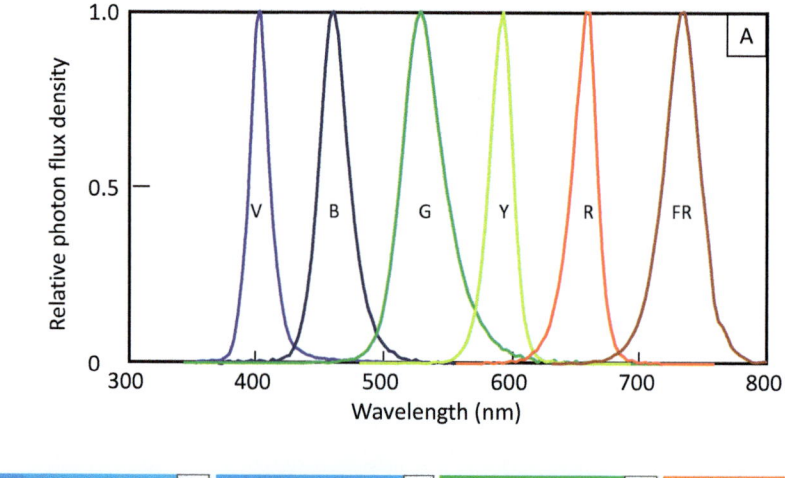

Fig. 31.7 Green cos lettuce plants illuminated under different colors of LEDs. Spectral distributions of LED light that illuminate cos lettuce plants (**a**). V, B, G, Y, R, and FR in the panel (**a**) denote *violet*, *blue*, *green*, *yellow*, *red*, and *far red*, respectively. The spectral distributions of photon flux density were measured at the top of the plants. Changes in the apparent color of the plants under different LED light (**b–i**). The panels, (**b**), (**c**), (**d**), (**e**), (**f**), (**g**), (**h**), and (**i**) were photographed under *violet*, *blue*, *green*, *yellow*, and *red* monochromatic LED light; *blue* and *green* dichromatic LED lighting; *blue*, *green*, and *red* trichromatic LEDs; and *violet*, *blue*, *green*, *yellow*, *red*, and *far-red* polychromatic LED light, respectively (The graph and photographs were provided by Prof. K. Fujiwara)

31.2.5 Color Appearance of Green Plants Under LED Lighting

Green plants appear green under sunlight because they contain chlorophyll, which absorbs most of blue (400–500 nm) and red (600–700 nm) regions of visible light spectrum. Green light (500–560 nm) is less absorbed when compared with blue and red light, but rather reflected by chlorophyll, making green plants appear green. If green plants are illuminated with either monochromatic blue or red LEDs, or dichromatic blue and red LEDs, there will be a significant change in their color (Fig. 31.7).

Changes in the color of plant tissue are a common symptom of plant disease. Abnormality in chlorophyll often causes yellowing of normal green tissue. Viruses cause mosaic patterns in green plants, characterized by leaves mottled with yellow, white, and green spots or streaks. The overdevelopment of anthocyanin causes anthocyanescence, resulting in a purplish color of the plant. Each pathological or abnormal plant color has a characteristic reflectance spectrum. In fact, many studies have revealed that plant diseases can be diagnosed using the reflectance spectrum (Sankarana et al. 2010). These characteristic spectra change color in the pathological region under monochromatic or dichromatic LED illumination.

For the purposes of quality assurance, special luminaires should be installed with LED in a plant factory and greenhouse with artificial lighting. These luminaires should have a color-rendering property sufficient to make the color of green plants appear natural. The color-rendering property determines the ability of a light source to provide visible color to *humans*. The higher the color-rendering property, the better the light source in work areas. Such luminaires have a relatively broad and flat spectral distribution, such as fluorescent, metal halide, and white LED lamps.

31.3 Possible Health Effects of Monochromatic LED Light in Plant Factories and Greenhouses with Artificial Lighting

31.3.1 Circadian Rhythm

Most of our daily activities are governed by the circadian rhythm. Circadian rhythm is approximately 24 h cycle oscillating in the brain. For example, we wake up in the morning and fall asleep at night (sleep-wake cycle). Blood pressure goes up in afternoon and down in early morning. Body temperature increases during daytime and decreases at night. Circadian variation can also be observed in psychological or cognitive functions. For example, attention and memory improve during the day and deteriorate at night. Color perception is reported to be governed by the circadian rhythm to a certain extent (Morita et al. 1994).

Impairment of the circadian rhythm causes some diseases. Epidemiological studies revealed that a larger proportion of shift workers suffer from diabetes mellitus in comparison to non-shift workers. Sugar metabolism is dependent on the circadian rhythm of the *human* body. Abnormalities in sugar metabolism are thought to cause diabetes mellitus in shift workers. Other epidemiological studies suggest an increased risk of breast cancer for the nurses on shift work (Hansena and Stevensb 2012).

Melatonin, a hormone secreted from pineal gland behind the third ventricle of the brain, regulates many *human* cells, tissues, organs, and organ systems including the brain, retina, cardiovascular system, liver, gallbladder, intestine, kidneys, immune cells, adipocytes, prostate and breast epithelial cells, ovary/granulosa cells, myometrium, and skin (Ekmekcioglu 2006). Although the precise intracellular signaling cascade is currently unknown, melatonin has possible antiproliferative effects on breast cancer cell proliferation (Sanchez-Barcelo et al. 2005). Melatonin is secreted exclusively at night with a robust circadian rhythm, but melatonin secretion is strongly inhibited by bright light. Absence of antiproliferative effects of melatonin may facilitate the development of breast cancer.

The phase response curve (PRC; Fig. 31.8) is a curve describing the relationship between a light stimulus and a corresponding shift in the circadian rhythm (phase shift). The PRC shows that the sleep and wake-up time shifts to earlier or later in the day. Light exposure in early morning advances the phase of circadian rhythm (phase advance), whereas light exposure in early evening delays the phase (phase delay). To have a comfortable sleep, we should avoid the strong light exposure in early evening and have enough light in early morning. In fact, the epidemiological studies reported that adolescents diagnosed with delayed sleep phase disorder were exposed to more ambient light between 10:00 pm and 2:00 am and less ambient light between 8:00 and 9:00 am compared with healthy adolescents (Auger et al. 2011; Gradisar and Crowley 2013).

In a plant factory with artificial lighting, laborers can work night shifts. Exposure to excessive light at night increases the possibility of disorders such as insomnia, diabetes mellitus, and breast cancer, as shown above. The PRC indicates that light exposure should be considered when determining a working time for the laborers who work the night shift in such places.

In 2002, my colleagues and I provided the evidence that intrinsically photosensitive retinal ganglion cell (ipRGC) contains a special photopigment, melanopsin, which transduces the light to electrical signals in the mammalian retina and sends out these signals to the brain in order to reset the circadian rhythm (Hattar et al. 2002; Berson et al. 2002; Lucas et al. 2003). The *human* ipRGC has peak photosensitivity at around 480 nm (Bailes and Lucas 2013). This wavelength corresponds to the blue light. The LED lights often used for plant cultivation have maximum wavelengths of around 460 nm (blue) and 660 nm (red). The BL may influence the laborers' circadian rhythm during early evening in plant factories. It delays the circadian phase, inducing insomnia and other circadian rhythm disorders such as delayed sleep phase syndrome (Smith and Eastman 2009).

Of course, polychromatic light containing BL also affects the *human* circadian rhythm (Revell and Skene 2007; Revell et al. 2010). Interestingly, a recent physiological study showed that BL-enriched fluorescent lamps suppress melatonin secretion more strongly than regular white fluorescent lamps (Brainard et al. 2015).

If the laborers work the night shift in the facilities, blue-enriched lighting may worsen the diseases caused by circadian rhythm impairment such as diabetes mellitus. It is reported that eyeglasses with BL filter protects against these diseases (Burkhart and Phelps 2009). Many plastic materials can filter out UV-A (380–315 nm) and UV-B (314–280 nm), which are also harmful to the eyes due to their molecular structures. However, clear polycarbonate cannot filter out BL. - Yellow-tinted plastics are the preferred material, because this filters out BL as well

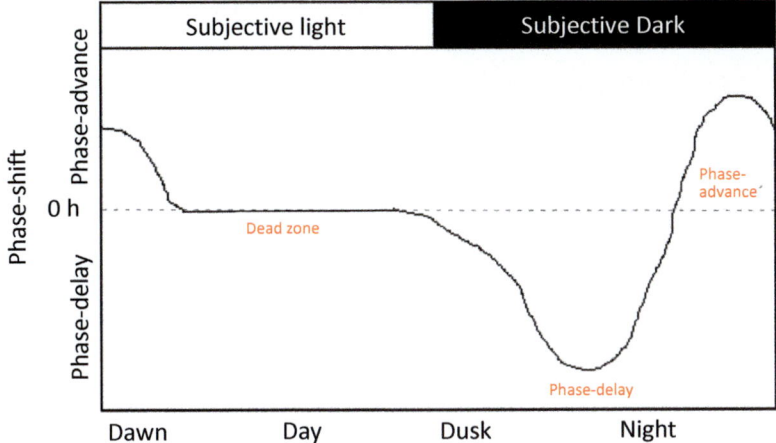

Fig. 31.8 The schematic diagram of phase response curve of *human* circadian rhythm. Subjective light and subjective dark are a time-of-day sensations to recognize nighttime and daytime, respectively

as UV-A and UV-B. Scheduled use of such eyeglasses in the evening or at night may also improve occupational health in the plant factories.

31.3.2 Hazardous Effects of BL on the Eye

In 1980, a research group (Ham et al. 1980) reported that strong BL exposure causes severe chronic retinal damage. *Rhesus monkeys* were exposed to high-intensity BL at 441 nm for 1000 s. Two days after the BL exposure, lesions were formed in the retinal pigmented epithelium (RPE) possibly by oxidative stress (Narimatsu et al. 2015). RPE is a black-pigmented tissue that is located just behind the retina of the eye (Fig. 31.2). The melanin, a pigment component in the RPE, strongly absorbs BL. Therefore, the RPE is subject to BL. The RPE has several important functions such as light absorption, epithelial transport, spatial ion buffering, visual cycle, phagocytosis, secretion, and immune modulation. Extensive damage to RPE results in severe chronic visual impairment.

Rhesus monkeys' eyes are similar to *humans* and ideal for ophthalmological experiments. Of course, the same experiments cannot be performed in *human* eyes due to the ethical restrictions. However, the same damage is plausibly caused to *human* RPE by the intense BL.

The natural lens of the eye strongly absorbs BL as well. BL induces possible clouding or opacification of the lens. Clouded lens is a common eye disease called cataract (Roberts 2011). Cataract significantly decreases the performances and quality of life of the laborers. The eyeglasses with yellow-tinted plastic lenses will prevent the laborer from such chronic eye diseases (Kernt et al. 2012).

31.3.3 Hazardous Effects of UV-C on the Eye

UV-C light (279–200 nm) from the sun does not usually reach the surface of the earth, because the ozone layer in the earth's atmosphere absorbs and filters out virtually all of UV-C. In our daily life, UV-C is irradiated solely by artificial lighting. UV-C is germicidal in nature and is used in some plant factories for sterilization of water.

The cornea and the lens of the eye are known to absorb most UV-C (Boettner and Wolter 1962; Wickert et al. 1999; Meyer-Rochow 2000). Therefore, exposure to UV-C results in corneal damage and opacification of the lens (Doughty et al. 1997). Hazardous effect of UV-C on the eye is greater than BL and must be avoided in working places such as plant factories. UV-C lamps and luminaires must be kept out of sight in such places.

31.4 Glare and Plant Cultivation Works

Glare is a visual sensation caused by an excessive luminance in the visual field. It can disable laborers' visual function temporally (disability glare) or simply cause discomfort (discomfort glare). Disability glare reduces visual performance and visibility thereby inhibiting the laborers from performing their plant cultivation work, resulting in a decrease in productivity in the plant factories and greenhouses with glaring luminaires.

Discomfort glare is sometimes accompanied by disability glare and reduces visibility. It causes stress in occupational environment. It is well known that the lower the background luminance, the larger the size of glare source, and the lower the angle in the visual field, the more discomfort glare is produced (Fig. 31.9; Luckiesh and Moss 1927–1932; Sanders and McCormick 1993).

Glare is also caused by reflectance from the surface of appliances such as floor, shelves, benches, and inner walls of the plant factories and greenhouses. Such glare is called reflected glare, while the glare caused directly from the luminaire is named direct glare. Reflected glare can be specular (as from smooth, polished, mirror-like surfaces), spread (as from brushed, etched, or pebbled surfaces), and diffuse (as from flat-painted or matte surfaces; Sanders and McCormich 1993).

To control glare, we should improve the offending light source, attach a douser to the luminaire, and color the appliances that do not reflect the light. The luminaire should not be installed in the field of view of the laborers. The anti-glare eyeglasses will be effective in some cases. To reduce reflected glare, we should make the proper work place layout and choose the surfaces of the appliances and the inner wall that diffuses the light. The workplace should be designed to avoid the glaring reflected light from the appliances. Flat painted wall, nonglossy walls, and crinkled finish on the surfaces of appliances are desirable, while bright metal, glass, and glossy wallpaper should be avoided.

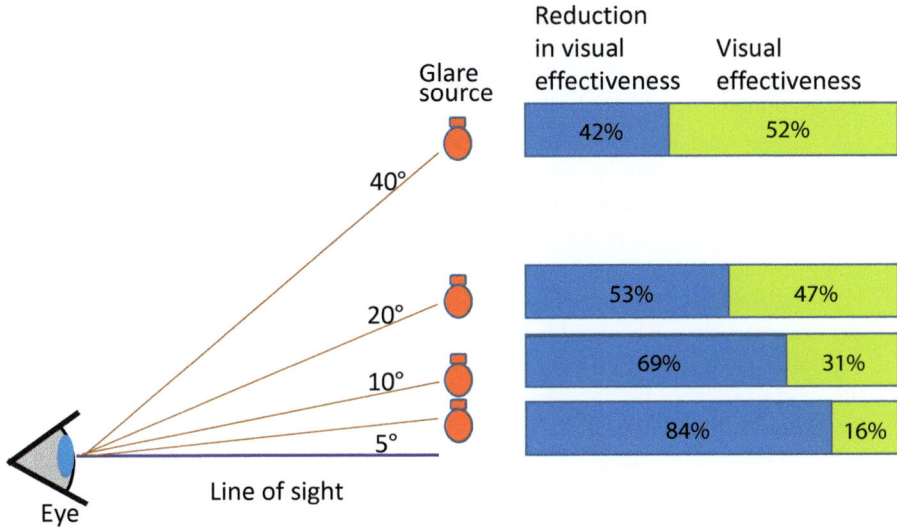

Fig. 31.9 Effects of direct glare on visual effectiveness. The disability of glare becomes greater as the glare source gets closer to the line of sight (Luckiesh and Moss 1927–1932)

31.5 The Desirable Lighting Environment for the Laborers and the Occupational Health Regulations of Lighting

The LEDs revolutionized the plant cultivation in plant factories and greenhouses. They have a possibility to raise the productivity of horticultural and medical plants and to lower the electric bill conspicuously. However, no study exists for the optimization of the lightings in such places. In the last section, I summarize the desirable lighting environment for plant cultivation works as well as the occupational health regulations of lightings to avoid health problems in laborers of plant factories and greenhouses:

1. We should install the LED lamps or fluorescent lamps the spectral distributions of which are broad and uniform enough to illuminate the plants in a similar manner to the daylight. We cannot discriminate the colors of plants under the illumination containing the markedly strong narrow band lights from LED lamps.
2. BL has a possible harmful effect on the circadian rhythms of the laborers. It is pointed out that some severe diseases such as diabetes mellitus and breast cancer can be related to the abnormal regulations of circadian rhythm by BL. The *human* circadian system is highly sensitive to BL at night. The yellow-tinted plastic lenses filter out BL. The scheduled use of the eyeglasses with the yellow-tinted plastic lenses can have a beneficial effect on the laborers who work at night.

3. It is reported that exposure to high-intensity BL tears the retinal pigmented epithelium and clouds the lens, resulting in severe chronic visual impairment. In the working places with strong BLs, the laborers should put on eyeglasses with yellow-tinted plastic lenses throughout their working time.
4. UV-C lamps are mainly used for germicidal purposes and sterilization of water in the plant factories. It is reported that UV-C also damages *human* eyes. Hazardous effect of UV-C on the eyes is greater than BL. UV-C lamps and luminaires must be kept out of sight in the working places.
5. Glare reduces visual performance and visibility, resulting in a decrease in laborer productivity in the plant factories or greenhouses with artificial lighting. To control glare, we should attach a douser to the luminaire and color the appliances that do not reflect the light. The luminaire should not be installed in the field of view of the laborers. The anti-glare eyeglasses can reduce the glare. The inner wall of working place should diffuse the light instead of reflecting it in order to avoid the glare.

We so far have no data to optimize the lighting environment for the laborers in the plant factories or the greenhouses with artificial lighting. In fact, I could not locate a study on the lighting environment in such places. It will be required to survey the current status of working environments to prepare the occupational regulations concerning the lightings and the luminaires. In the future, such lighting regulations will improve feasibility and comfort in work places as well as raise productivity and healthiness of the laborers.

References

Auger RR, Burgess HJ, Dierkhising RA et al (2011) Light exposure among adolescents with delayed sleep phase disorder: a prospective cohort study. Chronobiol Int 28:911–920

Bailes HJ1, Lucas RJ (2013) Human melanopsin forms a pigment maximally sensitive to blue light ($\lambda_{max} \approx 479$ nm) supporting activation of G(q/11) and G(i/o) signalling cascades. Proc Biol Sci 280(1759):20122987, 10.1098

Berson DM, Dunn FA, Takao M (2002) Phototransduction by retinal ganglion cells that set the circadian clock. Science 295:1070–1073

Boettner EA, Wolter JR (1962) Transmission of ocular media. Invest Ophthalmol 1:776–783

Bowmaker JK, Dartnall HJ (1980) Visual pigments of rods and cones in a human retina. J Physiol 298:501–511

Bowmaker JK, Dartnall HJ, Mollon JD (1980) Microspectrophotometric demonstration of four classes of photoreceptor in an old world primate, Macaca fascicularis. J Physiol 298:131–143

Brainard GC, Hanifin JP, Warfield B et al (2015) Short-wavelength enrichment of polychromatic light enhances human melatonin suppression potency. J Pineal Res 58:352–361

Burkhart K, Phelps JR (2009) Amber lenses to block blue light and improve sleep: a randomized trial. Chronobiol Int 26:1602–1612

Doughty MJ, Cullen AP, Monteith-McMaster CA (1997) Aqueous humour and crystalline lens changes associated with ultraviolet radiation or mechanical damage to corneal epithelium in freshwater rainbow trout eyes. J Photochem Photobiol 41:165–172

Ekmekcioglu C (2006) Melatonin receptors in humans: biological role and clinical relevance. Biomed Pharmacother 60:97–108

Gradisar M, Crowley S (2013) Delayed sleep phase disorder in youth. Curr Opin Psychiatry 26:580–585

Ham WT Jr, Ruffolo JJ Jr, Mueller HA et al (1980) The nature of retinal radiation damage: dependence on wavelength, power level and exposure time. Vision Res 20:1105–1111

Hansena J, Stevensb RG (2012) Case–control study of shift-work and breast cancer risk in Danish nurses: impact of shift systems. Eur J Cancer 48:1722–1729

Hattar S, Liao HW, Takao M et al (2002) Melanopsin-containing retinal ganglion cells: architecture, projections, and intrinsic photosensitivity. Science 295:1065–1070

Kernt M, Walch A, Neubauer AS et al (2012) Filtering blue light reduces light-induced oxidative stress, senescence and accumulation of extracellular matrix proteins in human retinal pigment epithelium cells. Clin Experiment Ophthalmol 40:e87–e97

Lucas RJ, Hattar S, Takao M et al (2003) Diminished pupillary light reflex at high irradiances in melanopsin-knockout mice. Science 299:245–247

Luckiesh M, Moss FK (1927–1932) The new science of seeing. In: Interpreting the science of seeing into lighting practice, vol 1. General Electric Co, Cleveland

Meyer-Rochow VB (2000) Risks, especially for the eye, emanating from the rise of solar UV-radiation in the Arctic and Antarctic regions. Int J Circumpolar Health 59:38–51

Morita T, Ohyama M, Tokura H (1994) Diurnal variation of hue discriminatory capability under artificial constant illumination. Experientia 50:641–643

Narimatsu T, Negishi K, Miyake S et al (2015) Blue light-induced inflammatory marker expression in the retinal pigment epithelium-choroid of mice and the protective effect of a yellow intraocular lens material in vivo. Exp Eye Res 132:48–51

Okahara S, Kataoka M, Shima M et al (2014) The possibility of the elderly and individuals with disabilities working in a plant factory with artificial lighting – the activity of the upper limbs and trunk muscles in the standing and sitting positions. Jpn J Occup Med Traum 62:38–43, in Japanese

Revell VL, Skene DJ (2007) Light-induced melatonin suppression in humans with polychromatic and monochromatic light. Chronobiol Int 24:1125–1137

Revell VL, Barrett DC, Schlangen LJ et al (2010) Predicting human nocturnal nonvisual responses to monochromatic and polychromatic light with a melanopsin photosensitivity function. Chronobiol Int 27:1762–1777

Roberts JE (2011) Ultraviolet radiation as a risk factor for cataract and macular degeneration. Eye Contact Lens 37:246–249

Rodieck RW (1998) The first steps in seeing. Sinauer Associates, Sunderland

Sanchez-Barcelo EJ, Cos S, Mediavilla D et al (2005) Melatonin-estrogen interactions in breast cancer. J Pineal Res 38:217–222

Sanders MS, McCormich EJ (1993) Human factors in engineering and design. McGraw-Hill Education, New York

Sankarana S, Mishraa A, Ehsania R, Davisb C (2010) A review of advanced techniques for detecting plant diseases. Comput Electron Agric 72:1–13

Sekuler R, Randolph B (2002) Perception. McGraw-Hill, New York

Smith MR, Eastman CI (2009) Phase delaying the human circadian clock with blue-enriched polychromatic light. Chronobiol Int 26:709–725

Wandell BA (1995) Foundations of vision. Sinauer Associates, Sunderland

Wickert H, Zaar K, Grauer A et al (1999) Differential induction of proto-oncogene expression and cell death in ocular tissues following UV irradiation of the rat eye. Br J Ophthalmol 83:225–230

Chapter 32
Moving Toward Self-Learning Closed Plant Production Systems

Toyoki Kozai and Kazuhiro Fujiwara

Abstract A general scheme of next-generation closed plant production systems (CPPS) with self-learning ability (s-CPPS), consisting of virtual and real CPPS as the two main components, is presented. In virtual CPPS, various models to show the environmental effects on plant growth and mass and energy balance in the CPPS are embeded and then used for predictive simulations of real CPPS. The values of coefficients (parameters) in the models are adjusted by a parameterization process to fit better with those in real CPPS. Once an s-CPPS is developed, it can be utilized, after some revisions, in open and semi-closed greenhouse crop production systems. Recent global technologies are advantageous in s-CPPS. For example, information and communication technology (ICT); artificial intelligence (AI) with big data mining, deep learning, pattern recognition, etc.; the Internet of Things (IoT); and bioinformatics can be combined to make s-CPPS smarter. s-CPPS are integrated plant production systems for next-generation urban agriculture.

Keywords Closed plant production system (CPPS) • Predictive simulation • Self-learning • Real CPPS • Virtual CPPS

32.1 Introduction

The aim of this book is to contribute to developing next-generation plant factories and greenhouses with LEDs for urban agriculture. Plant factories and closed greenhouses are one type of closed plant production system (CPPS). In a CPPS, the consumption of fossil fuel and water and the emission of waste are reduced considerably (Kozai 2013). However, to develop a CPPS as a sustainable plant

T. Kozai (✉)
Japan Plant Factory Association (NPO), Kashiwa-no-ha, Kashiwa, Chiba 277-0882, Japan
e-mail: kozai@faculty.chiba-u.jp

K. Fujiwara
Graduate School of Agricultural and Life Sciences, The University of Tokyo, 1-1-1 Yayoi, Bunkyo-ku, Tokyo 113-8657, Japan
e-mail: afuji@mail.ecc.u-tokyo.ac.jp

production system, electricity consumption for lighting must also be decreased significantly, for example, with intelligent LED lighting. Such smart CPPS will require innovative methodology incorporating advanced global technologies under an appropriate clear vision, mission, goals, and concepts leading to sustainability.

32.2 Mission

The world is facing four interrelated issues involving agriculture/food, the environment/ecosystems, resources/energy, and the quality of life as urban populations increase, agricultural populations decrease with a greater percentage of elderly farmers, and global climate change creates abnormal local weather patterns with more droughts, flooding, spread of disease, etc. Since these four issues are closely interrelated globally and locally, they must be resolved simultaneously based on a common concept, methodology, and technology under varying social, economic, and energy constraints (Kozai et al. 2015a).

Our current mission is to design, develop, construct, and operate specialized food and ornamental plant production systems for high-yielding, high-quality produce. Those systems must contribute to a better quality of life, while conserving the environment and resources. Specifically, to create a healthy, sustainable eco-society, we need to develop resource-efficient, economically feasible plant production systems that allow maximum scheduled yields of high-quality plants at the optimal timing using minimum resources. This will result in lowered costs, greater benefits, and fewer emissions of environmental pollutants and waste under given social, environmental, and resource conditions, contributing to human welfare and social sustainability.

With that mission, we set goals for different specialized food and plant production systems: aquaculture, mushroom culture, and other biomass production. Among the various food and plant production systems, this book focuses on plant factories and greenhouses with LED lighting, which will play an increasingly important role in solving the four interrelated issues simultaneously. This volume discussed the characteristics, roles, opportunities, and challenges of CPPS from different practical aspects.

32.3 Next-Generation CPPS: s-CPPS

A general scheme of next-generation s-CPPS is shown in Fig. 32.1. The two main components of s-CPPS are real (or actual) and virtual CPPS. Virtual CPPS evolve by self-learning during commercial production using the data collected from the real CPPS. The software and hardware requirements of CPPS in operation are then revised based on messages from the virtual CPPS.

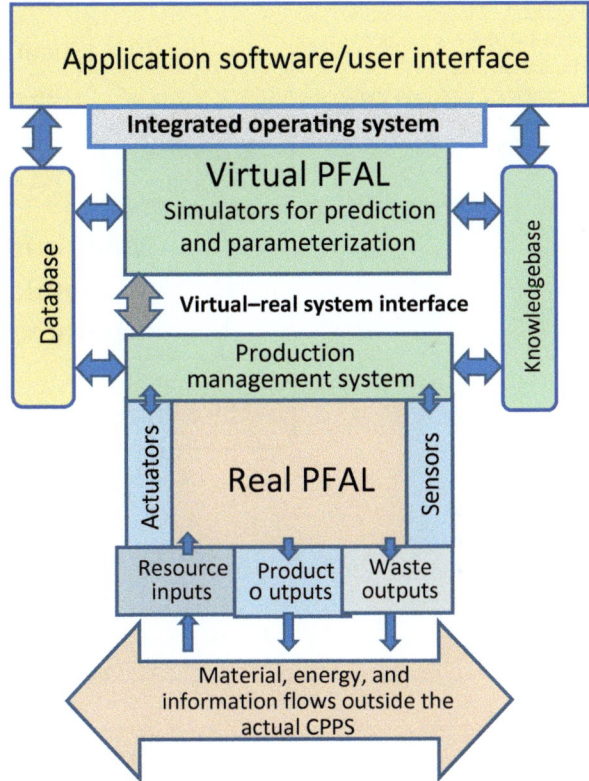

Fig. 32.1 Scheme showing the next generation closed plant production systems (CPPS) with self-learning ability, referred to as s-CPPS in this chapter

s-CPPS are connected via the Internet with other CPPS and different open/closed databases (Harper and Siller 2015). Figure 32.1 illustrates how databases store information on products and waste, the environment, resource inputs, product outputs, market prices, costs, weather, coefficients, etc. "Knowledgebases" contain the models, know-how, rules, dictionaries, basic characteristics of plants, etc. needed to utilize those databases. s-CPPS rely on recent global technologies: information and communication technology (ICT); artificial intelligence (AI) with big data mining, deep learning, pattern recognition, etc.; the Internet of Things (IoT); and bioinformatics (genomics, transcriptomics, proteomics, metabolomics, phenomics, genome editing).

During the past half-century, numerous papers and books have been published on the modeling and simulation of environmental effects on plant growth, mass and energy balance, and environmental and image sensors for use in plant factories and greenhouses (Fig. 32.2). These models, as expressed by equations, are an integral part of virtual CPPS and used for predictive simulation. Since coefficient (parameter) values in the models change with different conditions, the values are adjusted by parameterization processes to fit real conditions in CPPS.

All of these changes in conditions are stored in the database and used for the evolution of the s-CPPS via self-learning. The model structure and configuration are modified often, with partial support from human experts, and stored in the

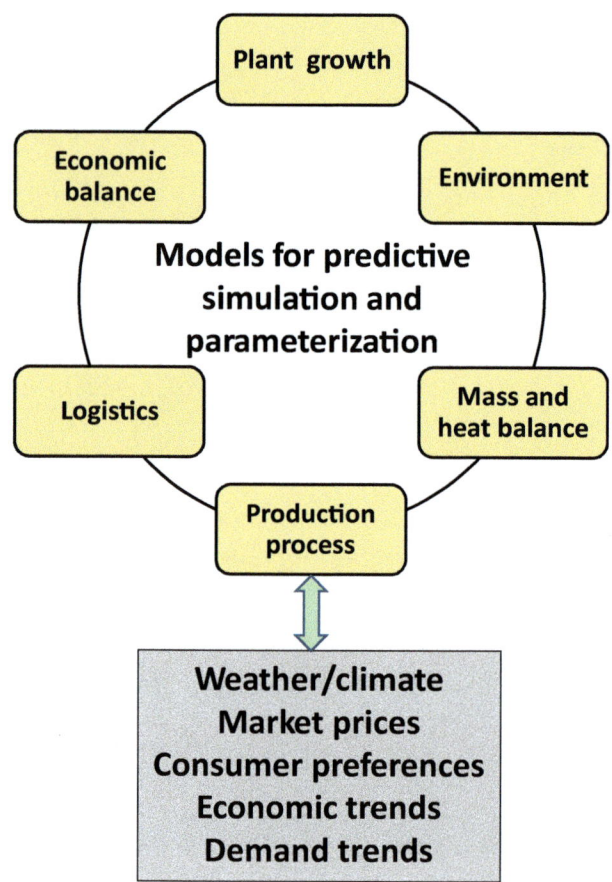

Fig. 32.2 Models used in virtual CPPS for predictive simulation and parameterization and then applied to culture spaces in real CPPS

knowledgebase. Once an s-CPPS is developed, the methodology and most models can be applied after specific revisions for use in open or semi-closed greenhouse crop production systems (Kozai et al. 2015b). Thus, s-CPPS will be key components in next-generation urban agriculture.

References

Harper C, Siller M (2015) OpenAG: a globally distributed network of food computing. Pervasive Comput 14(4):24–27

Kozai T (2013) Resource use efficiency of closed plant production system with artificial light: concept, estimation and application to plant factory. Proc Jpn Acad Ser B 89(10):447–461

Kozai T, Niu G, Takagaki M (eds) (2015a) Plant factory: an indoor vertical farming system for efficient quality food production. Academic, London, p 423

Kozai T, Kubota C, Takagaki M, Maruo T (2015b) Greenhouse environment control technologies for improving the sustainability of food production. Acta Hortic 1107:1–14

Index

A

Abscisic acid (ABA), 63
Absorbance, 116
ACESys. *See* Automation, culture, environment and systems (ACESys)
Adenosine triphosphate (ATP), 119
Africa, 307
Aging of the population, 360
AI. *See* Artificial intelligence (AI)
Air current speed, 178–183
Air movement, 178
Allene oxide synthase (AOS) genes, 262
Analytics, 32, 33
Angular distribution, 399
Angular light distribution, 102–103
Anthocyanin, 214, 241
Anthocyanin concentration, 255
Anti-florigen, 76–78
Anti-florigenic FT/TFL1 family protein (CsAFT), 77
Antioxidant activity, 240
Anxiety level, 355, 357
AOS. *See* Allene oxide synthase (AOS) genes
Artificial intelligence (AI), 43
Ascorbic acid, 248
Asia, 295–308, 319
ATP. *See* Adenosine triphosphate (ATP)
Automation, 34
Automation, culture, environment and systems (ACESys), 26, 27
Auxin, 65
Awareness, 353, 354, 357

B

B radiation, 195
Baby greens, 311
Belgium, 304
Blue light, 283
Boundary layer conductance, 165
Broad-spectrum light, 197
Business categories, 344

C

Canopy, 125
Canopy gap fraction, 141
Carbon assimilation, 120
Carbon dioxide (CO_2), 178, 180–186
β-Carotene, 249
Carotenoids, 116, 248
Cataract, 439
CCP. *See* CO_2 compensation point (CCP)
CEA. *See* Controlled environment agriculture (CEA)
CEPPs. *See* Controlled environment plant production systems (CEPPS)
Central and South America, 307
CFU. *See* Colony-forming unit (CFU)
China, 307
Chlorogenic acid, 242
Chlorophyll (Chl), 116, 205
Chlorophyll concentration, 255
Chloroplast, 118
Circadian rhythms, 67–69
City Farm, 302

Climate change, 360
Climate management, 313
Closed greenhouses, 13
Closed plant production system (CPPS), 12–18
CO_2 compensation point (CCP), 172
Colony-forming unit (CFU), 16
Color constancy, 435
Color perception, 432
Commercial plant factories, 295–298, 300, 302, 305–308
Comparing lighting options, 341
Components of the light environment, 50
Concept of CPPS, 13
Concurrent science, engineering and technology (ConSEnT), 31
Cone photoreceptors, 430
Consumer confidence, 358
Consumer understanding, 353
Consumers' image, 352, 353
Consumers' perception, 348
Consumers' understanding, 356
Controlled environment agriculture (CEA), 20–35
Controlled environment plant production systems (CEPPS), 22–23, 25, 26
Conversion efficiency, 417–418
Corynespora leaf spot disease, 266
CO_2 saturation point (CSP), 172
Cost performance (CP), 15
Cost reduction, 297, 308
C-PPFD, 96
CPPS. See Closed plant production system (CPPS)
CSP. See CO_2 saturation point (CSP)
CRDs. See Current-regulating diodes (CRDs)
Crowdsourced cataloging, 42
Cryptochromes, 61, 195
CsAFT. See Anti-florigenic FT/TFL1 family protein (CsAFT)
CSP. See CO_2 saturation point (CSP)
Cucumbers, 221
Culinary herbs, 311
Cuphea, 276
Current, powered by GE, 338, 339
Current-regulating diodes (CRDs), 408
Custom-made LEDs, 300
Cyber-physical systems, 31, 39

D

Daily light integral (DLI), 104, 204
Dark respiration, 168
Day-neutral plants (DNP), 76
Decision support, 32–33
De-etiolation, 64–65
Democratization of food production, 45
DIALux, 92
Digital plant recipes, 38
Dimming, 409
Disease resistance, 261–264, 266–268, 270, 272, 273
DLI. See Daily light integral (DLI)
DNP. See Day-neutral plants (DNP)
Dormancy, 272
DPPH free-radical scavenging capacity, 240
3D printing, 11
Drive circuits, 407–408
Duration of life, 17
Duty ratio, 387

E

Economic analysis, 199, 341–343
Economic feasibility, 341
Edema, 276
Education, 356, 359, 361
Efficacy, 235
Efficiency, 308
Electrical energy, 418
Electrical energy use efficiency, 419
Electron transport, 119, 120
Ellipsoidal leaf angle density function, 139
End-of-day FR light (EOD-FR), 66
End-of-day lighting, 155
Energy-photon conversion efficacy, 392
Energy transfer, 418
Energy use efficiency, 419
England, 303
EOD. See End-of-day FR light (EOD-FR)
Erectophile, 139
Ergosterol, 248
Europe, 295–308, 329
Evolution, 21–24
Expolinear model, 158
Extension growth, 194
External coincidence model, 78
Extinction coefficient, 127

F

Farmbox greens, 311
Far-red (FR, 700–800 nm) radiation, 193
Far-red light, 284
FLC. See FLOWERING LOCUS C (FLC)
Flavonoids, 116
Flexibility, 308

Florigen, 76
Flowering, 75–76, 192
FLOWERING LOCUS C (FLC), 84
FLOWERING LOCUS T (FT), 76
Fluence bioengineering, 314
Food computer, 39
Food education, 45
Forward current, 381
FR-dependent, 195
Free-word association, 349
FR-neutral, 195
Fruit quality, 227–228
Fruit yields, 219
FT. *See* FLOWERING LOCUS T (FT)
Functional–structural plant models (FSPMs), 159
Functional substances, 270
Functional vegetables, 360, 361

G
GA. *See* Gibberellin (GA)
Gamma correction, 145
Gating effects, 69–70
GE, 300, 314
Gibberellin (GA), 63
Glare, 440
Global technology, 8
Global trends, 318
Glucose concentrations, 253
G radiation, 197
Green LED lighting sources, 273
Green light, 262–268, 272, 273
Greenhouses, 21–22
Gross photosynthetic rate, 171
Grower-driven LEDs, 300
Growth recipe, 308
Growth-related gene, 268

H
Half width, 381
Heat dissipation, 409–410
Heliospectra, 314, 330–332
High-irradiance responses (HIR), 59
High-pressure sodium (HPS), 220
Hong Kong, 307
Hort Americas, 315
Hue angle, 213

I
Illuminance, 370
Illumitex, 303, 314, 339
Information and communication technology (ICT), 6–8, 12

Information processing speeds, 9
Intelligence-empowered CEA, 28–30
Intelligent lighting, 344
Intensity, 308
Internal coincidence model, 78
Internet of Food (IoF), 38
Intracanopy lighting, 222
Intravision, 314, 333, 334
Intrinsically photosensitive retinal ganglion cell (ipRGC), 438
Intumescence, 275–285
Investment costs, 229
Irradiance, 369
Iterative experimentation, 39
Ivy geraniums, 276

J
Japan, 296–300, 303, 306–308

K
Keystone technology, 300, 319, 320
Knowledge, 352, 353, 358
Kyocera, 300, 320

L
Labor, 313
LCOE. *See* Levelized cost of electricity (LCOE)
LCP. *See* Light compensation point (LCP)
LDP. *See* Long-day plants (LDP)
Leaf area density (LAD), 139
Leaf area index (LAI), 126, 182
Leaf area ratio (LAR), 134
Leaf boundary layer resistance, 178
Leaf senescence, 53
Leafy greens, 311
LED. *See* Light-emitting diode (LED)
Levelized cost of electricity (LCOE), 10
LFR. *See* Low-fluence response (LFR)
Lidar, 142
Light compensation point (LCP), 173
Light-emitting diode (LED), 4, 403, 429
　control gear, 379
　grow light, 296
　lamp, 379
　lighting business, 344
　lighting players, 317–335, 337–345
　lighting system, 380
　lighting trends, 300
　light source, 379
　luminaire, 380
　module, 379

Light-emitting diode (LED) (*cont.*)
   package, 378
   PFAL market, 296
   sulfurization, 388
Light energy, 421
Lighting book, 234
Lighting companies, 314
Lighting metrics, 233
Lighting technologies, 234
Light intensity, 184–185
Light saturation point (LSP), 173
Light uniformity, 228
Light use efficiency (LUE), 152
Lipoxygenase (LOX) genes, 262
Local technologies, 7
Long-day plants (LDP), 76, 192
Low-fluence response (LFR), 59
Low light, 311
LSP. *See* Light saturation point (LSP)
L-system, 159
LUE. *See* Light use efficiency (LUE)
LumiGrow, 314
Luminous efficacy, 4, 392, 417
Luminous flux, 370
Luminous intensity, 369, 383, 399

M
Maltose (malt sugar) concentration, 253
Marginal costs
   of DNA sequencing, 9
   of information, 7
Market, 296–300
Marketing, 361
Marketing activities, 356, 358, 359
Market size, 297
Melatonin, 437
Metamers, 434
Microgreens, 311
Microterroirs, 44
Middle East, 307
MIRAI, 299, 300
Monosaccharide concentration, 253
Multiple reflections, 98

N
NADPH. *See* Nicotinamide adenine dinucleotide phosphate (NADPH)
NB. *See* Night-break (NB)
NDVI. *See* Normalized difference vegetation index (NDVI)
Net assimilation rate (NAR), 134

The Netherlands, 302
Net photosynthetic rate, 171
Next-generation CPPS, 446–448
Next-generation urban agriculture, 12
Nicotinamide adenine dinucleotide phosphate (NADPH), 119
Night-break (NB), 77
Nihon Advanced Agri, 300, 321
Nisshinbo, 297
Normalized difference vegetation index (NDVI), 143
North America, 310–315, 337
Nutraceuticals, 45
Nutrition, 313
Nutritional value, 351
Nutritious plants, 297

O
Occupational regulations, 442
Oedema, 276
Open-ended question, 350
Open Phenome Project, 42
Optimal lighting direction, 106
Optimal light quality, 107
Optimal photo- and dark periods, 106–107
Optimal PPFD, 104–106
Ornamentals, 192
Ornamental sweet potato, 276
Osmotin-like proteins, 262
Other regions, 295–308
Overhead lighting, 220

P
PAL. *See* Phenylalanine ammonia-lyase (PAL)
Panama, 307
Panasonic, 314
Parallel layout, 101
Peak absorption, 204
Peak wavelength, 384
Percent loss (%L) of photosynthetic photons, 96
Perpendicular layout, 101
PFALs. *See* Plant factories with artificial lightings (PFALs)
PFALs business, 344
Pharmaceutical products, 240
Phase response curve, 438
Phenylalanine ammonia-lyase (PAL), 270
Philips, 300, 314
Philips lighting, 335, 336
Photomorphogenesis, 64

Index

Photomorphogenic lighting, 203
Photon efficiency, 229
Photon flux, 370
Photon flux density, 371
Photon intensity, 370
Photon irradiance, 371
Photon number, 371
Photoperiod, 75
Photoperiodic control, 198
Photoperiodic efficacy, 196
Photoperiodic flower bud differentiation, 272
Photoperiodic lighting, 192
Photoperiodism, 75
Photoreceptors, 57–62, 204
Photorespiration, 169
Photosynthesis, 178, 181
Photosynthetic capacity, 173
Photosynthetic photon flux density (PPFD), 396–398
Photosynthetic photon flux fluence rate (PPFFR), 138
Photosynthetic photon number efficacy, 392, 400–401
Photosynthetic photon use efficiency, 420
Photosynthetically active radiant energy efficiency, 400
Photosynthetically active radiation, 375
Photosystem, 119
Phototropins, 61
Phototropism, 65
Phyllotaxis, 158
Phytochrome, 205
Phytochrome photoequilibrium (PPE), 193, 205
Phytochromes, 59–61, 193
Phytomation, 23
Pixel value, 145
Planophile, 139
Plant canopy analyzer, 141
Plant density, 311
Plant factories with artificial lightings (PFALs), 3, 295
Plant factory, 23
Plant growth regulator, 211
PlantLab, 304
Plant morphology, 5
Plant phenomics, 42
PL light systems, 314
p–n junction, 405
Potato, 277
PPE. *See* Phytochrome photoequilibrium (PPE)
PPFD. *See* Photosynthetic photon flux density (PPFD)

PPFFR. *See* Photosynthetic photon flux fluence rate (PPFFR)
Protected cultivation, 21
Pulsed light, 387
Pulse width modulation (PWM), 387, 409

Q
Quality, 308
Quantity of light, 370
Quantum sensor, 138
Q_{10} value, 169

R
R:FR ratio, 66, 199
Radiant energy, 369
Radiant flux, 369, 384
Radiant intensity, 368
Rate variable control, 14–16
Ratio of blue/red light, 242
Real CPPS, 446
Red light, 284
Reflectance, 92
Reflected glare, 440
Relative annual productivity, 16
Relative growth rate (RGR), 134
Relative leaf growth rate (RLGR), 134
Resource use efficiency (RUE), 15
Resource-saving, 17
Russia, 304

S
Saccharide concentrations, 254
Sample spreadsheet, 342
SAR. *See* Shade-avoidance response (SAR)
s-CPPS, 446
SDPs. *See* Short-day plants (SDPs)
Seed germination, 62
Seed technology, 313
Self-learning systems, 108
Semiconductor, 405
Shade-avoidance response (SAR), 66–67
Shinetsu Kagaku, 300
Short-day plants (SDPs), 76, 192
Showa Denko, 300, 322
SIA. *See* Systems informatics and analytics (SIA)
Side reflector, 98
Skotomorphogenesis, 64
Software businesses, 308
Solar spectral irradiance, 371

Sole-source lighting, 205
Solid angle, 368
Solidlite, 326, 327
South Korea, 306
Spectral distribution, 193, 371, 399
Spectral distribution curve, 395–396
Spider mite, 268
S-PPFD, 96
SPREAD, 301
Standard chromaticity diagram, 433
Standards, 234
Stanley, 300
Stanley electric, 323
State variable, 14
Stimulus words, 349
Stomatal conductance, 165
Strawberry anthracnose, 263
Strawberry PFAL, 297
Stroma, 119
Sucrose concentration, 250
Sugar concentration, 250
Sustainable PFAL business models, 345
Sweet peppers, 221
Systems analysis, 20, 30
Systems informatics and analytics (SIA), 30, 34

T
Taiwan, 305
TAS. *See* Total antioxidant system (TAS)
TERMINAL FLOWER 1 (TFL1), 77
Terrestrial laser scanners (TLS), 142
Thylakoid, 118
TLS. *See* Terrestrial laser scanners (TLS)
α-Tocopherol, 248
Tomato, 221, 276
Toshiba, 300, 302, 314, 324, 325
Total antioxidant system (TAS), 242
Total flavonoid concentration, 246
Total phenolic compounds, 240
Total phenolic concentration, 246
Traceability, 17
Transistor, 408
Transition from FLs to LED lightings, 298

Transition of built-up cash, 343
Translocation, 122
Transparent specifications, 308
Transpiration, 121, 122, 178–183
Transpiration rate, 165
True-or-false questions, 354, 355

U
Ultraviolet light, 278
Understanding, 357
Upward-directed LED lighting system, 53
Urban agriculture, 3
Urban food and agriculture systems, 24–25
Urbanization, 360
USA and Canada, 309
Ushio lighting, 325, 326
UV-B receptor (UVR8), 62

V
Valoya, 314
Varieties, 296
Vernalisation, 84–85
Vertical farming, 295, 309–315
Very-low-fluence responses (VLFR), 59
Viewing half angle, 384
Virtual CPPS, 446
VLFR. *See* Very-low-fluence responses (VLFR)
Volatile elicitors, 44

W
Water, 313
Wavelength conversion phosphor, 389

X
Xanthophylls, 249

Z
Zeitlupe family proteins, 62

344, 49, ac Air Slit & Side Reflector 95 Fig 7.4 1/2 Angle
06 7.4.1.1 Spherical PPFD, DLI 7.4.2, 107 7.4.2.1 UV/FR
16 UV Flavinoids 8.2.2, Light Attenuation 105
173 Light Saturation, 1 mole CO2 fixed 8 moles Absorbed photons
174 Graph, 184 Airspeed + Leaf temperature 193 FR Flowering
MState Flowering Light, 194 Fig 14.1 B450 DR660 FR735
193 Phytochrome Photoequilibrium .63 - .72 2Lext
196 Comparative Chart HPS - LEDs 193 CraigRunkle 2015 Rapid
Flowering LD plants Meng & Runkle 2016

Deedre Shannon Craig "Determining Effective
Ratios of Red and Far-red Far-red light from
Light-Emitting Diodes that Control flowering
of photoperiodic ornemental crops"
(Light + temperature)

   Chinese Hibiscus flowered 9 days sooner
   @ 420 umol < 840 umol, plants grew
   under 300 umol 2x dry weight over
   100 umole irradiance, plants also flowered
   20 days sooner ↑ DLI +dry weight
   + ↑ flower bud # + lateral branching, +Flower
   Diameter + ↑ root growth, reduced time
   to flower (Runkle Heins 2006)
   June MI FL 50 moles/day ave January
   MI 15 mol-m-d DLI 15 moles
   More light Reduce #of Nodes that develope before
cryptochromes 1st Flower bud Impatiens flowered 17% faster
   UV A - (320-380) + Blue can inhibit stem
   Elongation B:R 1.6 ↓ 17-28%
   B:R 1.05 ↓ Reduced Elongation 25 to
UV 50 umole/m2/sec Nat Day light R:FR Ratio 1:1.15
   High R inhibits stem elongation
   (over FR)

R - more branching LD plants flower more rapidly when get FR light at end of DAY FR-absorbing filter inhibits Elongation - Gallium Nitride chips - 440 NM R+FR

LED chips Gallium Al chips - 630-940 peak 650 + 720
Cree 231 lm/watt p62 (Craig) Ratios R-FR
p209 70%R/30%blue p222 "1% Rule" 1% more light Produces/results 1% more yield p223 Intracanopy p229
7.8. Basic good Info p234 Runkle New Book Light Mgnt in controlled environments p235 ASABE LED standards p244 Flashing Yellow 596nm ↑Antioxidents
247 Flavor - Blue LEDs 250 UV chart p253 Table 18.7
Red light before Harvest ↑sugar content (Brix - watermellon)
p255 Table 18.9 sugars ? nm 455nm 535nm 470nm lesser 505
p257 chlorophyll spectrum   chapter 19 Green light Disease Resist.
p267 Green light - fruit enlargement control spider mites
p298 Philips + Showa Denko site 660nm optimum wavelengths
Photosynthesis p300 Ag Led cos p314 - Growlight Co. p318
p330 Heliospectra - Early bloom recipe 470 & 660 + uv Emmerson Effect
pulsed light p387) Sulfur p388 p389 root development timing
blue + Red 399 Comparable Led blue Red White table p407
LED circuits How to calculate resistance p412 Dark Respiration + Photorespiration p424 Co2, temp, vaporpressure deficit, PAR light spectrum 438 460nm B 660nm R + UV dang danger UV ABC yellow tinted glasses UV + Blue light protection 438 + 439

Heliospectra turn of BR White - Run FR End of 12 hr Day
cuts 10-14 days off Flowering Cycle
Short day plants Flower when there is Adequate Pfr

666 synthesis
vegitative
Pr →(red)→ Pfr
←(Far Red)←
740nm ↓ Destruction
Reproductive Flowering

← Dark Reversion
slow

Mich State 2016 Meng Runkle
Flowering 600-700nm

~~Regulate~~
FR Flowering only 1-2 umol needed

Feit Electric Plant Grow Light
Total Grow Night & Day Mgt Light
Philips GreenPower LED
Flowering Lamps
DR/W   DR/W/FR

CPI Antony Rowe
Chippenham, UK
2017-07-20 09:44